W0227810

KEY TOPICS
IN BRAIN
RESEARCH

*Edited by A. Carlsson,
P. Riederer, and H. Beckmann*

H. Przuntek and P. Riederer (eds.)

Early Diagnosis and Preventive Therapy in Parkinson's Disease

Springer-Verlag Wien GmbH

Prof. Dr. Horst Przuntek
St. Josef Hospital, University of Bochum, Bochum, Federal Republic of Germany

Prof. Dr. Peter Riederer
Clinical Neurochemistry, Department of Psychiatry, University of Würzburg, Würzburg, Federal Republic of Germany

Originally published by Springer-Verlag/Wien in 1989

With 59 Figures (1 in color)

ISSN 0934–1420

ISBN 978-3-211-82080-3 ISBN 978-3-7091-8994-8 (eBook)
DOI 10.1007/978-3-7091-8994-8

Preface

Recent studies show an age-dependence of a variety of neurotransmitters, their metabolites and the receptor systems. Similar research strategies are in line with the suggestion that Parkinson's disease cannot be regarded as a simple pronunciation of aging. Rather the underlying pathophysiology is independent of or superimposed to normal aging processes. It has thus been hitherto assumed that Parkinson's disease, if diagnosed early, could be treated in such a way that any progression could be halted or at least minimized. This idea prompted us to organize a workshop dedicated to the question of an early diagnosis and possible therapeutic strategies. A variety of methodological possibilities are available. However, uncertainity exists about the specificity and sensitivity of such methods. In fact, our current knowledge and today's availability of circumscribed methods of different origin do not put us in a position to be able to come to a conclusive diagnosis of Parkinson's disease by the use of only one of these strategies. On the contrary, a high rate of diagnostic accuracy can be achieved only by the use of several methods including imaging techniques, biochemical, electrophysiological and clinical assessments.

A second item concerned the possible therapeutic strategies of early Parkinson's disease. Here, overall agreement was achieved between all participants: a patient with early signs of Parkinson's disease should be treated from the outset. Furthermore, it became evident that not all MPTP-antagonizing substances are of therapeutic value in Parkinson's disease. It also transpired that a variety of animal species are more useful for simulating the early phase of Parkinson's disease (e.g. rats, mice) while others are more suitable for covering the symptoms of advanced phases (e.g. monkeys).

These proceedings give a survey of the most recent aspects of early diagnosis and preventive therapeutical strategies. By including the most important parts of the discussions, we hope to give the reader a lively insight into the problems behind these important new vistas and stimulate the scientific community into developing methods and diagnostic possibilities to help prolong both the life-expectancy and the quality of life of those rendered succeptible to Parkinson's disease.

This workshop would not have been possible without the financial support of Hoffmann-La Roche, Grenzach-Wyhlen. In particular, we would like to thank Mr. J. Wanzlik for his help in organizing this meeting.

Furthermore we are grateful to Dr. K. Blau who improved the english style of a number of manuscripts and the "Discussions". The secretarial help by Mrs. J. Philipp, Mrs. H. Bigge, Mrs. I. Riederer, and Dr. R. Steinberg is gratefully acknowledged.

Last but not least, we thank Springer-Verlag Wien-New York for the excellent production and the smooth cooperation over the last months.

October 1988 **H. Przuntek** and **P. Riederer**

Contents

Listed in Current Contents

Welcome and introduction

H. G. Mertens

Neurological Clinic, University of Würzburg, Federal Republic of Germany

Ladies and Gentlemen,

It gives me great pleasure to welcome you to the Symposium on "Early Diagnosis and Preventive Therapy in Parkinson's Disease".

The Neurological Clinic of Würzburg University is the oldest Neurological clinic in Bavaria, and right from its foundation the investigation and treatment of extrapyramidal disease has been a significant aspect of its work, starting with Professor Schaltenbrandt. In this connection we may recall his stereotactic atlas, which even today is still in regular use in the stereotactic department.

The early diagnosis of extrapyramidal disease seemed in the first place to be of particular value is genetically determined diseases, since it offered the best chance of discovering and treating affected patients in the families afflicted with the disease. Thus in the seventies our attention in Würzburg was above all focussed on the early diagnosis and treatment of Wilson's disease.

After we had established the possibility of early diagnosis of Wilson's disease, and the resulting early treatment, we turned increasingly to attempts at an early diagnosis of Huntington's chorea, where psychometric measurements, biochemical determinations and gene probes found their application. It turned out that the psychometric procedures as well as attention to fine motor performance showed clear evidence of the disease before it became evident through its clinical manifestations.

It was precisely the results of these investigations which led us to suspect that it might be possible that Parkinson's disease too might be detectable at an earlier stage than that at which the diagnosis becomes evident from the clinical signs. In contrast to the genetically determined diseases, however, the early diagnosis of Parkinson's disease represents a much more difficult problem, since, as was the case earlier with tuberculosis, there was no way of avoiding serial investigations.

If we start out from the fact that the disease manifests itself subclinically for years before frank clinical symptoms appear, then it should ideally be possible to recognise it by a carefully directed early diagnostic methodology. And if animal experiments with the MPTP-model (1-methyl-4-phenyl-1,2,3,6-tetrahydropyridine) were to come up with drugs that were prophylactically effective it might then be possible to investigate such drugs for their effectiveness in preventing the progression of the disease in Parkinsonian patients.

It therefore seems particularly important to investigate whether psychometric procedures might not be suited, as in Huntington's chorea, to provide evidence of the presence of Parkinson's disease. Various studies indicated that the pre-morbid personality in Parkinson's disease was characterised by depression, introversion, rigidity and inflexibility. These finding were confirmed in twin studies, from which it became evident that the twin who developed Parkinson's disease was less extroverted, more introverted and more depressive than the unaffected twin. In more detailed experiments this was shown by higher perseverance-scores e. g. in the Wisconsin card-sorting test. It remains to be established whether such disturbances are ascribable to the degeneration of the mesocortico-limbic system.

Poewe and Gerstenbrand in particular noted early on that the relatives of patients who subsequently developed Parkinson's disease observed distinct early personality changes, so we have to ask ourselves whether such changes might be quantifiable and could be made more specific by suitably designed additional investigations. We also need to enquire whether dementia is a typical feature of Parkinson's disease or whether it is rather occurring within the framework of a multi-infarct syndrome or a dementia of the Alzheimer type.

Patients with Parkinson's disease very early on complain of subjective and autonomic disturbances, which tend to be ignored because they are overshadowed by the more prominent motor disturbances. These subjective symptoms generally take the form of deafness, paresthesia, crawling of the skin, and aches and pains. While it may be difficult to quantify such subjective disturbances, it might be relatively simple to measure autonomic disturbances that manifest themselves by defects in temperature regulation and by seborrhoea.

In Huntington's chorea a pathological posture is seen early on, before other hyperkinetic signs appear which then generally raise the suspicion of a choreatic hyperkinesia. In Parkinson's disease we ought perhaps to consider the possibility that video-recordings might help us early on to document deviations from the patient's normal state. In particular this approach should enable us to make early recordings and to detect and evaluate changes in the patient's postural reflexes.

One of the things one notices is that stress or emotional tension can provoke tremor in Parkinsonian patients. It might be worth seeing whether it is possible, analogous to the effect of curare in myasthenia, by using a stress-inducing substance e. g. β-carboline prematurely to elicit a resting tremor, or by administering a short-acting dopamine agonist perhaps to discover some kind of threshold dose at which a healthy person would not show any movement deficit but a Parkinson patient would respond by showing akinesia and rigor.

Any attempts at early diagnosis procedures must in the first place be based on early-diagnostic procedures which are simple and which can eventually be applied routinely by medical ancillary personnel. For this reason we should use batteries of tests which focus on motor performance so that any noticeable deviations from normal can be followed by diagnostic procedures specific for Parkinson's disease. Because it is the complex motor performance which is affected in Parkinson's disease it should be possible, by refining the testing procedures currently available, to detect motor disturbances earlier than is possible at present. We might, for example, submit patients who have come to our notice because of defects in fine motor performance to some investigation like positron-emission tomography ("PET-scanning"), although it is at this point in time not clear at what clinical stage quantifiable PET-scan results would appear, and what kind of band–width we might expect from age–correlated investigations.

Since we must assume that the pathological changes in Parkinson's disease are not restricted only to the central nervous system, we may expect that beyond psychometric, kinesiological, electrophysiological and pictorial-recording modalities we may find detectable effects in clinical chemical investigations, perhaps on blood cell receptors or changes in enzyme activities. Whether we should extend these to studies on urinary excretion or cerebrospinal fluid changes will be discussed during the symposium.

Parkinson's disease has indeed become a paradigm of the way the pathogenesis and therapy of a neurological disease should be investigated. The last few years particularly have seen us pass a number of important new milestones with regard to our understanding of factors affecting motor function: there has been the MPTP-model, the development of transplantation techniques and the study of Lewy-bodies, as well as the biochemical investigation of tyrosine hydroxylase, of iron metabolism and of the investigation of neuropeptides. There have been fresh insights into the function of the differentiated dopamine-receptors and the reciprocal GABA-ergic and dopaminergic innervation. These are all factors which cannot fail to influence the treatment of Parkinson's disease. We have to take a fresh look at the treatment approaches used hitherto particularly from the viewpoint of preventive treatment: for example we should look for an answer to the problem, bearing in mind that monoamine oxidase B is not found in the dopaminergic nerve-endings of the striatum, just how MPTP-antagonists might be suited to the provision of a preventive therapy. Further, we have to ask ourselves whether our present pharmacokinetic strategies, particularly with oral administration, where only a fraction of the drug (less than 1% as a rule) reaches the brain following intermittent administration, are not vastly inferior to alternative methods by which the drugs are continuously instilled into the brain. It is my wish for this symposium that a substantial number of these and other questions will find answer or the beginnings of answers. I also sincerely hope that this symposium will be the occasion for closer collaboration between the European researchers active in studies on Parkinson's disease and their counterparts in the friendly neighbouring countries.

Correspondence: Prof. Dr. H. G. Mertens, Neurological Clinic, University of Würzburg, Josef-Schneider-Strasse 11, D-8700 Würzburg, Federal Republic of Germany.

The premorbid personality of patients with Parkinson's disease

An early sign of their disease?

W. Poewe, F. Gerstenbrand, E. Karamat, and B. Schmidhuber-Eiler

Department of Neurology, University of Innsbruck, Austria

Summary

A total of 85 patients with idiopathic Parkinson's disease have been tested for characteristic premorbid personality features. In three different studies with varying test methods patients consistently appeared more introverted, depressed, rigid and inflexible than age-matched controls. It is suggested that together with impairment of conceptual shift and mental slowing found in early stages of Parkinson's disease these premorbid personality traits might represent early signs of brain dopaminergic dysfunction.

Introduction

Underterred by the lack of a reliable method permitting the retrospective assessment of permorbid personality traits neurologists and psychiatrists have since the beginning of this century been impressed with certain peculiar features in the personality of patients with Parkinson's disease being present long before the onset of their illness (Todes and Lees, 1985, for review). With remarkable consistence the patients' previous personality is reported to be dominated by traits of moral rigidity and inflexibility combined with a tendency towards depression and introversion. Many of the earlier authors have taken a psychodynamic view of their findings and have postulated a causal role of lifelong suppression of aggressive tendencies and emotional deprivation for the eventual development of Parkinson's disease (Cohen-Booth, 1935; Sands, 1942; Mitscherlich, 1960) and this view has more recently been emphasized by Todes

(1984). More in line with our current understanding of this illness others have argued that a characteristic premorbid parkinsonian personality might be an early clinical manifestation of the underlying dysfunction of brain dopaminergic systems (Poewe et al., 1983; Ward et al., 1983) and that it might be somehow related to the phenomenon of bradyphrenia later observed in patients (Rogers et al., 1987).

In this paper a summary of the authors' results in studies of premorbid characteristics in the personality of Parkinson patients will be given and discussed with respect to their possible significance as an early disease sign.

Patients and methods

A total of 85 patients with idiopathic Parkinson's disease have been studied in three separate investigations of premorbid personality traits. All were receiving chronic L-dopa treatment and the first series comprised 14 males and 14 females (mean age 67 years; mean duration of disease 9 years) who where tested by the Giessen test (GT) administered in a retrospective fashion to the patients and a close relative or companion (Poewe et al., 1983). The GT was selected as a personality inventory for this investigation because it provides for both a self and outside assessment of the patients personality features so that the bias of retrospective judgement could be kept as low as possible. This was supplemented by a second set of studies in 12 male and 12 female patients (mean age 65 years; mean duration of disease 7,5 years) where in addition to the Giessen test the Freiburg Personality inventory (FPI) was administered to assess both the patients' present personality status and–in a modified version–retrospectively also their previous personality. In both series the results could be compared to age specific standard values from control populations available for both the GT and the FPI and statistical significance was calculated by means of Student's t-test.

In our most recent ongoing study 38 patients (22 males, 16 females; mean age 61,4 years; mean duration of disease 5,2 years) and 17 age-matched healthy controls (11 males, 6 females; mean age 59 years) have so far been investigated. Tests included a mini mental state (MMS) examination and the WAIS to exclude demented patients or controls, the geriatric depression scale (GDS) and Cattell's 16 PF personality inventory. Statistical differences of scores between patients and controls were calculated using a one-way analysis of variance.

To assess premorbid character and behavioural features patients and controls underwent a one-hour semi-standardized interview by an experienced psychologist (E.K.) designed to cover as extensively as possible biographical data, premorbid habits, hobbies, family life, professional career and social activities. These interviews were repeated with a close relative or companion of the patient or control person (mostly spouses) to obtain their version of the proband's premorbid biography and character. The interviews were stenographically recorded and subsequently evaluated independently by the interviewer and a second psychologist blind to the diagnosis and unaware of the background of the study. For evaluation of the interview material seven "bipolar" scales were defined ("introverted/depressed vs. outgoing/happy"; "workoholic vs. easy going"; "pedantic vs. generous"; "rigid vs. flexible"; "loner vs. sociable"; "smoker vs. non-smoker"; "teetotaller vs. keen on alcohol") and ratings were done by scoring one on either pole if the respective traits were prominent in the material or zero when there was no obvious trend into either direction.

Results

The results of our first study have been previously published (Poewe et al., 1983). The most striking trends towards a deviation from score values of a control population were found in scales 3 ("control") and 4 ("basic mood") of the GT. Parkinsonians appeared to have been overcontrolled in their premorbid personalities, which in this test is defined by features as "talented in dealing with money", "overorderly", "overambitious", "timid", "overly self-reflective", "swallowing anger", "dependent". These trends were evident both on self- and foreign-assessment.

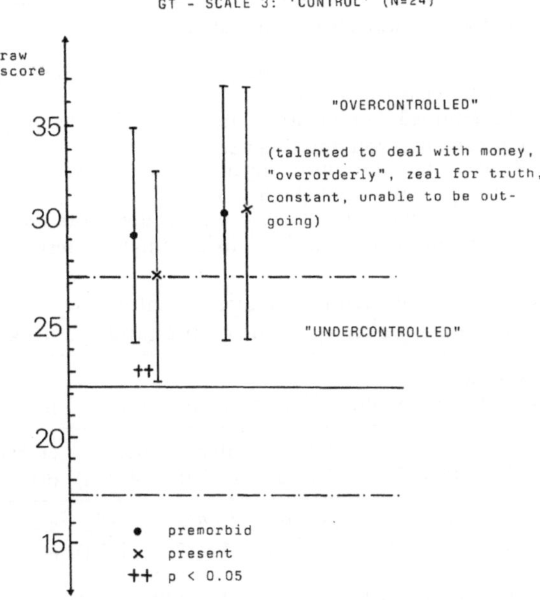

Fig. 1

Table 1. Percent of patients and controls scoring as normal, mildly or severly depressed on the GDS

Score	Controls (N = 17)	Patients (N = 33)
≤ 10 (normal)	88.2% (N = 15)	31.6% (N = 12)
11–15 (mildly depressed)	5.9% (N = 1)	31.6% (N = 12)
≥ 16 (severely depressed)	5.9% (N = 1)	36.8% (N = 14)

Table 2. Results of assessment of present personality of patients and controls by Cattell's 16 PF

Factor	Mean scores (± std. deviation)	
	Patients (N = 33)	Controls (N = 17)
N	5.4 (± 1.9)[a]	3.9 (± 1.6)
O	6.5 (± 1.6)[a]	4.9 (± 2.7)
Q4	5.1 (± 2.0)[b]	4.1 (± 1.5)
QII	5.1 (± 1.6)[a]	6.3 (± 2.3)

[a] $p \leq 0.01$, [b] $p = 0.015$, one way analysis of variance

Factor N: Forthright, natural, genuine (low scores) vs. Shrewd, calculating, socially alert (high scores)
Factor O: Unperturbed, confident, secure (low scores) vs. Apprehensive, self reproaching, worrying (high scores)
Factor Q4: Relaxed, tranquil, unfrustrated (low scores) vs. Tense, driven, restless, overwrought (high scores)
Factor QII: Low adjustment, skeptical, cautious (low scores) vs. high adjustment, confident, low anxiety (high scores)

Table 3. Evaluation of semi-standarized interviews into premorbid behaviour and personality (percent of probands scoring on item)

	introverted depressed	workoholic	pedantic	rigid	loner	non-smoker	"teeto-taller"
Patients (N = 33)	49% (48/50)	71.5% (50/85)	75% (74/75)	50% (42/58)	47.5% (45/50)	66.5% (61/72)	28% (27/29)
Controls (N = 17)	17.5% (11/24)	55.5% (41/70)	29.5% (24/35)	14.5% (12/17)	17,5% (11/24)	49,5% (41/58)	29.5% (24/35)

Percentages given as means of two ratings with individual ratings in brackets (for explanation see text)

In the second study with the GT very similar results were obtained and the mean scores and their standard deviations are depicted in Fig. 1 for scale 3 of this test. In contrast to our first series patients and their relatives/companions this time went through a retrospective as well as the original version of the GT. While there was again a marked deviation from normal controls towards a rigid and overcontrolled type of personality in all modes of assessment there was a significant difference in the patients' self-assessment of their previous and present personality in that they appeared less rigid in their present as compared to their premorbid state (see Fig. 1). In the FPI patients' scores showed overall less differences from the standard scores of a control population but there were remarkable trends towards a difference in scales 3 (depression), 4 (arousal), 7 (dominance) and 8 (inhibition), where Parkinsonians appeared more depressed, tranquil, less dominant and more tense both in their actual and previous personalities.

In the third set of studies there were no statistically significant differences between patients and controls in the WAIS or MMS scores, but patients were significantly more depressed on the GDS (Table 1). Statistically significant differences between patients and normal controls were also evident in several of the primary and second-order personality factors of Cattell's 16 PF and they are summarized in Table 2. Overall patients appeared more socially alert, apprehensive, worrying, tense, anxious and introverted.

Evaluation of the interview material relating to the patients' and healthy controls' premorbid habits and character yielded the results summarized in Table 3. Both raters found a considerably greater percentage of patients to score on the items of "introverted/depressed", "workoholic", "pedantic", "rigid", "loner" and "non-smoker".

Discussion

In three different studies with several methodological approaches a total of 85 patients with idiopathic Parkinson's disease have shown a consistent premorbid personality pattern characterized by features of depression, introversion, rigidity and inflexibility. There is a remarkable accordance between these findings and those reported in the literature since the beginning of the century (Todes and Lees, 1985). Similar observations have also been made in a recent twin study of Parkinson's disease where in a mutual premorbid personality assessment between twins the parkinsonians generally appeared less outgoing and more introverted and depressed than their healthy co-twins (Ward et al., 1983). This consistence among the findings of a number of reports lends support to their possible significance despite the reservations one must have towards retrospective assessment of personality features. There is also evidence that the well documented tendency of Parkinsonians to have been premorbid non-smokers does not reflect an unknown protective factor against the disease associated with cigarette smoking but may rather be an expression of the patients' premorbid character (Poewe et al., 1983; Golbe et al., 1986).

It seems possible that the reported peculiaritis in the premorbid character of patients are linked to certain psychological abnormalities found in early and untreated stages of the disease. These include difficulties of conceptual shift as expressed in a significantly greater tendency towards perseverative errors in the Wisconsin Card Sorting Test when compared to age-matched controls (Lees and Smith, 1983) as well as slowing of mental processing time similar to that of depressives in a digit symbol substitution task (Rogers et al., 1987). Such psychometric findings have been linked to the degeneration of dopaminergic meso-cortico-limbic projections in Parkinson's disease (Agid et al., 1984), so that a speculative interpretation of premorbid personality changes in this illness could view them as an early manifestation of dopaminergic dysfunction.

Further studies of the premorbid personality changes in Parkinson's disease are clearly needed and should include control groups of other chronically disabling diseases to test the specifity of the postulated parkinsonian personality. It might also be worthwhile to examine introverted and withdrawn depressive patients by positron emission tomography to clarify whether such psychic states are associated with functional alterations of brain dopaminergic systems.

Acknowledgement

The authors are grateful to G. W. Kemmler, Department of Biostatistics, University of Innsbruck, for valuable assistance with statistical evaluation of test results.

References

Agid Y, Ruberg M, Dubois B, Javoy-Agid F (1984) Biochemical substrates of mental disturbances in Parkinson's disease. Adv Neurol 40: 211–218

Cohen-Booth G (1935) Paralysis Agitans. Entstehungsbedingungen und Beeinflussungsmöglichkeiten. Nervenarzt 8: 69–83

Golbe LI, Cody RA, Duvoisin RC (1986) Smoking and Parkinson's disease. Search for a dose-response relationship. Arch Neurol 43: 774–778

Lees AJ, Smith E (1983) Cognitive deficits in the early stages of Parkinson's disease. Brain 106: 257–270

Mitscherlich M (1960) The psychic state of patients suffering from parkinsonism. Psychosom Med 1: 317–324

Poewe W, Gerstenbrand F, Ransmayr G, Plörer S (1983) Premorbid personality of Parkinson patients. J Neural Transm [Suppl] 19: 215–224

Rogers D, Lees AJ, Smith E, Trimble M, Stern GM (1987) Bradyphrenia in Parkinson's disease and psychomotor retardation in depressive illness. Brain 110: 761–776

Sands IR (1942) The type of personality susceptible to Parkinson's disease. J Mt Sin Hosp 9: 792–794

Todes CJ (1984) Idiopathic Parkinson's disease and depression: a psychosomatic view. J Neurol Neurosurg Psychiatr 47: 298–301

Todes CJ, Lees AJ (1985) The pre-morbid personality of patients with Parkinson's disease. J Neurol Neurosurg Psychiatr 48: 97–100

Ward CD, Duvoisin RC, Ince SE, Nutt JD, Eldridge R, Calne DB (1983) Parkinson's disease in 65 pairs of twins and in a set of quadruplets. Neurology 33: 815–824

Correspondence: Dr. W. Poewe, Department of Neurology, University of Innsbruck, Anichstrasse 35, A-6020 Innsbruck, Austria.

Parkinson's disease: development of dementia in aging

W. Danielczyk[1] and P. Fischer[2]

[1] Neurological Department, Geriatric Hospital Lainz, Vienna, and Ludwig Boltzmann-Institut für Altersforschung: Arbeitsgruppe Alzheimer-Demenz

[2] Neurological Institute, University of Vienna, Vienna, Austria

Summary

Mental changes in Parkinson's disease have various etiologies which are separable on clinical and neuropsychological grounds. Dementia in Parkinson's disease have been diagnosed much to frequently in the last two decades without using criteria for dementia as given in the DSM III. Reversible pharmacotoxic psychoses, depression, and isolated cognitive impairments may be misdiagnosed as dementia. At our department the mean death-age of 575 Parkinsonian patients has grown for about 12 years since 1950. Even very old patients with pure idiopathic Parkinson's disease (proven at autopsy) were not found to be demented during life. But a high percentage of Parkinsonian patients were suffering from additional cerebrovascular or -degenerative disease during the final stages of their diseases.

Introduction

An increasing number of people in industrialized countries today reaches an age which formerly had been unattainable for most individuals. In parallel, the incidence of cerebropathies associated with advanced age, above all senile dementia of Alzheimer's type (SDAT) and cerebrovascular disease and Parkinson's disease (PD), has increased. Today the majority of patients suffering from PD are in the 8th decade of their life (Maier-Hoehn, 1986). During the period 1950 to 1959 the average death age of parkinsonian patients at the Neurological Department of our Geriatric Hospital was 65.6 ± 9.4 years for women and 66.2 ± 9.0 years for men; this increased to 77.9 ± 6.9 years and 78.1 ± 6.7 years, respectively, for the period 1980 to 1985. Psychiatric complications occur by far more frequently in older parkinsonian patients than in younger ones, although also in juvenile parkinsonian patients dopamine (DA) level is reduced below 30% of the normal value. It thus appears that aging plays an important role in psychiatric complications in PD, which is mediated by intra- and extracerebral multimorbidity. In the first few years of idiopathic PD most patients are certainly not demented (Marsden, 1980), but may show "isolated cognitive impairments" (Growdon and Corkin, 1986; Lees and Smith, 1983).

The prevalence of dementia in Parkinson's disease (PD) differs enourmously between various research groups, which is partly due to different definitions of dementia. Another important discrepancy arises from varying selection criteria for PD patients, which will be demonstrated in this article regarding age and agedependent coincidental cerebropathies. A first comparison of mental changes in patients with Alzheimer's disease, multi-infarct dementia and Parkinson's disease was given by Danielczyk (1982). Recently, the prevalence of dementia in idiopathic PD was estimated as 10.9% (Mayeux et al., 1988).

Isolated cognitive impairment

The distinction between dementia on one hand and isolated cognitive impairment on the other hand does not only depend on the definition of dementia but is further complicated by particular problems in interpreting low scores in (neuro)psychological tests of PD patients. Two such problems have been extensively discussed in the literature and concern i) primary sensorimotor deficits and ii) depression. It is impossible to test patients without cotesting sensorimotor functions and it is hardly possible to differentiate between depression and bradyphrenia (Rogers et al., 1988).

Frequently reported isolated cognitive deficits concern visuo-spatial functions (Boller et al., 1984), memory functions (Mohr et al., 1987; Huber et al., 1987) and socalled "frontal lobe functions" (Lees and Smith, 1983; Bowen et al., 1975; Taylor et al., 1987). Investigation of visuo-spatial functions is not only contaminated by central sensorimotor processing but is also influenced by primary sensory alterations concerning somatosensory and especially visual functions. Studies on the retinal role of DA, for instance, suggest specific retinal dysfunction in PD patients (Bodis-Wollner and Onofrj, 1986).

The existence of memory impairments and frontal symptoms in PD patients of early stages is important for the understanding of cognitive deficits in PD, because of DAergic innervation of brain structures involved in these cognitive functions. There is DAergic input to nucleus amygdalae and hippocampal formation and frontal lobes via the ascending DAergic mesocorticolimbic pathway (Marsden, 1980). Thus, specific cognitive impairments of memory and frontal lobe function may be closely associated to pathology of the DAergic system, which is also supported by the fact, that these impairments are improved by dopaminergic therapy (Lees and Smith, 1983; Mohr et al., 1987; Bowen et al., 1975; Taylor et al., 1987).

According to DSM III the coincidence of memory disturbances and at least one other cognitive impairment allows the diagnosis of dementia, if the loss of intellectual abilities is of sufficient severity to interfere with social or occupational functioning. This again is a specific problem of diagnosing dementia in PD, because it is an uncritical interpretation of a patient's disability, if one calls a patient demented whose impaired "functioning" is obviously due to motor impairment. Thus, the problem of definition of dementia remains, even if standardized clinical criteria are used. In our opinion these isolated cognitive impairments are not responsible for impaired social functioning. PD patients at our department are mainly helpless because of motor symptoms at the time of admission (Danielczyk, 1983).

In our opinion circumscribed cognitive deficits do not represent the onset of a dementing process as an integral part of PD. They may be stable for a long time and do not progress, as it is the case in SDAT. In our ongoing prospective longitudinal study on dementia in PD there are some patients without dementia even in stage 5 of the disease, who have been showing mild and stable deficits in memory or constructional abilities for 3 years of testing and more than 10 years after the onset of PD.

Pharmacotoxic psychosis, delirium

Another specific problem of epidemiology of dementia in PD, which is frequently neglected, is the occurrence of pharmacotoxic psychoses in PD of advanced stages, which may be longlasting if medication is not changed. These reactive psychoses frequently present as confusional states with disorientation even in nondemented PD patients, causing reversible EEG-changes (Danielczyk, 1983). They are certainly not part of dementia in PD but belong to receptor mechanisms in treated PD. In our present sample of 20 PD patients, who have shown at least one reactive psychosis since their admission to our department, only 50% (10 patients) suffered also from dementia, according to DSM III, extensive psychological testing, and a Mini-Mental State score of less than 24.

Neuropathological correlations

The theoretically most important question concerning PD and dementia is, which correlations exist between cognitive disturbances on one hand and neuropathological changes on the other hand. Are cognitive changes in PD due to pathology of the DAergic system? But not only DA deficiency has been described in typical idiopathic PD. The other monoaminergic ascending systems are also regularly affected and this may influence cognition. Cognitive impairments due to noradrenergic deficits found very little attention in the discussion of dementia in PD, but would also explain temporary improvement of impairments after L-dopa therapy. Neuropathological changes in the raphe nuclei and in locus coeruleus may also be associated with clinical sign of depression or bradyphrenia, respectively. Besides monaminergic pathology recent reports suggest lesions of the nucleus basalis Meynert to be correlated with severity of dementia in PD (Jellinger, 1986). But in our opinion these multisystem involvement in idiopathic PD (excluding diffuse Lewy body disease, progressive supranuclear palsy, nigrostriatal degeneration, ...) do not account for severe dementia.

Dementia and aging

It is evident to the clinician that some PD patients develop severe dementia with rapid progress of cognitive deterioration. All these patients are old ones, leading to the conclusion that severe dementia in PD is a problem of the old patient and has little to do with subcortical degenerative changes of monoaminergic systems. (The association of primary degeneration of nucleus basalis Meynert neurons without cortical Alzheimer pathology and severe dementia needs further examination.)

Normally the number of Alzheimer type lesions and the risk of ischaemic cerebral lesions increase with age as does age-dependent DA-deficiency (Hornykiewicz, 1985). In about 30% of old patients Lewy-bodies in the substantia nigra and locus coeruleus have been described, which is about 15 times higher than is the prevalence of PD in this old population. In correlation to these subliminal neuropathological changes the incidence of clinically diagnosed SDAT, multiinfarct dementia (MID), and PD increases with age (Rajput et al., 1984).

Thus, one can expect to find an age-associated increase of dementia in PD, which stems from two sources. Firstly, there is simple addition of PD and DAT or MID, respectively. This would explain, that up to 15% of old PD patients are demented, even without assuming an elevated incidence of DAT in PD (Quinn et al., 1986).

Secondly, there ist the possibility of functional interaction between age-dependent DA-deficiency (unable to cause parkinsonian symptoms alone) and cerebral polypathy (also unable to cause parkinsonian symptoms alone, but causing cognitive symptoms) (Quinn et al., 1986). This cerebral polypathy includes neurofibrillary changes of neurons in substantia nigra, locus coeruleus, hippocampal areas and cortex and/or vascular changes in the latter regions or, more frequently, in the basal ganglia. (The importance of white matter lucencies for dementia in aged patients, increasingly detected by MRI, needs further examination.)

That implies, that there is a certain proportion of old patients, diagnosed as "cerebral disease and Parkinson's syndrome" in the aged, which will not show massive degenerative pathology of the substantia nigra. The expected discrepancy between neuropathologically defined PD and clinically found PD, which increases with age, shall be demonstrated in detail. Its forms an important bias of investigations of dementia in advanced PD.

Clinico-pathological correlations

At our institution 29 patients died in the years 1986/87, who had the diagnosis "Parkinson's syndrome" (mean age 79.1 years). Because we only investigated inpatients from a chronic care hospital, we expected a high percentage of dementia in our PD population, which reached 55 percents. 27 of these 29 patients were autopsied. Of these 27 patients only 10 patients clinically suffered from idiopathic PD in the beginning of their illness. 17 patients showed vascular accidents, cognitive deficits or even dementia before "Parkinson's syndrome" was diagnosed. Thus, there was a large proportion of old patients with the diagnosis "Parkinson's syndrome", who suffered from another cerebropathy prior to parkinsonism.

These results were confirmed by neuropathology, which was carried out by Prof. Jellinger. 14 patients (out of 27) suffered primarily from Alzheimer's disease (n = 8; all demented) or vascular pathology (always including basal ganglia; n = 6; 2 demented) and showed only subliminal degenerative changes in the substantia nigra. Lewy-bodies were described in every patient, but cell loss did not reach severity of PD-pathology (Jellinger, 1986). These patients obviously showed parkinsonian symptoms (including response to DAergic therapy) due to interaction of subliminal pathology of the DAergic system and other cerebropathies.

9 patients showed massive parkinsonian pathology with co-incidence of cerebro-vascular pathology (n = 6; 2 demented) or Alzheimer's type pathology (n = 2; all demented) or both (1 case; demented). Only 4 patients showed pure parkinsonian pathology with loss of pigmented neurons in locus coeruleus and especially in substantia nigra. These 4 patients were not demented according to clinical exercise (DSM III) and Mini-Mental State examination. One female patient (79 years) of these 4 patients with pure PD pathology revealed impairment of remote memory. PD was first diagnosed 15 years ago. Substantia nigra and locus coeruleus were severely devastated. In addition to these changes there was severe neuronal loss in the nucleus basalis of Meynert (with Lewy-bodies in some remaining neurons). Thus, this patient's memory deficit could be caused by severe PD pathology, which extended to the basal nucleus.

If one only regards the clinically defined subsample of pure idiopathic PD in the beginning of the disease (n = 10), 6 patients showed additional pathology (causing dementia in 4 patients). If one regards the neuropathologically defined subsample of idiopathic PD (n = 13), 3 patients clinically developed PD at the time or after other cerebropathies occurred and 6 patients obviously developed cerebral polypathy secondary to PD-pathology.

In the past the mixture of clinically and neuropathologically defined PD was one source of discrepancy regarding dementia in PD. A discrepancy which increases with increasing age of patients investigated.

Single case study

Another interesting patient (90 years old female) died some months ago and did not belong to the latter study. She took part in our prospective longitudinal neuropsychological study of dementias in PD. She was demented as defined by a Mini-Mental State score of 17. Her memory was only slightly reduced, which weakened our diagnosis of dementia. She had great difficulties in attention and concentration; she had frequent short episodes of sleeping during the day and even fell asleep during a vocabulary test. She lacked any motivational capacity. Verbal IQ was average, but there were problems in testing spatial intelligence with Raven's coloured progressive matrices. Neuropathologically this patient revealed severe degenerative pathology in DAergic and noradrenergic systems, especially. The locus coeruleus was completely devastated, as were great parts of the substantia nigra. Nevertheless there were no degenerative changes detectable in the nucleus basalis of Meynert and surprisingly nearly no Alzheimer lesions in the hippocampal formation and in cortical regions, respectively. There were also no cerebrovascular changes. That could mean, that nucleus basalis pathology is no part of the PD pathology, but represents additional pathology. It further shows that parkinsonian pathology per se may cause dementia in the very late stage of the disease. But this "dementia" is amalgamated with non-cognitive symptoms such as depression, lack of motivation, disturbance of biological rhythms et cetera, possibly linked to noradrenergic deficiency.

Conclusion

It is evident, that cerebral polypathy in the aging causes a high percentage of dementia in neuropathologically verified PD. These secondary dementias in PD of advanced stages and age may be favoured by PD-associated noradrenergic pathology and by primary loss of cholinergic neurons from the nucleus basalis Meynert. Certainly, dementia in PD is not an homogenous syndrome, but arises from different interacting cerebral pathologies. The exact frequencies and severities of cognitive symptomatology remain to be elucidated in prospective longitudinal clinicopathological studies with quantification of various cognitive functions and quantification of neuropathological-neurochemical changes.

Acknowledgements

The authors gratefully acknowledge the neuropathological diagnoses carried out by Prof. Dr. K. Jellinger. This study was supported by grants from the City of Vienna and from the Austrian Academy of Sciences.

References

Bodis-Wollner, Onofrj M (1986) The visual system in Parkinson's disease. In: Yahr MD, Bergmann KJ (eds) Advances in neurology, vol 45. Raven Press, New York, pp 1–18

Boller F, Passafiume D, Keefe NC, Rogers K, Morrow L, Kim Y (1984) Visuospatial impairment in Parkinson's disease. Role of perceptual and motor factors. Arch Neurol 41: 485–490

Bowen FP, Kamienny RS, Burns MM, Yahr MD (1975) Effects of levodopa treatment on concept formation. Neurology 25: 701–704

Danielczyk W (1982) Dementive Veränderungen bei der Langzeitbehandlung von Parkinson-Kranken – ihre Bedeutung im Vergleich zu anderen chronisch Nervenkranken. In: Fischer P-A (ed) Psychopathologie des Parkinson-Syndroms. Editiones Roche, Basel, pp 115–127

Danielczyk W (1983) Various mental behavioural disorders in Parkinson's disease, primary degenerative senile dementia, and multi infarction dementia. J Neural Transm 56: 161–176

Danielczyk W (1986) Akinetische Krisen, akinetische Endzustände und Sterbealter bei hospitalisierten Parkinson-Patienten. In: Fischer P-A (ed) Spätsyndrome der Parkinson-Krankheit. Editiones Roche, Basel, pp 89–98

Fahn S (1988) An estimate of the prevalence of dementia in idiopathic Parkinson's disease, Arch Neurol 45: 260–262

Growdon JH, Corkin S (1986) Cognitive impairments in Parkinson's disease. In: Yahr MD, Bergmann KJ (eds) Advances in neurology, vol 45. Raven Press, New York, pp 383–392

Hornykiewicz O (1985) Brain dopamine and ageing. Interdisc Topics Geront 19: 143–155

Huber SJ, Shulman HG, Paulson GW, Shuttleworth EC (1987) Fluctuations in plasma dopamine level impair memory in Parkinson's disease. Neurology 37: 1371–1375

Jellinger K (1986) Overview of morphological changes in Parkinson's disease. In: Yahr MD, Bergman KJ (eds) Advances in neurology, vol 45. Raven Press, New York, pp 1–18

Lees AJ, Smith E (1983) Cognitive deficits in the early stages of Parkinson's disease. Brain 106: 257–270

Maier-Hoehn MM (1986) Parkinson's disease: Progression and mortality. In: Yahr MD, Bergmann KJ (eds) Advances in neurology, vol 45. Raven Press, New York, pp 457–461

Marsden CD (1980) The enigma of the basal ganglia and movement. TINS 3: 284–287

Mayeux R, Stern Y, Rosenstein R, Marder K, Hauser A, Cote L, Fahn St (1988) An estimate of the prevalence of dementia in idiopathic Parkinson's disease. Arch Neurol 45: 260–262

Mohr E, Fabrini G, Ruggieri St, Fedio P, Chase ThN (1987) Cognitive concomitants of dopamine system stimulation in parkinsonian patients. J Neurol Neurosurg Psychiatr 50: 1192–1196

Quinn NP, Rossor N, Marsden CD (1986) Dementia and Parkinson's disease–pathological and neurochemical considerations. Br Med Bull 42/1: 86–90

Rajput AH, Offord KP, Beard CM, Kurland LT (1984) Epidemiology of parkinsonism: incidence, classification, and mortality. Ann Neurol 16: 278–282

Rogers D, Lees AJ, Smith E, Trimble M, Stern GM (1987) Bradyphrenia in Parkinson's disease and psychomotor retardation in depressive illness. Brain 110: 761–776

Taylor AE, Saint-Cyr JA, Lang AE (1987) Parkinson's disease–cognitive changes in relation to treatment response. Brain 110: 35–51

Correspondence: Prof. Dr. W. Danielczyk, Neurological Department, Geriatric Hospital Lainz, Versorgungsheimplatz 1, A-1130 Wien, Austria.

Psychometric assessment of early signs of dementia in special consideration of Parkinson's and Alzheimer's disease—an update

R. Steinberg and **H. Przuntek**

Department of Neurology, St. Josef University-Hospital,
Bochum, Federal Republic of Germany

Summary

The combination of subtests for assessment of higher brain functions reported in this paper, serves not only to identify the degree of dementia, but additionally is able to detect the special impairment of parkinsonians. Bradyphrenia is shown by unimpaired AW, GF, ZN and sum performance (untimed), and impaired timed subtests.—In many cases test protocol of BE and MT shows correct answers, but outside of the time limit.—In DAT subgroup—on the contrary—we see a general tendency of decreasing test performance.

When selecting practicable measurements, there appeared grouping criteria of utility.

The first group consists of scores, involved by different higher brain functions, german transformed versions available: trail making, digit symbol substitution, MMS, Benton test, and FRT.

These tests allow an economical assessment of a general demential process and a rough graduation. They are an useful possibility of "short-screening" in nonclinical area.

A deeper analysis of cognitive impairment is obtainable by the scores reported in the present paper. They are employable in the areas of clinical use and research, because of additional possibility of scoring isolated cognitive functions (cognitive speed, recent auditory memory etc.).

Concerning the measurement of every-day-life functions the presented subtests are not yet sufficient. But on the one hand they fulfil the formal assumptions of identifying dementia, and on the other hand measurement of "every-day-life" functions should not be misinterpreted as scoring of unique or peculiar capabilities of individuals.

Finally we have to be aware of the possibilities of computer-aided testing as a method of fast quantification of test scores.

For better comprehension of typical parkinsonians' patterns of impairment, we have to enlarge our test battery by a personality inventory and by reaction-time methods to assess cognitive speed.

Introduction

In patients with Huntington's disease early signs of intellectual impairment were found by means of Wechsler adult-intelligence-scale up to twenty years before onset of motor impairment (Butters et al., 1978). So it may be possible to find early signs of intellectual dysfunction or personality changes long before the classical symptoms of Parkinson's disease are apparent. Today there is widespread agreement of akinesia, rigor, tremor, depression and dementia being part of Parkinson's disease.

Dementia in Parkinson's disease has been classified as bradyphrenia as early as 1922 by Naville (Rogers et al., 1987). He described bradyphrenia, synonymous with subcortical dementia and psychic akinesia, as slowing of cognitive processing associated with impairment of concentration and apathy.

The following criteria are suggested for diagnosis of bradyphrenia (Mayeux, 1986): Persistent impairment of attention and vigilance, no evidence of impaired consciousness associated with dementia, or depression, insiduous onset with little daily fluctuations.

But as one finds dementia in Alzheimer's disease too; from the neuropsychological point of view it is necessary to differentiate dementia of Alzheimer's type (DAT) and dementia in Parkinson's disease. Our ability to differentiate DAT and PD with dementia still remains questionable.

Dementia or global cognitive impairment, is defined by the criteria of DSM III (abbrevations see appendix):

Loss of intellectual ability with social and occupational dysfunction.

Memory impairment.

One or more of the following:

1. Impaired thinking, semantics
2. Impaired judgment
3. Aphasias, apraxias, etc.
4. Personality changes

Consciousness not clouded

Either:

1. Related organic cause,
2. Non-organic mental disorders ruled out

An important observation is impairment occurring after intellectual maturity.

So one is impressed that on the one hand in Parkinson's disease with dementia, bradyphrenia is accompanied by little global cognitive impairment, and on the other hand in DAT global impairment is associated with little bradyphrenia. So one has to find a psychometric test battery for assessing the specific impaired of these higher brain functions.

Methodological requirements for selection of suitable subtests of a dementia test battery

1. Economy in assessment and evaluation

This permits repeated clinical employment and careful treatment of patients, and maintains homogeneous performance level during the whole test session.

2. Selection of standardized psychometrical measurement and subtests with normalized scores

Ensuring test quality (objectivity, validity and reliability) renders comparison of former results, and leads to availability of valid parameters, taken from previous research.

3. Selection of a catalog of relevant and isolated measurable higher brain functions

This permits assessment of low contaminated impairments, ensures test scores being measurements of well-known single factors, diminishing unknown common variance in test results, avoids excessive amounts burden of data without additional claryfying variance. Preferable is hypothesis guided measurement in "probe-technique" in contrast to widespread "screening technique" prefered when wishing an overview of all possible data. It is intended to assess a configuration of impairment characteristics, a pattern resulting from factor-analyzed subtests.

4. Examination of every-day-life functions of the patient sample

Relevant for quality of life of many patients and co-variant with every-day-life satisfaction.

5. Computer aided testing

Of increasing importance for selection of subtests is the question of possible assessment and quantitative evaluation by computers.

For the present state the last point is to be seen with regards to future development, so for the daily routine of today we need a "paper and pencil" version.

Regarding the above permises different considerations for the selections of a suitable subtest for assessment of early signs of dementia are becoming evident. Another conclusion is the association to the present state of research, we must be open to the fesh arrival of better methods.

In the present examination of dementia a psychometric test battery is used, consisting following timed and untimed subtests:

WIP (Dahl, 1968) a short form of Wechsler adult intelligence scale (WAIS), part of this are the subtests information, similarities, picture completion and block design. The KAI a short test of analytical intelligence (Lehrl et al., 1980), contains subtests of letter reading and digits and letter short memory span.

The d 2 (Brickenkamp, 1978) an attention-stress-test.

The AAT-naming-subtest (Huber et al., 1982) from Aachener Aphasie-Test. And the Beck Depression Inventory (BDI) (Beck et al., 1981) by means of which we assess depression caused memory or intellectual or motor impairment.

The test scores obtained from the subtests of this measurement lead to a pattern of information about dementia in general and bradyphrenia, in particular by the timed subtests: Picture completion, block design and letter reading.

Aspects of the pre-morbid personality of Morbus Parkinson

Parkinsonians (Todes and Lees, 1985) are generally industrious with a moralistic attitude to life. A chronic intrapersonal conflict has been remarked: the suppressed aggressiveness. The typical parkinsonian is making a religion of success by constantly striving for independence from outside interference and seeking freedom from authority, albeit within a framework of social conformity. The personality is hidden behind a social mask, where we can find strong hostile and sadistic impulses, an excessive degree of selfcontrol, repressing emotional and instinctive drives. On the one hand there was found a personality of parkinsonians, with illness of short duration, as heterogeneous–while the personality of patients with a long duration history showed a constriction, towards mental rigidity and inertia, attributed to living with the illness. On the other hand we find the opposit view, finding the most severe psychological consequences in the beginning of impairment and an adaptative effect when living with the disorder.

The compulsive neurotic character with lack of self-assertiveness finds expression in being industrious, exact and of great ambition. Their mental welfare was dependent on job achievement. M. Mitscherlich (1965) "believed that these 'behavior patterns' were laid down in early childhood at around the age of two years, when affective life was profoundly influencing the motorsystem". Here lack of self-assertivness was regarded as central to the parkinsonian patient. These individuals had a permanent readyness for motor activation, but inhibition led to motor blocking and ambivalence.

A limited range of emotional expression, mental inflexibility, moral rigidity and a perfectionistic approach to life was found in patients, with additional difficulty in coping with emotional stresses. Parkinsonians are described as having considerable introverted aggressive drive but avoiding harming others, as having no education in "savoir vivre" and inflexibility and difficulty in expressing aggression.

Psychological methods to assess such personality are the Giessen psychometric personality inventory, a psychoanalytically orientated test, and the FPI (Freiburger personality inventory), and with such means they were found to be introverted, reliable, responsible, subordinate, loyal, and with lack of flexibility. To get such data patients are instructed to fill out the questionnaire in a retrospective manner.

Ward et al. (1984) studied identical twins in whom behavioral differences were found "as early as the first decade of life between the affected and nonaffected sibling. The affected twin tended from early childhood to be less "usually the leader" and "more self controlled" and by the age of sixteen they were also more "nervous". Ten years before the development of Parkinson's disease they had become "less aggressive, quieter and less confident and light hearted", than their unaffected sibling.

Additionally it is found that parkinsonian patients tend to be non-smokers by habit.

20% of 176 parkinsonian patients had required psychiatric attention for major depressive illness before the appearance of motor disabilities. This depressive predisposition together with an over-controlled personality seems to be typical.

Early signs of impairment of higher brain functions

Two variables of demential impairment are well known and examined. One of these is called "visuospatial deficit", the other one "impairment of cognitive processing".

Two patterns of neuropsychological performance were found (el-Awar et al., 1987): Six out of twelve have "focal" abnormalities and the other 50% have generalized impairment of cognitive functions.

These patterns of learning deficits were assessed by a dementia rating scale (DRS), the Boston naming test (BNT) a test of verbal fluency, and the Wechsler memory scale.

The DRS contains subtests opf visual construction memory and language. The results show that 50% of the Parkinson's disease patients have rather focal impairment in memory functions. Two subgroups of Parkinson's disease were studied in another publication (Mortimer et al., 1982): one group with predominant tremor and relatively intact intellectual function and the other with more marked bradykinesia and neuropsychologic, impairment.

The neuropsychologic-test, administered to the patients by Mortimer et al. (1982), contained language production, comprehension, oculomotor scanning, visual spatial perception, psychomotor speed, concept set shifting, graphomotor transcoding, depression, short-term memory for prose passages, paired verbal associates and the following tests: WAIS; vocabulary; information; digit symbol; and block design subtests. The Węchsler memory scale with logical memory paired associated subtests were assessed in addition to the trail making test; Bender gestalt motor-test; memory; and a visual discrimination test. Further, the patients were examined by a finger-tapping-test and by the Zung self-rating depression scale.

Four clusters of neuropsychologic tests were identified:

Cluster 1 fund of information (WAIS vocabulary and information subtests).

Cluster 2 verbal memory (Wechsler memory scale: logical memory subtest, paired associates subtest (difficult items)–delayed and immediate performance).

Cluster 3 psychomotor speed (cancellation test, finger-tapping-test).

Cluster 4 visual spatial performance (WAIS block designs and digit symbol subtests; Bender gestalt motor test (untimed), trail making tests: forms A and B, visual discrimination test (untimed).

Perceptual motor or visuo-spatial, executive and higher order motor control tasks (Stern and Mayeux, 1987) are impaired, and choice selection time test performance cannot be improved by using advanced information as cue.

Three subtypes of Parkinson's disease were studied by Ransmayr et al. (1987) the akineto-rigid (RA type), tremor dominant, (T-type) and RAT-type (equivalance type) with equal expression of akinesia, rigidity and tremor. The neuropsychological test battery consisted of two parts. In part one most of WAIS subtest were assessed (block design subtest without time limit). Additionally there was the Benton visual retention test of visual memory and the Zung self rating scale of depression. In a second session, part 2, several untimed tests were assessed: The line cancellation test, the figure test of Rybakoff, the IQ tasks of the intelligence structure test, (IST) and a self-designed test for visual estimation of varying distances (black, straight lines). A positive correlation of akinesia with number of errors in the Bender gestalt test and of rigidity with depression score wa found. No consistent pattern of neuropsychological impairment among the three subgroups of idiopathic Parkinson's disease has been found, the authors admitted a lack of success in detecting a specific association of akineto-rigid-Parkinson's disease (AR-type) with cognitive and visuo-spatial deficits as suggested earlier.

The neuropsychological profile of primary parkinsonism among Filipinos (Lourdes, 1987) was assessed by means of WAIS, Wechsler memory scale, Boston naming test, trail making, and spontaneous written expression. 21% of patients were found to be demented as defined by DSM III. All obtained subaverage scores on comprehension, object assembly, picture arrangement, and part B of trail making. Arithmetic and block design ranked second in difficulty for this group.

Table 1. Subjects with subaverage peer scores on the WAIS [15]

Test	Demented %	Nondemented %
Verbal-tests		
comprehension	100	15
information	57	23
digit span	57	19
arithmetic	86	8
vocabulary	57	4
similarities	43	8
Performance tests		
object assembly	100	19
digit symbol	86	19
picture arrangement	100	12
block design	86	12
picture completion	57	4
	n=7	n=26

In a follow-up study one year after first test session, two of the patients displayed significant gains in neuropsychological test profile. One of the two patients seemed to be depressed at the first test session, and the second increased his scores of similarities, digit span, picture completion, block design and picture arrangement subtests, but incidentally now had difficulties in written expression.

The authors sum up the neuropsychological profile as revealing deficits in visuo spatial processing, social judgment, attention and mental tracking, recent memory and word retrievel.

Using the Sternberg paradigma, bradyphrenia or cognitive, mental slowing, has been found as a specific deficit (Wilson et al., 1980), by independent measuring of speed and accuracy of short term memory retrievel. Procedure: 1. Memory set size: The subject is given a variable number of digits to remember. 2. Probe digits, one at a time are presented. 3. Subjects had to classify a probe digit as belonging or not belonging to 1. 4. Response speed and accuracy are recorded.

An increased scanning time for older patients has been found, reflecting a slowing of the search processes, this is due to other dementias too. But no correlation has been found with "accuracy".

Possible involvement of attention processes in Parkinson's disease (Girotti et al., 1987) have been found by means of Zazzo's attention test and Wechsler adult intelligence scale. Reaction times were analyzed by a computed examination of motor reaction time (RT) and movement times (MT). Reaction time with predicted stimulus was significantly slower for parkinsonians in comparison with controls. The speed and inaccuracy aspects of Zazzo's attention test were both significantly different from controls too. A good correlation was also found among RT and between RT and MT. RT was the span of time that elapsed from the moment when the central light was switched off and when contact was lost between the subject's forefinger and the central sensor. MT was the span of time that elapsed from the loss of contact between the subject's forefinger and the central sensor to when the peripheral sensor was touched.

Additionally WAIS and movement time—with and without predicted stimulus—were significant. So data show PD patients being impaired in attention. Maybe the rate of visual information processing is slowed in PD patients and the deficit is clearly manifested when a time limit is set. On the other hand visual elaboration processes seem to be normal in PD patients when no time limit is given.

In 1978 five factor-analyzed cognitive main components were proposed (Jakobi et al., 1978): fluid and crystallized intelligence, visualisation, cognitive speed, and cognitive flexibility (with reservation).

Fluid intelligence has been defined as inductive, deductive and conclusional thinking, assessed by Figure Reasoning Test.

Crystallized intelligence, defined as enduring knowledge of words and conceptions and their availability, tested by vocabulary subtest of WAIS for example.

Visualization or visual spatial orientation, assessed by Figure Reasoning Test, Benton visual retention test, and block design of WAIS. Test scores of the latter were additionally influenced by cognitive speed. This contains cognitive productivity, quick changing of perspectives, speed of perception, and shifting aptitude.

To sum-up, it could be shown that there was impairment of cognitive functions in fluid intelligence of parkinsonians and in measurement of cognitive speed, as had been expected before, whereas crystallized intelligence and visualization were rather intact. This is in accord with a view of Parkinson's disease containing akinesia, tremor, bradyphrenia, vegetative disturbances, and additionally (Fischer and Jakobi, 1978) cerebral atrophy, dementia, and pathological EEG changes for example, each additional symptom stressing a negative course.

Psychometric differentiation of dementia in parkinsonism and Alzheimer's disease

With regard to test economy (see above) and with limitation to comparable, transcoded german normalized versions, significant differences were found (Mildworf, 1986) in several subtests; the term "significant" difference is used, if $p < 0.01$ and the term "high significant" if $p < 0.001$.

In general, neuropsychological test performance of parkinsonians is better than performance of DAT patients. In comparison with normal controls PD patients show poor performance.

Two forms of Benton tests (0'/45') recent visual memory test differentiate the three groups at a highly significant level.

Remote memory is significantly impaired only in DAT group.

Block design differentiates the three subgroups at a highly significant level.

Trail-making shows the same differences.

Naming or categorial naming differentiates DAT and dementia in PD in a highly significant fashion, and also normal controls and PD significantly as well.

Super span (recent auditory memory) and digit span are able to differentiate PD with dementia and DAT highly significantly, and PD and normal controls significantly. In the next step parkinsonians were divided into subgroups with dementia (mini mental scale <21); and without dementia (MMS >21).

None of the correlations between motor symptoms (bradykinesia, rigidity and tremor) reached statistical significance, suggesting that the motor symptoms of Parkinson's disease consist of relatively independent dimensions. A significant association was found between bradykinesia and psychomotor speed, timed and untimed visual spatial performance and spatial orientation memory (-0.34 to -0.45, $p<0.01$).

Performance on subtest of WAIS: vocabulary and information seemed to be influenced relatively little by the disease.

The visual spatial deficit found in most parkinsonian patients has a more specific relationship to the motor symptoms.

Another study (Rogers et al., 1987) found significant slowing related in the parkinsonian patients to structural brain disorder and affective impairment. The patients were assessed neuropsychologically by the national adult reading test (NART), by full scale WAIS, by Webster rating scale for Parkinson's disease, by Hamilton rating scale for primary depressive illness and a digit symbol substitution test (DST), a computer aided testing procedure.

There were significant differences between the three groups for response time on the digit symbol-test ($p<0.001$) and also for response time on the symbol-test ($p<0.001$) and movement time ($p<0.001$).

The "response time" of the digit symbol test is the time the patients needed to answer a symbol on the screen with the correct number and the "response time" of the symbol test is the time the patients needed to answer symbol and number on screen with the appropriate number; "matching time" is called the difference of these two scores indicating the cognitive processing. Movement time is the motor performance to press the number keys on the keyboard.

Response times for both tests and movement time were significantly prolonged in both parkinsonian and depressed patients compaired with the normal controls, but the differences between the two patient groups were not significant. Differences in matching time did not achieve statistical significance either. The parkinsonian patients with structural disorders, reported on CT-scan, had a significantly slower DST response time ($p<0.001$) than the patients without. In contrast to the whole parkinsonian group these nine patients had a significantly longer matching time ($p<0.001$) than the normal controls. In the 18 parkinsonian patients without structural brain disorder. DST response time was no longer significantly different from that in the normal control group. Additionally in both groups of patients slowing of response was found with no significant impairment of general intellectual function, consistent with reports of subtile cognitive changes in early untreated Parkinson's diseases. The slowing of response in cognitive tests is only a significant feature in a proportion of patients with parkinsonism, and a proportion of patients with depressive illness.

In MPTP (1-methyl-4-phenyl-1, 2, 3, 6-tetrahydropyridine) induced parkinsonism were found (Mayeux, 1986) visuo spatial and executive function impairments, as well as perceptual motor dysfunction (Stern and Langston, 1985) assessed by drawing, puzzle assembly from WAIS, line orientation, tracing, tracking of tactile tasks. The pattern of intellectual changes seen in the MPTP patients was similar to that reported in indiopathic Parkinson's disease. Performance on general measures of intellectual function, such as a modified MMS is also usually lowered in PD. Other measures, specially performance IQ of the WAIS, are also lower in patients than age-matched controls. But the abnormalities were relatively subtle, and attention, memory, digit span, calculation, and overall language test performance was comparable to controls. This study examined six MPTP induced parkinsonism patients with significantly lower school education in comparison with normal controls.

Super span and digit span are measurements for capacity of recent auditory memory. We use the KAI (Lehrl et al., 1980), containing information-psychological subtests "reading speed of letters", immediate "repeating of digits" and of "letters" in sets of increasing numbers, as a completed index of recent auditory memory, or of cognitive speed.

Remote memory is involved in the information scale of WAIS.

Naming and categorial naming are assessed by the naming subtest of AAT (Aachener aphasia test) (Huber et al., 1982).

Trail making is transcoded to ZVT (Reitan, 1958). However this score is contaminated by different functions: visuo spatial perception, visuo motor coordination, and motor execution of unknown degree.

Psychometric differences of neuropsychological performance

For exemplification of the suggested differences we show data out of our current research:

Table 2

Test	? DAT	Parkinsonism	
1* AW	94.0	101.5	(means)
1 GF	87.0	105.2	
1 BE	61.8	66.4	
1 MT	60.5	70.9	
1 sum	77.0	90.7	
1 BuL	83.7	88.1	
1 ZN	85.0	105.8	
2 GZ	86.5	80.5	
2 GZ-F	82.5	81.8	
3 AAT ben.	56.3	–	

* Parameters: 1 = IQ, sd = 15; 2 = SW, sd = 10; 3 = T, sd = 10. *AW* information; *GF* similarities; *BE* picture completion (timed); *MT* block design (timed); *sum* sum IQ (parameters of WIP); *BuL* letter reading (timed); *ZN* digit and letter repeating (KAI); *GZ* total number of signs; *GZ-FR* total number–mistakes (d 2); *AAT ben* AAT subtest naming; DAT group n = 03, mean age = 57, PD group n = 13, mean age = 65. Excluded were all patients with BDI > 15

Appendix

Repeated used abbreviations

AAT	Aachener aphasia test
AW	Allgemeinwissen, information subtest of WAIS
BDI	Beck depression inventory
BE	Bilder ergänzen, picture completion subtest of WAIS
BNT	Boston naming test
DAT	dementia of Alzheimer's type
DRS	dementia rating scale
DSM III	diagnostic and statistical manual of mental disorders, American Psychiatric Association
DST	digit symbol substitution test, subtest of WAIS
d 2	Aufmerksamkeits-Belastungs-Test, attention-stress test
FRT	figure reasoning test
GF	Gemeinsamkeiten finden, similarities subtest of WAIS
KAI	Kurztest für Analytische Intelligenz
MMS	mini mental scale
MT	Mosaiktest, block design subtest of WAIS
NART	national adult reading test
PD	Parkinson's disease
WAIS	Wechsler adult intelligence scale
WIP	short form of HAWIE, german transcoding of WAIS
WMS	Wechsler memory scale
ZVT	Zahlen-Verbindungs-Test, trail making test

References

Beck AT, Mendelson M, Mock J, Erbaugh J (1981) Beck depression inventory, BDI

Brickenkamp R (1978) Test d 2. Verlag f Psychologie Dr CJ Hogrefe, Göttingen Toronto Zürich

Butters N, Sax D, Montgomery K, Tarlow S (1978) Comparison of the neuropsychological deficits associated with early and advanced Huntington's disease. Arch Neurol 35: 585–589

Dahl G (1968) WIP-1; Reduzierter Wechsler-Intelligenztest. Anton Hain-Verlag, Meisenheim am Glan

El-Awar M, Becker JT, Hammond KM, Nebes RD, Boller F (1987) Learning deficits in Parkinson's disease. Comparison with Alzheimer's disease and normal aging. Arch Neurol 44/2: 180–184

Fischer PA, Jakobi P (1978) Diagnostik hirnorganischer Störungen. In: Handbuch der klin Psychologie. Hogrefe, Göttingen, S 1756–1782

Girotti F, Grassi MP, Carella F, Soliveri P, Musicco M, Lamperti E, Caraceni T (1987) Possible involvement of attention processes in Parkinson's disease. Adv Neurol 45: 425–429

Huber W, Poeck K, Weniger D, Willmes K (1982) Der Aachener Aphasietest (AAT). Verlag Hogrefe, Göttingen

Jakobi P, Fischer P-A, Schneider E (1978) Kognitive Störungen von Parkinson-Patienten. Aus: Frankfurter Symposium 10/11. Schattauer, Stuttgart, S 219–230

Lehrl S, Gallwitz A, Blaha L (1980) KAI; Kurztest für allgemeine Intelligenz. VLESS Verlagsgesellschaft mbH, Vaterstetten

Lourdes K (1987) Neuropsychological profile of primary parkinsonism among filipinos. Adv Neurol 45: 421

Mayeux R (1986) Mental state. In: Koller WC (ed) Handbook of Parkinson's disease. Marcel Dekker, New York Basel, p 127

Mildworf B, Globus M, Melamed E (1986) Patterns of cognitive impairment in patients with DAT and Parkinson's disease. In: Fisher A, Hanin I, Lackman C (eds) Adv in behav biol, vol 29, p 135

Mortimer JA, Pirozzolo FJ, Hansch FC, Webster DD (1982) Relationship of motor symptoms to intellectual deficits in Parkinson's disease. Neurology 32: 133–137

Ransmayr G, Poewe W, Plörer S, Birbamer G, Gerstenbrand F (1987) Psychometric findings in clinical subtypes of Parkinson's disease Adv Neurol 45: 409–411

Reitan RM (1958) Validity of the trail making test as an indicator of organic brain damage. Percept Mot Skills 8: 271–276

Rogers D, Lees AJ, Smith E, Trimble M, Stern G (1987) Bradyphrenia in Parkinson's disease and psychomotor retardation in depressive illness. An experimental study. Brain 110/3: 761–776

Stern Y, Langston W (1985) Intellectual changes in patients with MPTP induced parkinsonism. Neurology 35: 1506–1509

Stern Y, Mayeux E (1987) Intellectual impairment in Parkinson's disease. Adv Neurol 45: 405

Todes CJ, Lees AJ (1985) The premorbid personality of patients with Parkinson's disease. J Neurol Neurosurg Psychiat 48: 97–100

Ward CD, Duvoisin RC, Ince SE, Nutt JD, Eldridge R, Calne DB, Dambrosia J (1984) Parkinson's disease in twins. In: Hassler RG, Christ JF (eds) Adv neurol, vol 40. Raven Press, New York, pp 341–344

Wilson RS, Kaszniąk AW, Klawans HL, Garron DC (1980) High speed memory scanning in parkinsonism. Cortex 6: 67–72

Correspondence: Dipl.-Psych. R. Steinberg, Department of Neurology, St. Josef Hospital, University of Bochum, Gudrunstrasse 56, D-4630 Bochum 1, Federal Republic of Germany.

Sensory and musculo-skeletal dysfunction in Parkinson's disease—premonitory and permanent

M. B. Streifler

Section of Neurology, Sackler Medical School, Tel Aviv University, Tel Aviv, Israel

Summary

In Parkinson's disease the impressive motor symptomatology has side-tracked attention from other important constituents, like those pertaining to psycho-mental, autonomous and sensory systems. Pain and dysesthesias are present in about 60% of parkinsonians and predate the overt disease in about 20%.

Predominantly of neurogenic, "primary" origin they express a disorderly processing along central sensory pathways, a defective activity of dopaminergic and other transmitter-mediator systems.

Changes affecting musculo-skelettal structures contribute a somatic, "secondary" contingent of sensory discomfort and may determine the outward appearance of the patient.

*

The very conspicuous motor disabilities and signs in Parkinson's disease (PD) have always overshadowed its other important manifestations, like the intellectual, autonomous and the sensory ones. Although clearly alluded to in the old texts (Parkinson, 1817; Charcot, 1877) this latter issue has attracted a renewed and more profound interest only in the last twenty-five years. Thus Matsumoto et al. (1963) reported on "Somato-sensory status of parkinsonian patients before and after chemo-thalamotomy". Of the, not too numerous publications dealing with this theme to be found in the vast and quickly burgeoning literature on PD, the recent papers of Snider et al. (1976), Koller (1984), and Goetz et al. (1986) deserve special interest.

Pain and disagreeable sensations, usually underreported or only vaguely mentioned, have been established in about 60% of

PD-patients and are believed to usher in the characteristic disease symptomatology in up to 20%.

The sensory symptoms can grossly be classified into aches and pains and those of numbness and paresthesias, like formication, burning and coldness. Often they are hardly definable. They may be mild to tormenting, intermittent to constant and change with disease progression and the efficacy of treatment. They were more frequently encountered in post-encephalitic than in idiopathic parkinsonism. There is a controversy about their relative prevalence in the tremor-dominant versus the rigor-akinesia variety, but in essential tremor they have not been found more frequently than in normal controls (Snider et al., 1976).

Factors like rigidity, brady-akinesia and dystonia may play aggravating and causal roles. Patients may complain simultaneously and/or consecutively of sensory discomfort of different types.

The early, pre-motoric appearance of sensory complaints oftentimes is the cause for evaluations for suspected clinical situations like joint, bone and muscle disease, spondylopathic disorders, angina pectoris, phlebothrombosis, etc.

According to Snider et al. (1976), sensory complaints can be of "primary", neurogenic or of "secondary" origin, stemming from somatic, mainly muscular and/or skeletal structures, outside the nervous system. Often one is dealing with their combined effects. Primary sensory symptoms appear independently from the motoric elements in PD.

The most common primary complaint is pain, mostly spasmodic, vaguely delineated and in bouts, with preference for "off" periods.

Primary sensory symptoms can be the expression of dysfunction of the central nervous system, where pathological changes are known to exist, as well as of the peripheral sensory nerves, in which no specific pathology has been demonstrated in idiopathic parkinsonism. In most PD-patients with sensory complaints no abnormal "peripheral" sensory findings can be elicited on examination. Some 15% of cases may, however, have a slight, inconstant decrease in superficial touch and pain sensitivity and of vibration sense in the legs. Such findings are preferentially elicitable on the motorwise more affected side.

The study of cortical sensory functions in PD-patients with primary sensory symptoms has, by clinical testing yielded deficitary findings like delayed tactile perception (Dinnekstein et al., 1964), impaired visuospatial orientation and distortion of body image (Bowen et al., 1972) and there is mounting evidence for an altered sensory processing in PD. The nigro-striatal dopamine system seems to be responsive to certain sensory informations, which, in turn, guide or modulate basal ganglia motor programming (see also later on).

The well established abnormalities of the autonomic nervous system, which can influence sensory thresholds (Appenzeller and Gross, 1971) may also play a role in sensory functions.

The customary anti-parkinsonian drugs seem to alleviate sensory discomfort but some patients experience paresthesias when on anticholinergic therapy. Minor tranquilizers and antidepressants may be helpful and so also narcotics.

Secondary sensory symptoms are of somatic origin, due to localized musculo-skeletal disorders, like arthritis, bursitis, spondylopathy, dystonia, immobility, muscular tension and cramps. They may also be caused by the impact on the peripheral nerves of pressure, of impaired arterial perfusion and abnormal venous and lymphatic drainage, conditions leading to edematous swelling.

Goetz et al. (1986) studied pain and its association with motor fluctuations in 95, chronically treated out-patients, of which 46% reported on pain due to their PD. The pain categories listed were musculo-skeletal, joint pain, dystonia, radicular/neuritic pain and akathisia (in one patient). The predominant parkinsonian symptom associated with pain was akinesia. Of the 12 patients with tremor-dominant PD only three had pain—a nonsignificant association. Painful dystonias were reported in 28%, usually in the feet. The dystonias, not included in the studies of Snider et al. (1976), occurred early in the morning or late at night, hours away from the last medication. Radicular or neuritic pain was found in 14% and diffuse, generalized pain was elicited from the one patient with akathisia.

Joint pains, in shoulders, hips, knees and ankles, occurred in 14%—similar to the percentage of radicular/neuritic pain.

Syndromes suggestive of vascular, bone or thalamic pain were not encountered by these authors. In most patients (89%) pain was maximal when PD was most severe. But while in musculo-skeletal and in dystonia pains (66% of patients) pain severity and PD-disability went parallel, radicular pains were reported present equally both at maximal and minimal PD-disability. Patients with pains of all sorts were statistically younger, but objectively not more disabled than patients without pain.

In spite of some methodological differences and the slightly incongruous findings, this study is comparable with that of Snider et al. (1976). This group of authors, however, hesitates to adopt Sniders conclusion "that most pain in PD relates to centrally mediated dysfunction in sensory pathways".

These researchers (Goetz et al., 1986) had also reported a positive correlation between pain, depression and sleep disruption in PD-patients and stress the pain-alleviating properties of antidepressants in PD.

The pathogenesis of sensory dysfunction, especially of the primary variety is still little understood. No specific structural pathology of central sensory pathways has been demonstrated in PD. It can, however, be speculated that the substantia nigra and other catecholaminergic nuclei with their rich connections to basal ganglia, thalamic nuclei and their projections to sensory cortex can play important roles in the processing of sensory information.

Electrophysiologic studies of striatal influences on sensory activity indicate an inhibitory role (Krauthamer et al., 1967). The presence of spontaneous sensory symptoms in parkinsonism may therefore be viewed as due to a "release" of sensory functioning from extrapyramidal influences (Snider et al., 1976).

Also infra-tentorial structures like sympathetic ganglia, dorsal roots and columns, periaqueductal reticular pathways and brain stem sensory nuclei may function abnormally and cause or contribute to the sensory perturbations suffered by PD-patients.

As for the neurotransmitters involved in the overall clinical pathology of Parkinson's disease it should be kept in mind, that catecholamine projections are widely represented in the central nervous system, including areas connected with sensation (Snider et al., 1976). Also opioid receptors and enkephalines, which are believed to modulate the activity of dopamine, have been found in the striatum as well as in the dorsal areas of the spinal cord. The degeneration of such neurons, esp. of the striatum, in PD will lead to an impaired opioid-dopamine inter-action which can induce sensory disturbances (Nutt and Carter, 1984).

Disorderly musculo-skeletal manifestations in PD are mainly determined by rigidity, the hypertonic intermittent or continuous tenseness of the muscles. Most of the subjective impediments in daily living activities and alterations in outer appearance of the PD-patient are due to rigidity and its effects upon the functioning of joints, limbs and the supporting skeletal structures of the body.

Their initially subtle signs and symptoms appear early in the clinical evolution of the disease, usually prior to the disclosure of its real nature. Subjective complaints such as tightness, tiredness, exaggerated fatiguability, "getting slow and clumsy" are soon joined by more objective ones, like difficulties in turning in bed, a faltering and indistinct voice and a deteriorating handwriting.

Clinical examination at that stage can reveal stooping, pro-retro-pulsion, a leg-drawing or festinating gait, reduction up to loss of associated movements, cog-wheeling and tremors, mostly of the resting type.

Cog-wheeling/rigidity can be brought out or enhanced already early in the disease by the Matsumoto procedure (Matsumoto et al., 1963): While the patient moves repetitively one hand the passive rotation or flexion-extension of the contralateral wrist by the examiner will elicit rigidity and/or cog-wheeling.

Scoliotic deviation of the spine is a skeletal deformity frequently observed in PD-patients, with a thoraco-lumbar curve convex to the side of the body which is affected earlier and more severely. This even led to the question whether degeneration of the nigro-dopaminergic system can cause scoliosis. Based on a recent x-ray survey of 110 mildly disabled parkinsonians, italian authors (Della Sala, 1987) concluded that scoliosis could not be verified as frequently as clinically claimed and that asymmetrical symptom-atology is not a risk factor for scoliosis or spine decompensation, since in unilaterally affected patients the appearance of spine decompensation contralaterally to the more affected side amounted to an only low degree of statistical significance.

The neurophysiological mechanism underlying rigidity still evades explanation. Recordings from muscle spindle afferents in rigid PD, reveal an activity not present in normal subjects at rest. The hitherto assumed primary hyperactivity of both alpha and gamma motor neurons is rejected by Burke (Burke, 1986), who found no convincing evidence of a primary defect in fusimotor function. He pointed out that PD-rigidity is sustained by segmental afferent inputs, but not through conventional spinal proprioceptive reflexes. It has been observed that rigidity correlates with increased amplitude of the long-latency, transcerebral responses to sudden stretch. One may be dealing with a central "release" phenomenon with regard to long-latency stretch reflexes, which are mediated by motor pathways that do not traverse the basal ganglia.

Dystonia, an early, sometimes initial feature of PD lists among the causes of primary pain. Goetz et al. (1986) reported painful dystonias in 28% of 95 PD-patients, studied for pain. It seems established that antiparkinsonian medication, especially levo-dopa administration, is capable of increasing the incidence of dystonia and its painful characteristics.

A peculiar, dystonically appearing, but sometimes permanent skeletal deformity, observed in progressive PD, is the "striatal hand", in which here the adducted fingers are dystonically kept flexed at various angles from the meta-carpo-phalangeal joints and downwards.

Another condition connected with dysesthesia, the "restless legs syndrome" (Ekbom, 1960) should be mentioned here, because parkinsonism has been found in these rare patients more frequently than in the general population. The dysesthesiae consist in a "crawling sensation beneath the skin" and can usually be relieved by walking.

Taking up again the issue of musculo-skeletal dysfunction in PD with its important role in the causation of postural deformities, the curtailment of daily living activities and the almost relentless subjective discomfort, it is very intriguing that PD-patients appear to be so little aware and so little articulate about their postural deficit. This, to some part, might be ascribed to a dysfunction in central sensory processing and to an impairment of body-scheme functions.

References

Appenzeller O, Gross J (1971) Autonomic deficits in Parkinson's syndrome. Arch Neurol 24: 50–57

Bowen F, Hoehn M, Yahr DM (1972) Cerebral dominance in patients with parkinsonism. Neurology 22: 32–39

Burke D (1986) Rigidity/dystonia. Triangle 25 [Suppl 1]: 13 (and personal communication)

Charcot JM (1877) Lectures on diseases of the nervous system, vol 1 (translated by Sigerson G). The New Sydenham Society, London, p 137 (cited ex 10)

Della Sala S, Franchignoni FP, Grioni G, Mazzini L (1987) Scoliosis in Parkinson's disease (PD): a reappraisal. In: Proceedings of the international symposium on parkinsonism and aging, Milan, p 41

Dinnekstein A, Lowenthal M, Blake G (1964) Tactile delay in parkinsonism. J Nerv Ment Dis 139: 521–524

Ekbom K (1960) Restless legs syndrome. Neurology 10: 868–873

Goetz CG, Tanner CM, Wilson RS, Garron DC (1986) Pain in Parkinson's disease. Movement Disorders 1: 45–49

Koller WC (1984) Sensory symptoms in Parkinson's disease. Neurology 34: 957–959

Krauthamer G, Felty P, Albe-Fessard D (1967) Neurons of the medial diencephalon II. Excitation of central origin. J Neurophysiol 30: 81–97

Matsumoto K, Rossomann F, Lin TH, Cooper IS (1963) Studies on induced exacerbation of parkinsonism rigidity. The effect of contralateral voluntary activity. J Neurol Neurosurg Psychiatr 26: 27–32

Nutt JG, Carter JH (1984) Sensory symptoms in parkinsonism related to central dopaminergic function. Lancet ii: 456–457

Parkinson J (1817) An Essay on the shaking palsy. Sherwood, Neely and Jones, London, pp 47–48

Proctor F, Riklan M, Cooper IS, et al (1963) Somato-sensory status of parkinsonian patients before and after chemothalamotomy. Neurology 13: 906–912

Snider SR, Fahn S, Isgreen WP, Cote LJ (1976) Primary sensory symptoms in parkinsonism. Neurology 26: 423–429

Correspondence: Dr. M. B. Streifler, 53, David Hamelekh Blvd., 64237 Tel Aviv, Israel.

Autonomic nervous system screening in patients with early Parkinson's disease

A. D. Korczyn

Sackler Faculty of Medicine, Tel Aviv University, Tel Aviv, Israel

Summary

Clinical autonomic dysfunction is rarely a problem in the initial presentation of patients with Parkinson's disease, with the notable exception of constipation. Rather, the existence of definite and conspicuous manifestations should be an important indication that the underlying disease producing the extrapyramidal features is multi-system atrophy rather than Parkinson's disease. The frequent occurrence of constipation may be suggestive of Parkinson's disease, but a more detailed analysis as to the underlying mechanisms and characteristics which may possibly distinguish this from the common constipation of old age remain to be established. Furthermore, it will be of importance to look systematically for Lewy bodies in autonomic ganglia since these peripheral structures may provide tissue diagnosis, so much desired in this disease.

Introduction

Of the various manifestations of Parkinson's disease, a disproportionately small amount of attention was given to those involving the autonomic nervous system. This is unfortunate because of the importance to a better understanding of the disease and since these are not only common problems but are frequently amenable to therapy. Patients with Parkinson's disease frequently complain of problems of salivation, micturition and gastrointestinal function, and may also have defective cardiovascular control and temperature regulation (Appenzeller et al., 1971; Rajput and Rozdilsky, 1976; Korczyn, 1987). Seborrhea, a classical feature of Parkinsonism, is also a vegetative symptom. Because Parkinson's disease primarily affects the brain, these symptoms are usually ascribed to defective central regulation. However, there is some evidence of changes in sympathetic ganglia and in the adrenal glands in this disease (Den Hartog Jager and Bethlem, 1960; Den Hartog Jager, 1970). This may be a hint that peripheral, rather than central abnormalities, may cause impairment of the autonomic nervous system.

It is of course important to differentiate manifestations which are an integral part of the clinical spectrum of the disease from those which are iatrogenic (usually side-effects of antiparkinsonian or other drugs).

As will be evident from the following review, autonomic manifestations in patients with Parkinson's disease are rarely specific. Changes in various functions, such as of the gastro-intestinal tract, are common in old age and bear similar characteristics in normal aging and in Parkinson's disease, and data on their respective frequencies are not available. It is a formidable question whether in a particular patient the constipation is a manifestation of age, disease, therapy or a combination of these factors. Another problem which may occur, particularly regarding patients in whom the autonomic problems predominate, is whether the disturbances are part of the Shy Drager syndrome. These questions are usually not difficult to answer but may require detailed understanding of not only the pathophysiology but also pharmacological issues.

Gastrointestinal function

Various disturbances of the gastrointestinal function occur in Parkinson's disease. Sialorrhea is a common late manifestation of the disease, although it is much less obvious (both in frequency and severity) in treated patients. Most probably sialorrhea arises from poverty of automatic swallowing, and should therefore be regarded as being due to hypokinesia rather than a purely autonomic manifestation. The rate of saliva production in Parkinson's disease is not increased, and the drooling is not due to excessive salivation. This symptom responds satisfactorily to levodopa and its derivatives and is relatively non-responsive to anticholinergics. In fact, administration of antimuscarinics may further impair swallowing by increasing the viscosity of the saliva.

A more advanced manifestation of the same problem causes clinically significant swallowing difficulties, particularly evident for solid foods. Frequently associated chewing problems coexist, with a lower amplitude of chewing movements.

The difficulties in swallowing saliva and solid foods, if not sufficiently responsive to levodopa derivatives, may be relieved by drugs liquifying the saliva. A chance observation by one of our patients who had this problem and responded well to bromhexine has led us to the use of the drug in additional patients with satisfying results.

Probably the most common gastrointestinal manifestation of Parkinson's disease is constipation. This phenomenon occurs very commonly (Table 1) and our data on the frequency among spouses is considerably lower. Patients with Parkinson's disease in many cases report that constipation predated the first motor manifestations (46%), while both startet concomitantly in a few (19%). In the remainder (35%), where constipation has started after Parkinson's disease was diagnosed, the question of relationship to drugs can be addressed. Our impression is that several antiparkinsonian drugs are incriminated by the patients as exacerbating pre-existing constipation, but they rather infrequently cause constipation.

The constipation in Parkinson's disease will usually respond to various dietary and drug manipulations, including mild laxatives, but several patients depend on enemas. Occasional patient may develop intestinal pseudo-obstruction, later on leading to paralytic or adynamic ileus. Several patients known to us have undergone explorative laparatory and sigmoidostomy for the treatment of their ileus. In view of a previous report (Kreek et al., 1983) we have given naloxone, 1–15 mg i.v. to two patients with severe constipation complicating Parkinson's disease on four different occasions. There was no response to any of these injections.

In the case with the most severe constipation in our service, leading to intestinal pseudo-obstruction, the only practical solution

Table 1. Constipation in patients with Parkinson's disease and controls

	Males	Females	Total
Patients with Parkinson's disease	106	72	178
Constipation preceding parkinsonian symptomatology	26 (43%) (73 ± 7 y)	24 (51%) (73 ± 7 y)	50 (46%) (73 ± 6 y)
Onset coinciding	11 (18%) (74 ± 6 y)	10 (21%) (69 ± 9 y)	21 (19%) (71 ± 7 y)
Constipation starting after diagnosis of Parkinson's disease	25 (41%) (72 ± 6 y)	13 (28%) (72 ± 8 y)	38 (35%) (72 ± 7 y)
Total with constipation	62 (58%)	47 (65%)	109 (61%)
Controls*	6/37 (16%)	27/81 (33%)	33/118 (28%)

* Controls were spouses who were not diagnosed as having Parkinson's disease. Altogether, 102 spuses of 138 successive patients with Parkinson's disease could be interviewed

The age of each group, mean ± SD, is given in brackets

was the use of cholinomimetic drugs. The patient had no response to oral prostigmine or betanechol. Intravenous betanechol stimulated intestinal motility, audible and even visible through the abdominal wall. However, this had not resulted in fecal evacuation. It therefore became clear that the intestinal pseudo-obstruction was the result of impaired motility in the most distal part of the intestinal tract. We have therefore tried giving the patient betachechol by enema, with surprisingly productive results. This patient required the administration of 20—50 mg betanechol by enema (dissolved in a small amount of saline) repeatedly for several years until his death. Post mortem examination revealed diminution in the number of fibers and loss of ganglion cells within the muscularis mucosa with picnotic changes in the remaining neurons, particularly in the sigmoid colon and cecum (Dr. B. Ilie). Whether these changes represent an extreme form of a common process or whether they are due to a separate coexistent disorder, remains to be investigated.

Weight loss frequently accompanies Parkinson's disease, although the loss is relatively mild (Vardi et al., 1976). Several factors may be thought to contribute to this phenomenon, including the swallowing impairment noted above as well as the anorexia or nausea from dopaminergic therapy. The negative nitrogen balance resulting from hypokinesia and inactivity may also contribute. Interestingly, in a recent study (Rabey et al., 1986) we have found that patients with Parkinson's disease treated with lisuride gain weight. This observation is consistent with animal data (Horowski and Wachtel, 1978).

Cardiovascular system

There is a clinical impression that patients with Parkinson's disease are less likely than the general population to develop cardiovascular diseases e.g. hypertension and coronary artery disease. Whether this impression is correct should be concluded from more detailed studies, e.g. case-control studies.

Cardiac arrhythmias are rare in patients with Parkinson's disease, but may occur as side-effects of levodopa therapy. The use of peripheral decarboxylase inhibitors probably reduces the frequency of such episodes.

Livedo reticularis, or cutis marmorata, frequently occurs among patients treated with amantadine and is particularly common in women. The mechanism underlying these phenomena is unclear but may be related to venular dilatation. Although this phenomenon is innocuous it is important to recognize since it may otherwise lead to unnecessary investigations.

Edema, usually dependent, is another common manifestation. It almost always is absent or less conspicuous when the patient wakes up in the morning and develops during the day. It is only observable in advanced stages of the disease and probably occurs in patients who are hypokinetic and sedentary from the loss of the propelling force on venous flow by muscular massage. It is likely that amantadine may increase the edema, particularly if high doses are being used. On the other hand, ergot alkaloids like lisuride, have an effect on venomotor tone and thus may decrease venous stasis and edema formation.

The most incapacitating cardiovascular symptom is undoubtedly orthostatic hypotension. The hypotension in Parkinson's disease is particularly common following meals (Micieli et al., 1987), although the patients may not be aware of the association with food intake. The occurrence of orthostatic hypotension and post-prandial hypotension may suggest that the patient may not be suffering from paralysis agitans but from progressive autonomic failure. However, in Parkinson's disease other cardinal symptoms of the Shy Drager syndrome are absent, e.g. pupillary response abnormalities, iris atrophy and pharmacologic evidence for supersensitivity to direct adrenergic agonists. This may suggest that the problem is central rather than peripheral. In fact, recent data demonstrate normal blood pressure responses to tilt and we have shown that these patients have normal responses to the cold pressor test (Korczyn, 1987).

An important factor in the evaluation of orthostatic hypotension in patients with Parkinson's disease is the possible contribution of drugs. Several antiparkinsonian agents may cause or exacerbate orthostatic hypotension including levodopa. Levodopa may act through a mechanism similar to that of the antihypertensive drug alpha methyl dopa, i.e. centrally, although the peripheral conversion to dopamine may also contribute to the hypotension. This component will be eliminated by the use of decarboxylase inhibitors.

Direct acting dopamine agonists like bromocriptine or lisuride are particularly potent in precipitating hypotension by causing peripheral vasodilatation. Activation of dopamine receptors located on sympathetic terminals powerfully inhibits stimulation-induced norepinephrine release. This action is abolished by treatment with the D-2 receptor antagonist domperidone.

Pupillary changes

Few reports are available mentioning pupillary abnormalities in Parkinson's disease, and these are all drug-related. However, even if not posing a clinical problem, investigation of pupillary changes may be important since the pupillary size and its responses can be measured accurately and repeatedly using non-invasive techniques. Moreover, the effects of physiological stimuli and of locally-applied pharmacologic agents can be studied. The importance of such observations may lie in the fact that using the pupil as a model, pathophysiologic mechanisms may be discovered which could be relevant to our understanding of other autonomic manifestations.

The size of the pupil is commonly believed to be reduced in Parkinsonism, but again reference is to be made to the common miosis in old age (Korczyn et al., 1976). In fact, if correction is made for this factor, pupil size appears normal in Parkinsonian patients (Korczyn et al., 1985).

A relatively new method to evaluate autonomic disorders is the edge-light pupillary cycle time (Miller and Thompson, 1978 a). This method employs the light reflex, the physiology of which is well understood, to the study of the conduction in autonomic nerves. The ELPCT is known to be impaired in optic nerve disease (Miller and Thompson, 1978 b) and in patients with autonomic neuropathy (Martyn and Ewing, 1986). We have measured the ELPCT in a small group of patients with Parkinson's disease and have found it to be abnormal in most, with a progressive slowing in more advanced stages of the disease (Table 2). The pathogenesis of this phenomenon is yet unknown. Although abnormal delays in visual evoked responses have been reported in Parkinson's disease (Bodis-Wollner and Yahr, 1978), the prolongation is small and unlikely to be responsible for the long delays observed by us. It is possible to count for these abnormalities by considering changes in the nucleus of Edinger-Westphal (Hunter, 1985). An alternative explanation is of

Table 2. Pupillary cycle time (PCT) in Parkinson's disease

	n	Age	PCT	PCT > 1000 ms
Normal controls	14	68 ± 14	880 ± 54	0/14 (0%)
Parkinson's disease				
Stage 1	7	72 ± 2	1057 ± 200	2/7 (29%)
Stage 2	21	66 ± 8	1221 ± 244*	17/21 (81%)
Stage 3	13	63 ± 12	1341 ± 345*	11/13 (85%)
All Park. patients	26	66 ± 9	1335 ± 322*	30/41 (73%)

Data are given in years (for age) and in milliseconds (for PCT). The staging of Parkinson's disease is according to the criteria of Hoehn and Yahr (1967). Asterisks indicate statistically significant differences from age-matched normal controls by Student's t test or by chi square test, as appropriate

defective sympathetic tone. We have recently shown that the ELPCT is prologed in patients with Horner's syndrome (Blumen et al., 1986), and the magnitude of the prolongation in patients with Parkinson's disease is suggestive of a sympathetic dysfunction.

Conclusions

Reviewing the literature as well as our own results have led us to conclude that patients with Parkinson's disease manifest impaired autonomic nervous system activity. The pattern of impairment, with prominent gastro-intestinal manifestations, is quite different from the abnormalities observed in progressive autonomic failure. Our understanding of the pathogenesis and pathophysiology of the autonomic dysfunction is still quite unsatisfactory.

An important facet of the autonomic dysfunction relates to drug-induced changes (Korczyn and Rubenstein, 1982). It is yet an open question whether the effects of the medication are "normal", i.e. whether the same alterations would be produced in non-Parkinsonian subjects when these drugs are given, or whether specific changes predispose patients with Parkinson's disease to develop these side effects. Be this as it may, recognition of these side effects would allow selection of appropriate drugs for the treatment of individual patients with Parkinson's disease.

From the preceding discussion it can safely be concluded that clinical autonomic dysfunction is rarely a problem in the initial presentation of a patient with Parkinson's disease (with the notable exception of constipation). Rather, the existence of definite and conspicuous manifestations should be an important indication that the underlying disease producing the extrapyramidal features is multi-system atrophy rather than Parkinson's disease. The frequent occurrence of constipation may be suggestive of Parkinson's disease, but a more detailed analysis as to the underlying mechanisms and characteristics which may possibly distinguish this from the common constipation of old age remain to be established. Furthermore, it will be of importance to look systematically for Lewy bodies in sympathetic ganglia since these peripheral structures may provide tissue diagnosis, so much desired in this disease.

References

Appenzeller O, Gross JE, Albuquerque NM (1971) Autonomic deficits in Parkinson's syndrome. Arch Neurol 24: 30–57

Blumen SC, Feiler-Ofry V, Korczyn AD (1986) The pupil cycle time in Horner's syndrome. J Clin Neuro Ophthalmol 6: 232–234

Bodis-Wollner I, Yahr M (1978) Measurements of visual evoked potentials in Parkinson's disease. Brain 101: 661–671

Den Hartog Jager WA, Bethlem J (1960) The distribution of the Lewy bodies in the central and autonomic nervous system in idiopathic paralysis agitans. J Neurol Neurosurg Psychiatr 23: 283–290

Den Hartog Jager WA (1970) Histochemistry of adrenal bodies in Parkinson's disease. Arch Neurol 23: 528–533

Hoehn MM, Yahr MD (1967) Parkinsonism: onset, progression and mortality. Neurology 17: 427–442

Horowski R, Wachtel H (1978) Direct dopaminergic action of lisuride hydrogen maleate, an ergot derivative, in mice. Eur J Pharmacol 36: 373–383

Hunter S (1985) The rostral mesencephalon in Parkinson's disease and Alzheimer's disease. Acta Neuropathol 68: 53–58

Korczyn AD, Laor N, Nemet P (1976) Sympathetic pupillary tone in old age. Arch Ophthalmol 94: 1905–1906

Korczyn AD, Rubenstein AE (1982) Autonomic nervous system complications of therapy. In: Silverstein A (ed) Neurological complications of therapy. Futura, New York, pp 405–418

Korczyn AD, Rubenstein AE, Yahr MD (1985) The pupil in Parkinson's disease. Proc, 8th international symposium on Parkinson's disease, New York, 1985, p P43

Korczyn AD (1987) Autonomic manifestations in Parkinson's disease. In: Nappi G, Caraceni T (eds) Morbo di parkinson e malattie extrapiramidali. Edizioni Mediche Italiane, Pavia, pp 201–205

Kreek JJ, Schaefer RA, Hahn EF, Fishman J (1983) Naloxone, a specific opioid antagonist reverses chronic idiopathic constipation. Lancet i: 261–262

Martyn CN, Ewing DJ (1986) Pupil cycle time: a simple way of measuring an autonomic reflex. J Neurol Neurosurg Psychiatr 49: 771–774

Micieli G, Martignoni E, Cavallini A, Sandrini G, Nappi G (1987) Postprandial and orthostatic hypotension in Parkinson's disease. Neurology 37: 386–393

Miller SD, Thompson HS (1978 a) Edge-light pupil cycle time. Br J Ophthalmol 62: 495–500

Miller SD, Thompson HS (1978 b) Pupil cycle time in optic neuritis. Am J Ophthalmol 85: 635–642

Rabey JM, Treves T, Streifler M, Korczyn AD (1986) Comparison of efficacy of lisuride hydrogen maleate with increased doses of levodopa in Parkinsonian patients. Adv Neurol 45: 569–572

Rajput AH, Rozdilsky B (1976) Dysautonomia in parkinsonism: A clinicopathological study. J Neurol Neurosurg Psychiatr 39: 1092–1100

Vardi J, Oberman Z, Rabey I, Streifler M, Ayalon D, Herzberg M (1976) Weight loss in patients treated long-term with levodopa. J Neurol Sci 30: 33–40

Correspondence: Prof. A. D. Korczyn, Sackler Faculty of Medicine, Tel Aviv University, 69978 Tel Aviv, Israel.

Clinical and biochemical characteristics of early depression in Parkinson's disease

W. Kuhn, G. Fuchs[1], **G. Laux**[2], **R. Rupprecht**[2], and **H. Przuntek**[3]

[1] Departments of [1] Neurology and [2] Psychiatry, University of Würzburg
[3] Department of Neurology, St. Josef University-Hospital, Bochum, Federal Republic of Germany

Summary

Depression is more prominent in untreated patients in the early phase of the disease than in treated patients who have suffered from Parkinson's disease more than six years. In depressive Parkinson patients low levels of urinary 3-methoxy-4-hydroxyphenylglycol (MHPG) and an abnormal Dexamethasone Suppression Test were found. Although a reactive component could not be excluded, both pharmacological and biochemical data support the hypothesis that alterations in norepinephrine metabolism could be the cause for depression in Parkinson's disease.

Introduction

The nature of depression in Parkinson's disease (PD) is presently not certain. On the one hand depression has been explained as a reaction to the functional disability in PD. This view has been supported by Mindham (1976), who found a significant positive correlation between depression and severity of motor disability. In contrast other authors have found no relation to the functional disability (Mayeux, 1984; Robins, 1976). They supposed a close association of depressive symptoms with pathological neurotransmitter disturbances that characterize PD. In patients with endogenous depression dysfunction of both catecholamine and serotonin metabolism has been postulated (Riederer, 1980). In the cerebrospinal fluid (CSF) of depressive Parkinson patients lower values of 5-hydroxyindoleacetic acid could be detected than in nondepressed patients (Mayeux, 1984).

These results support the hypothesis that reduced serotonin metabolism may cause depression in PD.

For testing the hypothesis that disturbed norepinephrine metabolism could be involved in the pathogenesis of depression in PD, we measured total urinary 3-methoxy-4-hydroxy-phenylglycol (MHPG) in depressive Parkinson patients. Furthermore, the Dexamethasone Suppression Test (DST) was performed in patients with PD because central noradrenergic dysfunction probably correlates with an abnormal DST (Kawamura, 1987).

Materials and methods

The demographic data of patients and controls are summarized in the legends of Fig. 1–3. Motor disability was evaluated by the Columbia University Rating Scale. Severity of depression was assessed by the Hamilton Rating Scale (HAMD) with 21 items and the Zung Scale with 20 items. Psychiatric diagnoses were derived according to the Diagnostic and Statistical Manual of Mental Disorders (DSM-III). After giving informed consent the DST was performed overnight. Plasma cortisol values were obtained at 8 a.m. and 4 p.m. after oral administration of 1 mg dexamethasone at 11 p.m. Cortisol non-suppression was defined as a failure to suppress both post-dexamethasone levels to below 5 μg/dl. One group of Parkinson patients (Fig. 2) was treated with Parkinson specific drugs (L-Dopa/Benserazid, Bromocriptine, Budipine, Deprenyl). Plasma cortisol was determined by a direct radioimmunoassay (Corning, Medfield, MA). Total urinary MHPG was measured by a gas-chromatographic method (Cagnasso, 1974). Statistical analysis was carried out using the U-test of Mann and Whitney.

Results

Early symptoms of endogenous and non-endogenous depression are summarized in Table 1.

Table 1. Early depressive symptoms

Sleep disturbances
Fatigue; hypersensitivity
Loss of energy, lack of drive (e)
Anhedonia (e, f)
Loss of interest (e, f)
Tension, anxious restlessness (m)
Panic attacks, agitation
Cardiac sensations, sweating

Significant more frequent in endogenous depression (e), males (m), females (f) (from Demel et al., 1980)

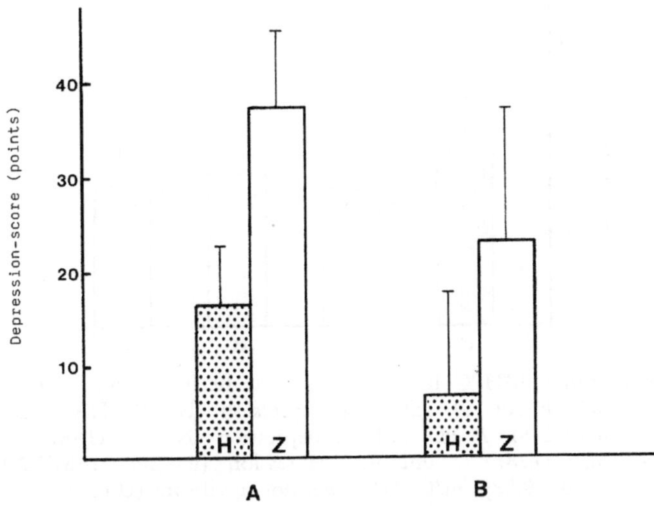

Fig. 1. Depression score: mean points $\bar{x} \pm$ SD. *A* untreated Parkinson patients (duration of disease: 1.2±0.9y; m/f=15/20; age: 61.1±13.8y; n=35). *B* treated Parkinson patients (dur. of dis.: 6.6±5.3y; m/f=24/14; age: 63.1±8.8y; n=38). *H* Hamilton rating scale; *Z* Zung rating scale

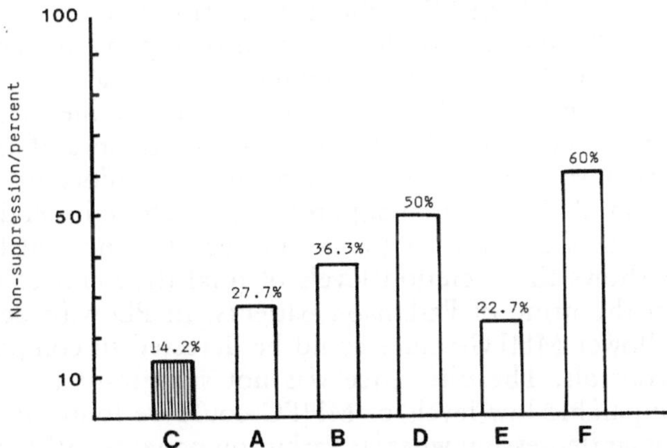

Fig. 2. Percentage of an abnormal DST in PD and depression. *C* controls (n=28; $\bar{x}$(HAMD)=2.0±2.9; age: 32±9.7y; m/f=14/14). *A* untreated PD without depression (n=18; $\bar{x}$(HAMD)=9.4±4.5; age: 64.3±10.8y; m/f=7/11). *B* untreated PD with depression (n=22; $\bar{x}$(HAMD)=21.3±4.5; age: 59.9±14.5y; m/f=6/16). *D* treated PD (n=18; $\bar{x}$(HAMD)=22.2±16.0; age: 63.2±9.5y; m/f=13/5). *E* non-endogenous depression (n=22; $\bar{x}$(HAMD)=23.5±8.2; age: 47±16y; m/f=7/15). *F* endogenous depression (n=28; $\bar{x}$(HAMD)=26.3±7.4; age: 49±14y; m/f=9/19)

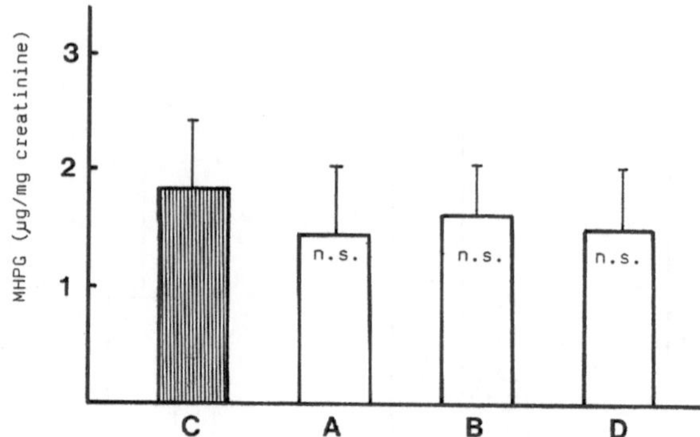

Fig. 3. Total urinary MHPG in PD ($\bar{x}\pm$SD). *C* healthy controls (n = 20; age: 37.9 ± 9.1y; m/f = 10/10). *A* PD with depression (N = 11; HAMD ≧ 17; age: 64.7 ± 8.5y; m/f = 3/8). *B* PD without depression (N = 13; HAMD < 17; age: 58.8 ± 12.8y; m/f = 6/8). *D* unipolar depression (n = 20; HAMD ≧ 17; age: 54.0 ± 9.58y; m/f = 8/12). *n.s.* not significant (U-test)

Fig. 1 demonstrates that depression is more prominent in untreated patients in the early phase of the disease than in treated patients—without specific antidepressant medication—who have suffered from PD more than six years.

Fig. 2 shows that 14.2% of the healthy controls were non-suppressors in the DST. In PD without depression five of 18 untreated patients (27.7%) showed no adequate cortisol response to dexamethasone. A higher degree of non-suppressors (36.3%) was found in untreated patients with depression. The highest value of non-suppression (50%) could be observed in treated patients suffering from PD more than six years. For comparison, in non-endogenous depression we found 22.7% non-suppressors, whereas in endogenous depression 18 of 28 patients (60%) showed an abnormal DST.

Fig. 3 shows the excretion levels of total (free and conjugated) MHPG in the urine of Parkinson patients. In PD with depression distinctly lower MHPG-values could be detected in comparison to healthy controls. The difference was not significant (p = 0.13, U-test). Comparable low levels of MHPG could be found in patients with unipolar depression whereas Parkinson patients without depression showed slightly higher values of total MHPG.

Discussion

About 50% of the untreated (de novo) Parkinson patients are depressed. The main depressive symptoms in PD are pessimism, hopelessness, decreased motivation and drive and increased concern with health (Gotham, 1986). Assessment of depression in PD with the HAMD showed significantly higher scores on items relating to suicide, work and interests, retardation, anxiety, general somatic symptoms and loss of insight in comparison to a control group with chronically disabled patients (Robins, 1976). Depressive symptoms were found in 37% of patients in the year preceding the initial evaluation of motor disability (Celesia, 1972). Treatment of de novo Parkinson patients with L-Dopa/Benserazid for six weeks reduces severity of depression by about 40% (unpublished observations). This lower level of depression could also be observed (Fig. 1) in patients suffering more than six years from PD. These patients received only antiparkinsonian drugs without any specific antidepressive medication. On the other hand significant improvement in depressive symptoms in PD has been reported with nortriptyline and imipramine (Andersen, 1980). In the case of nortriptyline no changes in motor disability could be observed, whereas for imipramine improvement of both depressive and parkinsonian symptoms was reported. Although a reactive component could not be excluded, these results support the hypothesis that depression in PD is mainly associated with disturbances of norepinephrine and/or serotonin metabolism.

The DST has been used to demonstrate dysfunction of the hypothalamic-pituitary-adrenal axis in depression. Reduced levels of central norepinephrine could cause an abnormal DST. In depressive de novo Parkinson patients a higher degree of non-suppressors (36.3%) was found than in patients without depression (27.7%).

Excretion of total MHPG in the urine of depressive Parkinson patients is distinctly lower than in healthy controls. Recently Kopin (1984) has demonstrated that about 20% of the total urinary MHPG is of central origin. Therefore by measuring total urinary MHPG in PD a maximum reduction of 20% could be expected as a reflection of central norepinephrine dysfunction. This may account for the non-significant differences between depressive Parkinson patients and healthy controls.

Both abnormal DST and lower urinary MHPG-values in depressive Parkinson patients support the hypothesis that alterations in norepinephrine metabolism could be the main cause for depression in PD.

References

Andersen J, Aabro E, Gulmann N, Hjelmsted A, Pedersen HE (1980) Antidepressive treatment in Parkinson's disease. Acta Neurol Scand 62: 210–219

Cagnasso M, Biondi PA (1974) Gaschromatographic estimation of urinary vanilmandelicacid and 3-methoxy-4-hydroxyphenylethyleneglycol after their purification on Sephadex LH 20. Ital J Biochem 5: 345–355

Celesia GG, Wanamaker WM (1972) Psychiatric disturbances in Parkinson's disease. Dis Nerv System 33: 577–583

Demel I, Schubert H, Unterthiner D (1980) Initialsymptome bei depressiven Erkrankungen. Neurol Psychiat 6: 263–266

Gotham AM, Brown RG, Marsden CD (1986) Depression in Parkinson's disease: a quantitative and qualitative analysis. Neurol Neurosurg Psychiatr 49: 381–389

Kawamura T, Kinoshita M, Iwasaki Y, Nemoto H (1987) Low-dose dexamethasone suppression test in Japanese patients with Parkinson's disease. J Neurol 234: 264–265

Kopin IJ, Jimerson DC, Markey SP, Ebert MH, Polinski RJ (1984) Disposition and metabolism of MHPG in humans: Application to studies in depression. Pharmacopsychiatry 17: 3–8

Mayeux R, Stern Y, Cote L, Williams JBW (1984) Altered serotonin metabolism in depressed patients with Parkinson's disease. Neurology 34: 642–646

Mindham RHS, Marsden CD, Parkes JD (1976) Psychiatric symptoms during L-Dopa therapy for Parkinson's disease and their relationship to physical disability. Psychol Med 6: 23–33

Riederer P, Birkmayer W (1980) A new concept: brain area specific imbalance of neurotransmitters in depression syndrome—human brain studies. In: Usdin E, Sourkes TL, Youdim MBH (eds) Enzymes and neurotransmitters in mental disease. Wiley, New York, p 261

Robins AH (1976) Depression in patients with parkinsonism. Br J Psychiatr 128: 141–145

Correspondence: Dr. W. Kuhn, Department of Psychiatry, University of Würzburg, Füchsleinstrasse 15, D-8700 Würzburg, Federal Republic of Germany.

Psychomotor investigations in depressed patients by comparison with Parkinson patients

G. Laux[1], W. Kuhn[2], and W. Classen[1]

Departments of [1]Psychiatry and [2]Neurology, University of Würzburg, Federal Republic of Germany

Summary

Patients with major depressive disorders show impairments of cognitive abilities and motor performance. Patients with Parkinson's disease (PD) also have slower reaction and movement times. Clinically, differential diagnosis between inhibited depression (D) and PD can be difficult. So, psychomotor functions of patients with D, PD and PD + D have been tested using the "Motorische Leistungs-Serie" (MLS) as objective assessment. Standard values of MLS in D were clearly higher than in PD. Furthermore, scores of PD + D were lower than those for PD (without depression). Under treatment an overall improvement of MLS performance in PD and PD + D (except Tapping) as well as in D (except Tapping and parts of line tracing) could be observed. Psychomotor slowing in D can be explained by central motivational deficits. Terminology of bradyphrenia as well as dopaminergic and/or noradrenergic dysfunctions being the possible neurochemical basis of this cognitive deficits are discussed.

Introduction

Patients with major depressive disorders show high thresholds to environmental stimuli, deterioration of mood often coincides with an impairment of cognitive abilities (Cohen et al., 1982). Severity of depressive symptoms and cognitive dysfunction are highly correlated in some studies (Rogers et al., 1987), cognitive and motor performance as well (motor retardation) is impaired in depressives (Abrams et al., 1981; Rogers et al., 1987; Guenther et al., 1988). The slowness of depressives in simple reaction time tasks may result from a deficit in their stimulus encoding and/or response execution, but is not caused by a dysfunction in the central information-identification process as in schizophrenia. Many investigators believe these findings do support the dysfunction hypothesis of nondominant hemisphere in major depressive disorders (Coffey, 1987).

Antidepressant medication influences cognitive abilities in depressives differentially: some antidepressants seem to improve performance, others hardly alter performance, others cause a deterioration of performance. On the other hand, a general amelioration of cognitive abilities has been postulated and attributed to the decline of depressive symptoms and concomitant improvement of memory and attention processes (Hobi, 1982).

Patients with Parkinson's disease also have slower reaction times (Bloxham et al., 1987); compared to reaction time movement time undergoes more substantial and consistent disturbance (Evarts et al., 1981). To improve objective assessment of Parkinson's disease the "Motorische Leistungs-Serie" (MLS) was used by Przuntek and Fischer (1986) and Kraus et al. (1987). Clinically, differential diagnosis between inhibited depression and Parkinson's disease can be difficult. So, we looked to psychomotor functions in MLS comparing depressed patients with Parkinson patients with and without depression for assessment of psychometric results supposing to be different.

Patients and methods

We investigated 29 inpatients (9 males, 20 females) with depressive syndromes diagnosed according to DSM-III criteria for major depressive episode and ICD-9 (15 unipolars No 296.1, 5 bipolars No 296.3, 9 reactive-neurotic No 300.4). Patients suffering from severe somatic, neurologic or further psychiatric pathology were excluded. Depressives were 48.9 (±12.2) years old on the average. At the beginning and end of the treatment, after 23 to 28 days, assessment of psychopathological symptoms was carried out using the Hamilton-Depression Scale (HAM-D). Under double blind conditions patients were treated with the tetracyclic antidepressant maprotiline (noradrenaline reuptake inhibitor; $n=14$) or with moclobemide, a new selective reversible MAO-A inhibitor ($n=15$).

Motor disability of Parkinson patients (mean age: 61.1 ± 13.8 years) was evaluated by the Columbia University Rating Scale. Severity of depression was assessed by HAM-D and the Zung-Scale. Parkinson patients ($n=14$; 3 males, 11 females) with depressive symptoms had a mean score (HAM-D) of 22.6 ± 5.7. In Parkinson patients without depression ($n=16$; 8 males, 8 females) a mean score (HAM-D) of 13.0 ± 3.8 was measured (Table 1). All Parkinson patients were free of drugs at t_1. In the following 12 weeks (t_2) all patients received a combination of L-Dopa/Benserazid and Deprenyl.

Table 1. Clinical data

	Age $\bar{x}$ (s)	HRSD $\bar{x}$ (s)	Columbia scores $\bar{x}$ (s)
Depressives (n = 30, 15 uni-, 5 bipolars)	48.9 y (12.2)	28.4 (5.1)	
Parkinson pts with depression (n = 14)		22.6 (5.7)	28.9 (10.2)
	61.1 y (13.8)		
Parkinson pts (n = 16)		13.0 (3.8)	30.1 (10.8)

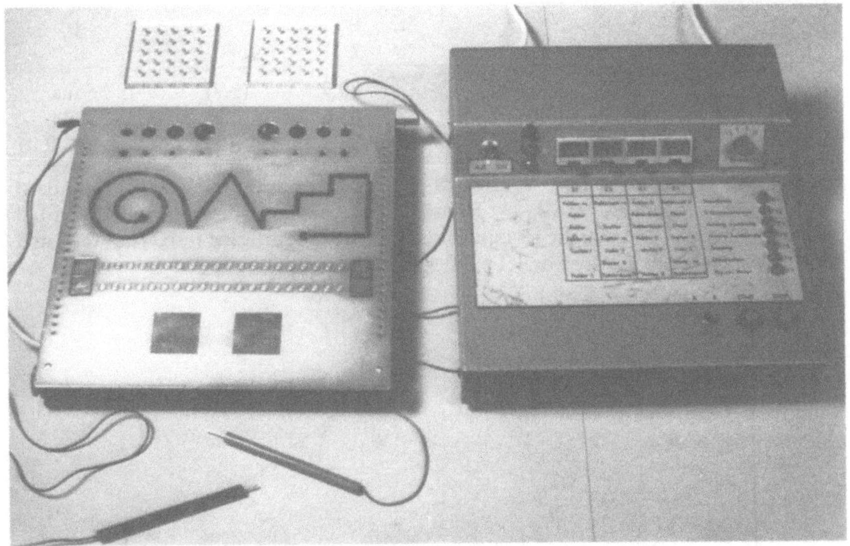

Fig. 1. Motorische Leistungs-Serie (MLS) composed of subtests steadiness, line tracing, aiming, tapping and pegboard

At the beginning (t_1) and end of the treatment (t_2) psychomotor performance examination was carried out.

Tests were done between 10–12 a.m. to control for diurnal variation of performance level.

In both sessions (t_1, t_2) examination comprised the "Motorische Leistungs-Serie" (MLS; Schoppe, 1974) with norms by Hamster (1980), measuring arm-hand-tremor (steadiness), examining the precision and speed of motor skills (line tracing) regarding duration of movement (total time: TT), time of mistakes (TM) and number of mistakes (NM), checking the accuracy of motor acts (aiming), the speed of imprecise movements (tapping) and accuracy as well as speed of movements (pegboard), each for the right (r) and left (l) hand (see Fig. 1).

Furthermore, acoustic and visual simple reaction time tasks using Bettendorff Reaktiometer T 96 were given, which are regarded as refined parameters of reaction. 30 visual and acoustic stimuli each were elicited at random with an interstimulus interval of 6 and 10 sec respectively. The patients were asked to press a button with a finger of their dominant hand when a defined stimulus appeared. Training periods preceded testing sessions in order to make the patients familiar with the tasks.

Original data were transformed to standard- and T-scores and repeated measures of analyses of variance (Manova) covariating age-effects, Fischer's t and Chi²-tests were computed.

Table 2. Visual and acoustic simple reaction times (T-values) using Bettendorff Reaktiometer T 96 in depressed inpatients (n = 30)

	t_1		t_2		p
	$\bar{x}$	s	$\bar{x}$	s	
vis RT	76.5	(9.2)	80.7	(9.7)	n.s.
ac RT	83.3	(13.5)	85.6	(14.2)	n.s.
HAM-D	28.4	(5.1)	15.2	(8.7)	≤ 0.01

t_1 before treatment, drug free. *t_2* after 28 days antidepressant treatment. *s* standard deviation $\bar{x}$ = means. *HAM-D* Hamilton Depression Scale. *vis RT* visual reaction time. *ac RT* acoustic reaction time. *n.s.* non significant (effects of repeated measures). Standard norms for RT: 100 ± 10 standard scores

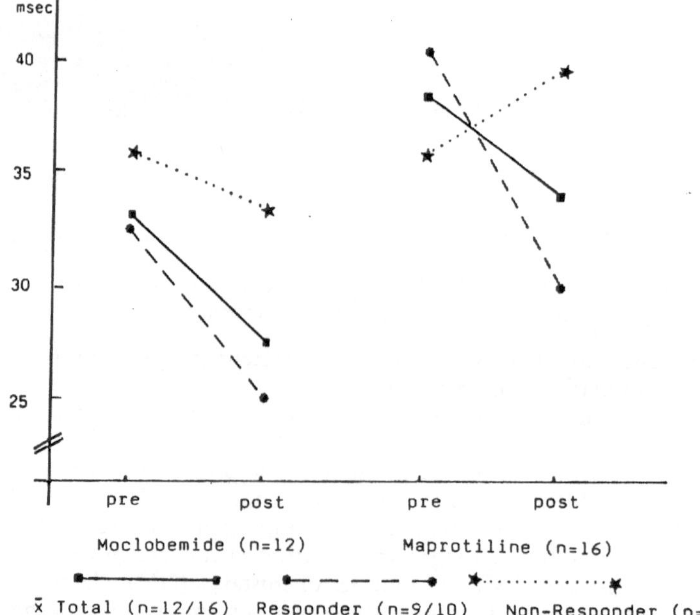

Fig. 2. Visual simple reaction times of depressed inpatients before and after treatment with different antidepressants

Results

Visual and acoustic simple reaction times in depressed patients are presented in Table 2.

Visual reaction times of depressives according to antidepressant treatment with moclobemide vs. maprotiline respectively and separated according to responders or non-responders are given in Fig. 2. Psychomotor performance test (MLS) results in depressives and Parkinson patients with and without depressive symptomatology are summarized in Table 3.

A significant overall improvement of depressive symptoms (HAM-D) could be demonstrated ($p < 0.01$). Considering visual and acoustic sensorimotor performance parameters improvement with antidepressant therapy could be seen in depressives, not reaching significance however. By separating treatment responders from non-responders, it was, however, possible to demonstrate divergent results with maprotiline: non-responders obtained deteriorated reaction times, responders improvement ($p < 0.05$).

Table 3. Psychomotor performance test (MLS) results (standard scores) in depressives and Parkinson patients before (t_1) and after (t_2) treatment

		Depressives n=29		Parkinson pts with depression n=14		Parkinson pts n=16	
	$\bar{x}$	t_1	t_2	t_1	t_2	t_1	t_2
Steadiness	r	94.3[a]	96.0	86.2[a]	90.1	91.3[a]	93.4
	l	96.0[a]	92.8	87.7[a]	90.9	88.3[a]	92.1
Line tracing							
TT	r	102.3	104.4*	95.4	100.5*	97.6	101.2*
	l	104.3	105.7	96.5	102.3	100.3	99.8
TM	r	108.2[b]	106.4	85.2[b]	92.9	91.0[b]	95.4
	l	109.7[b]	105.0	84.2[b]	90.6	88.6[b]	93.1
NM	r	111.4[b]	110.1	106.5[b]	111.5	100.3[b]	103.6
	l	110.9[b]	107.4	107.1[b]	114.1	103.3[b]	102.6
Aiming	r	92.9	95.0*	85.1	90.5*	90.5	93.2*
	l	91.3[b]	93.3**	81.5[b]	87.7**	82.6[b]	88.1**
Tapping	r	91.5	89.5	83.7	82.2	87.3	87.9
	l	86.7[a]	84.8	78.6[a]	78.6	81.3[a]	83.0
Pegboard	r	93.3[b]	91.9	81.5[b]	87.7	80.8[b]	87.6
	l	92.0[b]	92.8*	78.5[b]	82.7*	78.6[b]	83.5*

Norm: 100 ± 10 standard scores; significance of time course effects: * $p \leq 0.05$; ** $p \leq 0.01$. Signif. differences between groups: [a] $p \leq 0.05$; [b] $p \leq 0.01$

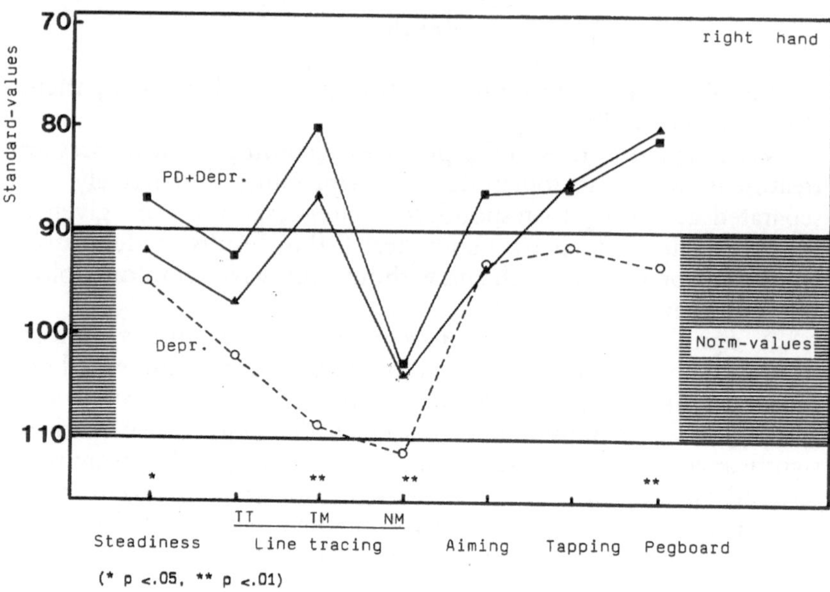

Fig. 3. MLS standard values of patients with depression and parkinsonian patients (PD) at t_1 (right hand). ○ Depression (n = 30); ■ PD without depression (n = 11; mean Hamilton score: 9.6 ± 2.8; mean Columbia score: 30.1 ± 10.8) ▲ PD with depression (n = 9; mean Hamilton score: 23.6 ± 4.0; mean Columbia score: 28.9 ± 10.2)

In the MLS performance of depressives was below the standard norm in tapping (left hand) only (Table 3). Under treatment significant amelioration in accuracy of performance (total time in the line tracing of the right hand) and aiming could be registered.

Standard values of MLS in depression are clearly higher than in Parkinson's disease. Furthermore, although there is no difference in motor disability–indicated by nearly identical scores in the Columbia rating scale–standard values of depressive Parkinson patients are lower than those of Parkinson patients without depression (Fig. 3). Under treatment an overall improvement of MLS performance in Parkinson's disease (except Tapping) both for patients with and without depression could be observed (Table 3).

Significant differences between depressed and Parkinson patients concerning MLS were found in nearly all subtests indicating higher performances of depressives before and after treatment and predominating lowest scores in Parkinson patients with depression.

Discussion

A wide variety of motor deficiencies has been reported in depressive disorders both in medicated and unmedicated patients yielding evidence that motor symptoms may be found in different motor subsystems. So alterations in gross motor activity and fine motor functions (retardation of simple and complex motor tasks) have been found (Guenther et al., 1988, for review). For example, Williams and Hemsley (1986) described how schizophrenics and depressives performed less well in complex choice reaction tasks. As compared to presentation of complex stimuli, tasks demanding complex reactions were shown to impair psychomotor functioning. This may be a hint of impaired information processing, and less disturbed stimulus perception in depressed patients. The results of our study showed a similar phenomenon. Scores below the normal range could be observed with respect to the tapping task, which comprises a pronounced speed factor. Not only precision and dexterity in motor acts, but also speed reactions were, and remained, impaired.

Regarding our results it seems astonishing that instead of mitigation of psychopathological symptoms in depressed patients taking antidepressants only aiming of the right hand showed improved performance, although high correlations between improvement of performance and psychopathological symptoms were reported (Cohen et al., 1982; Rogers et al., 1987). One could not postulate these effects to be influenced by drug action as impairment of performance is reported in drug free patients, too. Since simple motor acts require little cognitive processing and cognitive performance requires little in the way of motor activity, but both performances could be found being impaired with close relationship to depressive symptoms (Cohen et al., 1982), impairment in depressives may be explained by central motivational deficits. Hart and Kwentus (1987) observed cognitivebehavioural slowing in both depressive illness and Parkinson patients, a normal rate of processing information centrally appeared to distinguish depression from subcortical neurological disorders however. They also conclude that psychomotor slowing in the presence of normal information processing speed can be explained by a deficit in motivational state associated with depression.

In Parkinson's disease differences in MLS performance between depressive and non-depressive patients with comparable motor disability have been found (Fig. 3). In a similar way to endogenous depression central motivational deficits in depressive Parkinson patients could be the cause of those differences.

Guenther et al. (1988) recently suggested the existence of a psychotic motor syndrome (PMS) in endogenous depressed patients (and schizophrenics) involving disturbances of the lips and tongue, fine and gross movements of the dominant right hand and the complex motor coordination of the extremities. These authors saw clear improvement of the PMS with the amelioration of depressive symptoms suggesting PMS as a state marker for endogenous depression. Bradyphrenia (Rogers, 1986, for review), a syndrome including slowing of cognitive processing, more recently described as psychic akinesia and subcortical dementia, seems to be closely related, if not indistinguishable from the retardation of depressive illness (Rogers et al., 1987). Attempting to define its neuroanatomical and neurochemical basis it is a matter of debate if this cognitive deficits are due to dopaminergic and/or noradrenergic dysfunctions and which therapeutic trials—with dopaminergic or noradrenergic drugs—are treatment procedures being adequate.

References

Abrams R, Redfield J, Taylor MA (1981) Cognitive dysfunction in schizophrenia, affective disorder and organic brain disease. Br J Psychiat 139: 190–194

Bloxham CA, Dick DJ, Moore M (1987) Reaction times and attention in Parkinson's disease. J Neurol Neurosurg Psychiatr 50: 1178–1183

Classen W, Laux G (1988) Differences in sensorimotor performance between patients suffering from major depressive and schizophrenic disorders. Psychiatr Res (in press)

Coffey CE (1987) Cerebral lateralty and emotion: the neurology of depression. Compreh Psychiatr 28: 197–219

Cohen RM, Weingärtner H, Smallberg S, Pickar D, Murphy DL (1982) Effort and cognition in depression. Arch Gen Psychiatr 39: 593–597

Evarts EV, Tervainen H, Calme DB (1981) Reaction times in Parkinson's disease. Brain 104: 167–186

Guenther W, Guenther R, Streck P, Römig H, Rödel A (1988) Psychomotor disturbances in psychiatric patients as a possible basis for new attempts at differential diagnosis and therapy. Cross validation study on depressed patients. Eur Arch Psychiatr Neurol Sci 237: 65–73

Hamster W (1980) Die motorische Leistungsserie (MLS). Handanweisung. Dr Schuhfried GmbH, Mödling

Hart RP, Kwentus JA (1987) Psychomotor retardation and subcortical dysfunction in depression. J Neurol Neurosurg Psychiatr 50: 1263–1266

Hobi V (1982) Psychopharmaca, psychic illness, and driving ability: a contribution to debate. J Int Med Res 10: 283–305

Kraus PH, Klotz P, Fischer A, Przuntek H (1987) Assessment of symptoms of Parkinson's disease by apparative methods. J Neural Transm 25: 89–96

Przuntek H, Fischer A (1986) Beurteilung der feinmotorischen Fähigkeiten von Parkinson-Patienten mit der MLS (motorische Leistungsserie). In: Schnaberth G, Auff E (Hrsg) Das Parkinson-Syndrom. Basel, Editiones Roche, S 357–359

Rogers D (1986) Bradyphrenia in parkinsonism: a historical review. Psychol Med 16: 257–265

Rogers D, Lees AJ, Smith E, Trimble M, Stern GM (1987) Bradyphrenia in Parkinson's disease and psychomotor retardation in depressive illness. Brain 110: 761–776

Schoppe KJ (1974) Das MLS-Gerät. Ein neuer Testapparat zur Messung feinmotorischer Leistungen. Diagnostica 20: 43–46

Sternberg DE, Jarvik ME (1976) Memory functions in depression. Arch Gen Psychiatr 33: 219–224

Williams RM, Hamsley DR (1986) Choice reaction time performance in hospitalized schizophrenic patients and depressed patients. Eur Arch Neurol Sci 236: 169–173

Correspondence: Dr. G. Laux, Department of Psychiatry, University of Würzburg, Füchsleinstrasse 15, D-8700 Würzburg, Federal Republic of Germany.

Quantitative analysis of voluntary and involuntary motor phenomena in Parkinson's disease

H. Hefter, V. Hömberg, and H.-J. Freund
Department of Neurology, University of Düsseldorf, Federal Republic of Germany

Summary

Different approaches to quantify motor impairment in Parkinson's disease (PD) are presented which are helpful in the early diagnosis and therapy control: frequency analysis of tremor (involuntary finger and hand movements), frequency analysis of most rapid voluntary alternating finger movements, test of ability to move the fingers following a variable target frequency (hastening phenomenon), analysis of trajectories of most rapid single isometric indexfinger extensions, aming movements between two target zones of variable size (Fitt's law) and unrestricted pointing movements. With these methods the interaction between voluntary and involuntary movements can be quantified.

Introduction

As in other disorders of the extra-pyramidal motor system also in Parkinson's disease (PD), the motor dysfunction is a summation of both deficient voluntary motor activity, e.g. akinesia or bradykinesia (minus symptoms) and the intrusion of non-appropriate elements of involuntary motor activity into intended voluntary motor acts, e.g. dystonia or tremor (plus symptoms). The relative contribution of either component may vary from patient to patient.

The use of quantitative evaluation techniques of involuntary and voluntary movements may be helpful in delineating the relative contribution of each component and may help to understand the interaction of both. Furthermore, in addition to clinical scoring, quantitative motor analysis may be helpful in picking up motor dysfunctions at early stages in the course of a disease and in understanding similarities and differences between various extrapyramidal disorders which may give clues to differential diagnosis and treatment. In this chapter we will discuss several attempts to quantify various aspects of motor activity in Parkinson's disease.

I. Tremor

Tremor is usually recorded by accelerometers, which can easily be attached to various parts of the body. A very efficient way to process the accelerometer signal is frequency analysis by fast Fourier transformation (FFT) in order to yield information about the relative frequency content of the signal. Fig. 1 A shows a sequence of partially overlapping power spectra of tremor in a patient with Parkinson's disease. It is obvious that most of the energy of the spectrum is concentrated in a narrow frequency band around 4 Hz. It has been shown that this dominant frequency is the most reliable parameter to differentiate between various pathological tremors e.g. their variability to changes of the mechanical state of a limb (Hömberg et al., 1987). In contrast to the stability of dominant tremor frequency tremor amplitude is a much less reliable parameter with larger spontaneous variations (Hefter et al., 1987a).

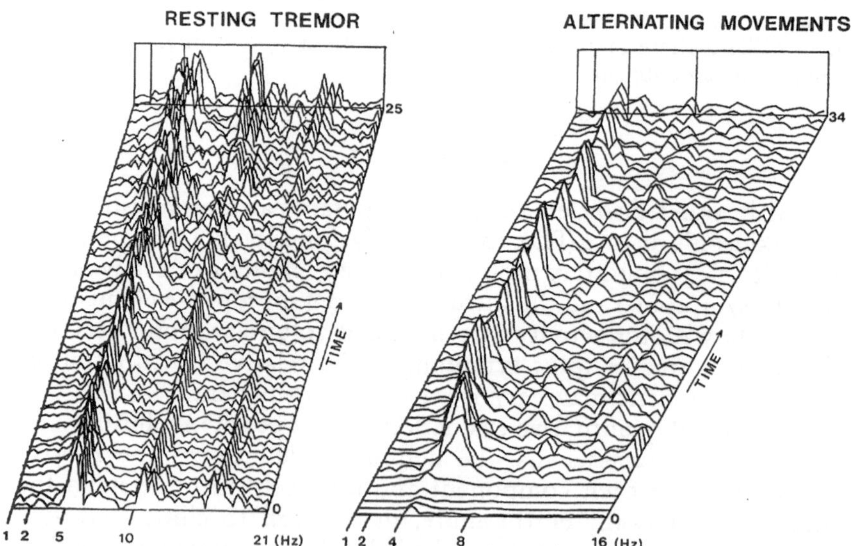

Fig. 1. Frequency analysis of tremor at rest (left side) and voluntary alternating finger movements (right side) in a PD-patient

II. Fastest alternating movements

Fastest index finger flexion-extension movements can be recorded in the same way as tremor by accelerometers. In this case the task is to move the index finger around the metacarpophalangeal joint at the maximal possible alternation frequency at small movement amplitudes. The accelerometer signal can be analysed by FFT and from the resulting power spectra the maximum alternation frequency can easily be determined. An example is shown in Fig. 1 B of a patient with Parkinsonian tremor at rest. He cannot move faster than the frequency of his resting tremor. In this case the maximal alternation frequency is around 4 Hz (normal 8 Hz). Such slowing of fastest alternating movements can also be found in other basal ganglia diseases especially in Huntington's disease (Hefter et al., 1987 b).

III. Hastening phenomenon

Although most patients with basal ganglia disease are not able to produce alternating movements faster than about 4 to 5 Hz, in some patients with Parkinson's disease especially in those without tremor at rest occasionally movement frequencies of 7 to 8 Hz can be found.

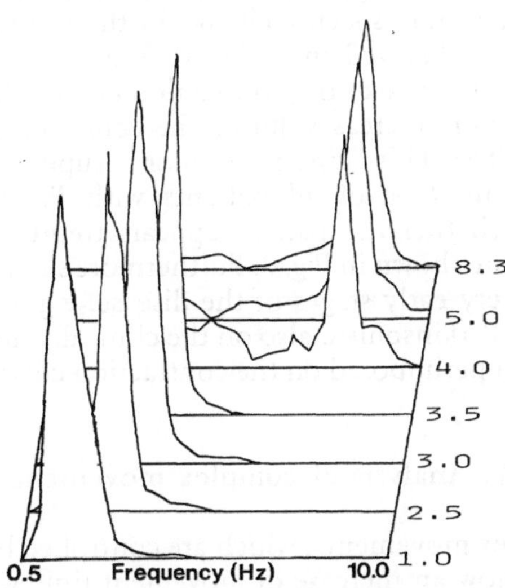

Fig. 2. Frequency analysis of finger movements of a PD-patient to be performed at a given target frequency as indicated by the numbers at the right side. There is a sudden jump to higher frequencies when the target frequency exceeds 4 Hz

When these patients perform voluntary alternating movements following given frequencies presented to them by a tone signal, they can easily follow this target up to 3 to 4 Hz. Above 4 Hz the frequency of these voluntary movements suddenly jumps up to 7 to 8 Hz although the target is still at lower frequencies. This lack in the range of modulation of movement frequencies has been called the hastening phenomenon. Fig. 2 gives an illustration of the spectral analysis of such a frequency following recording: with the target frequency increasing in steps of .5 Hz, above 3.5 Hz a Parkinsonian patient jumps up to an alternation frequency of around 8 Hz.

IV. Fastest isometric contractions

Brief isometric force pulses are a very elementary motor act. For non-targeted contractions in normals an independence of contraction time from contraction amplitude is observed, i.e. irrespective of the extent of the contraction the time needed for its completion remains almost the same (isochrony) (Freund und Büdingen, 1978). In both Parkinson's disease and Huntington's disease we showed that these contractions are slowed and the normal isochrony is disturbed or even lost depending on the degree of motor impairment (Hefter et al., 1987 b) whereas slowing seems to be a common feature in all basal ganglia diseases, a more specific change in the morphology of contraction curves is observed in Parkinson's disease: oscillations are superimposed on the rise of the contraction curve. The frequency of this "kinetic tremor" increases during the course of the contraction (Hefter et al., 1985). This "kinetic" tremor is superimposed on contraction curves in 95% of all patients with Parkinson's disease irrespective of whether the patient appears tremulous or akineto-rigid. Examples are shown in Fig. 3. Furthermore such oscillations are present in the very early stages of the disease, e.g. in patients with clinical "Hemi"-Parkinsonism also on the clinically normal side clear oscillations are superimposed on the contraction curves (Hömberg et al., 1985).

V. Analysis of complex movement

More complex movements, which are controlled by detailed sensory guidance show an increase of movement time with movement amplitude. Fitts in 1954 described a fundamental property of such aiming movements requiring patients to move a stylus to target zones different in width (W) and separated by various distances (D): In

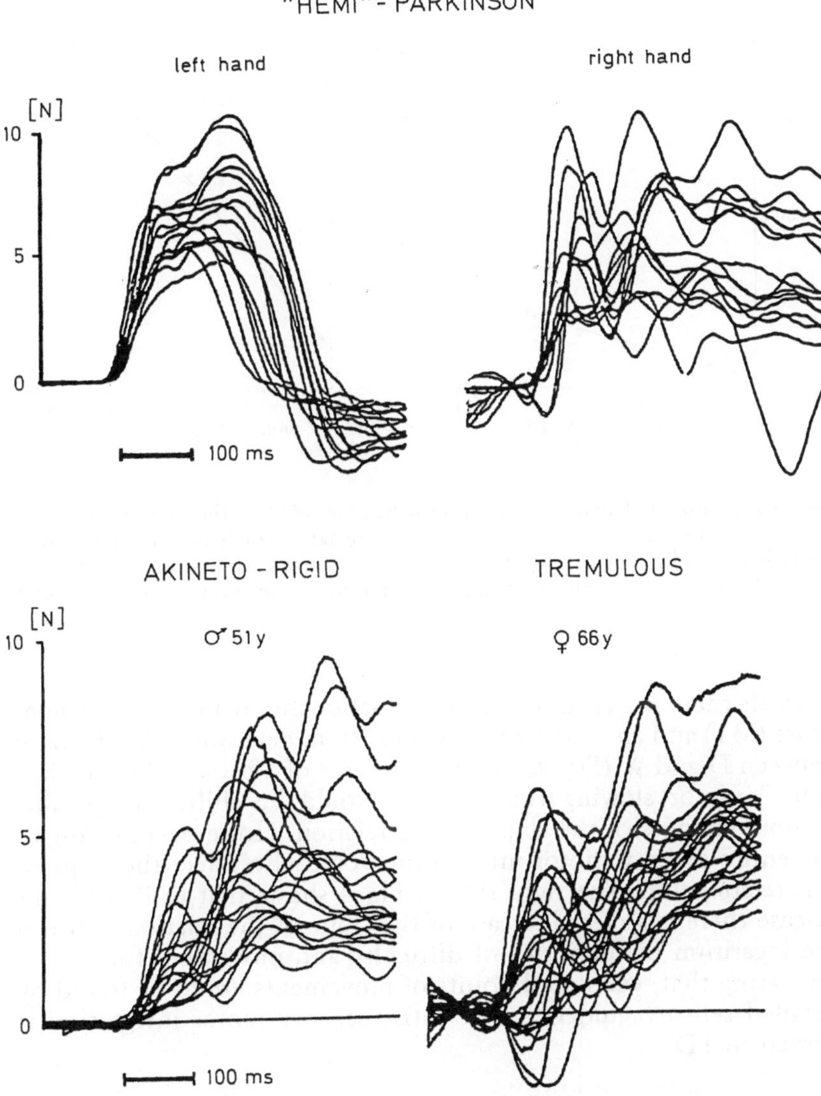

Fig. 3. Force trajectories of fastest isometric index finger extensions in an akineto-rigid PD-patient without tremor at rest (lower left), in a PD-patient with tremor at rest (lower right) and in a hemiparkinsonian patient (upper part). Also in the "normal" left hand a similar pattern of abnormalities is seen with prolongation of contraction time and superimposed oscillations

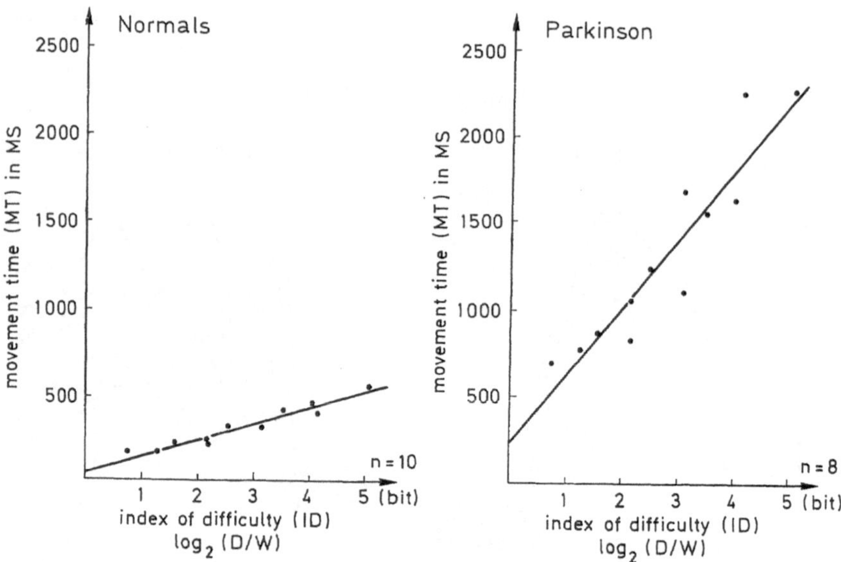

Fig. 4. Logarithmic dependence of movement time (MT) on the index of difficulty (defined as quotient of the distance D between the target zones and their width W in normals (left side) and PD-patients (right side). In both groups Fitt's law is valid, but in PD-patients the entire sensory motor information processing is markedly slowed

normals there is a clear logarithmic relationship between movement times (MT) and an "index of difficulty" (ID), determined by the ratio between D and W (Fig. 4, left side). In case of a patient showing just "pure" motor slowing this relation would be shifted in parallel without changes of the slope of this relation. However a slowing of the entire sensory motor integration would also affect the slope of this relationship. The right side of Fig. 4 shows that in Parkinson's disease there is a clear increase of the slope of the relation between the logarithm of the index of difficulty and movement time demonstrating that, when end points of movements are to be found by detailed sensory guidance, the entire sensory motor integration is slowed in PD.

VI. Interaction of voluntary and involuntary movements

Fig. 5 depicts a recording of a pointing movement in a Parkinsonian patient by an infrared measuring device. The task is to move

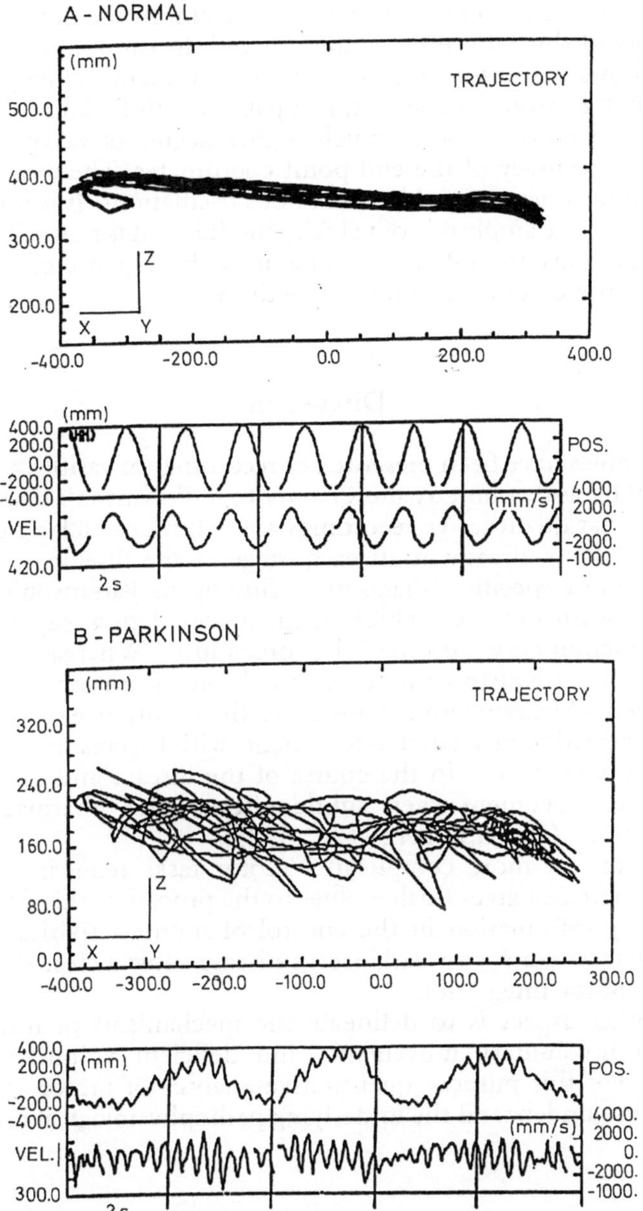

Fig. 5. Trajectory of the finger tip during a pointing movement between two targets 70 cm apart with arm outstretched and time course of position and velocity during the movement in a normal subject (the upper two parts) and in a PD-patient (the lower two parts)

the arm between 2 points separated by 70 cm. In the normal (A) the trajectory of this movement is smooth and the time course of movement shows a sinusoidal pattern with movement velocity varying between ±2 m/sec. In contrast, in a patient with Parkinson's disease (B) the trajectory reveals a much higher degree of variation and a decreased accuracy of the end point coordinates. The movement is slowed and superimposed by tremulous oscillations. It is interesting to note that the amplitude of velocity modulation is now taken up by the involuntary tremulous movements at the expense of decreased velocity modulation of voluntary activity.

Discussion

Examples have been presented of recordings of various aspects of involuntary and voluntary motor activity in Parkinson's disease. It is obvious that quantitative recordings are helpful in picking up some clinical deficits already at an early stage of the disease.

The most specific, "diagnostic" finding in Parkinson's disease seems to be the presence of high frequency oscillations superimposed on contraction curves during brief force pulses. Whereas slowing of contractions and disturbance of the isochronic features can be found in a variety of extrapyramidal disorders, the occurrence of such oscillations seemed to be restricted to patients with Parkinson's disease. It also occurs very early in the course of the disease and can also be helpful to document even subtle effects after pharmacological interventions (e.g. Hömberg et al., 1985).

The use of more complicated motor tasks requiring detailed sensory guidance gives further clues to the process of the integration of sensory information in the control of motor activities. Here it seems that patients with Parkinson's disease show a clear slowing of sensory-motor integration.

Another aspect is to delineate the mechanisms of interaction between involuntary movements and deficient voluntary motor activity. For this purpose quantitative analysis of motor activity is essential to understand the underlying pathophysiological processes.

References

Fitts PM (1954) The information capacity of the human motor system in controlling the amplitude of movement. J Exp Psychol 47: 381–391

Freund H-J, Büdingen HJ (1978) The relationship between speed and amplitude of the fastest voluntary contractions of human arm muscles. Exp Brain Res 31: 1–12

Hefter H, Hömberg V, Reiners K, Freund H-J (1985) Aktionstremor bei Parkinson-Patienten mit und ohne Ruhetremor. In: Schnaberth G, Auff E (Hrsg) Das Parkinson-Syndrom. Editiones Roche, Wien, Basel

Hefter H, Hömberg V, Reiners K, Freund H-J (1987 a) Stability of frequency during long-term recordings of hand tremor. Electroencph Clin Neurophysiol 67: 439–446

Hefter H, Hömberg V, Lange H, Freund H-J (1987 b) Impairment of rapid movements in Huntington's disease. Brain 110: 585–612

Hömberg V, Hefter H, Freund H-J (1985) Einfluß von Naloxon auf Bradykinese und Reaktionszeiten bei Parkinson-Patienten. In: Schnaberth G, Auff E (Hrsg) Das Parkinson-Syndrom. Editiones Roche, Wien, Basel

Hömberg V, Hefter H, Reiners K, Freund H-J (1987) Differential effects of changes in mechanical limb properties on physiological and pathological tremor. J Neurol Neurosurg Psychiatr 50: 568–579

Correspondence: Dr. H. Hefter, Department of Neurology, University of Düsseldorf, Moorenstrasse 4, D-4000 Düsseldorf 1, Federal Republic of Germany.

Freund HJ, Büdingen HJ (1978) The relationship between speed and amplitude of the fastest voluntary contractions of human arm muscles. Exp Brain Res 31: 1–12

Hefter H, Hömberg V, Reiners K, Freund HJ (1985) Akinesemessung bei Parkinson-Patienten mit und ohne Ruhetremor. In: Schneider C, [illegible] (Hrsg) Das Parkinson-Syndrom. Editiones Roche, Wien, Basel

Hefter H, Hömberg V, Reiners K, Freund HJ (1987) Stability of frequency during long-term recordings of hand tremor. Electroencephalogr Clin Neurophysiol 67: 439–446

Hefter H, Hömberg V, Lange HW, Freund HJ (1987) Impairment of rapid movement in Huntington's disease. Brain 110: 585–612

Hömberg V, Hefter H, Freund HJ (1985) [illegible] bei Parkinson-Patienten. In: [illegible] Das Parkinson-Syndrom. Editiones Roche, Wien, Basel

Hömberg V, Hefter H, Reiners K, Freund HJ (1987) Differential effects of changes in mechanical limb properties on physiological and pathological tremor. J Neurol Neurosurg Psychiatry 50: 568–579

Correspondence: Prof. Dr. H.-J. Freund, Neurologische Klinik, Universität Düsseldorf, Moorenstrasse 5, D-4000 Düsseldorf 1, Federal Republic of Germany.

Motor performance test

P. H. Kraus and **H. Przuntek**
Department of Neurology, St. Josef University-Hospital,
Bochum, Federal Republic of Germany

Summary

Looking for a simple tool for monitoring therapy and for early diagnosis of motor disturbances we used the motor performance test of Schoppe in a modified version.

In resolution and reliability tests using technical equipment for quantification of motor performance are superior to clinical examination. Quantification of movements and of movement disorders is the subject of a number of examinations using apparative tests. Most of them are designed to answer a specific scientific question, and therefore they only record selected parameters describing motion.

To obtain a better specifity in our testing we are using a battery of different apparative subtests in routine examination for control of course and to monitor therapy, with a base-line of age-corrected normal data. We found significant differences between patients and controls in the different subtests of our battery. To interpret the multidimensional test in its entirety we used discriminant-analysis, which gave an excellent distinction between patients and controls.

Introduction

The major motor symptoms of Parkinson's disease are motor deficiency (akinesia, bradykinesia, hypokinesia) and tremor. In clinical examination and in clinical scores (Duvoisin, 1971; Webster, 1968) motor deficiency is usually estimated in its entirety by complex movements (for example walking, diadochokinesia, turning around) or in anamnesis by estimation of disabilities in everyday life. To quantify the complex of symptoms in question there are different tests using technical equipment; most of them are meant to answer a specific scientific question, they use various technics to quantify different aspects of motion or different parts of movements.

Important examples are quantification of unvoluntary movements (especially tremor) with help of accelerometry, different tests measuring reaction times (Evarts et al., 1981), various methods of video processing as well as a number of examinations of complex movements (Barbeau, 1966; Duvoisin, 1971; Hallett and Khoshbin, 1980; Marsden and Schachter, 1981; Schwab et al., 1959; Teräväinen and Calne, 1980). Some of these results of apparative tests assess movements under the same aspects as they are described by clinical examination. Furthermore new aspects of motor disturbance become evident with the help of apparative quantification. On the one hand such methods give a noticeably better resolution than clinical examination, on the other hand they can really extend the spectrum of visible disturbances.

The results of apparative tests used in this way contribute to a better understanding of motor disorders. Some of these tests were used for monitoring therapy additionally to clinical estimation (Bowen et al., 1973; Schwab and Prichard, 1951). If a test gives results which correlate with the state of disease, and which are therefore relevant for estimation, it can be a useful resource.

Compared with clinical methods, standardized apparative tests also possess a greater objectivity. Apparative quantification also means a reduction of degrees of freedom. Clinical examination is a multidimensional decision; therefore it possesses a distinctly higher specifity in question of finding a diagnosis, because it considers every possible visible sign and symptom.

Clinical examination of Parkinsonian patients shows that especially in cases suffering from moderate symptoms there are clear differences of the clinical picture. An appropriate apparative assessment therefore needs a multidimensional battery of tests, which enables recording of all different aspects of the perturbation.

The basis of the battery we use is the "motorische Leistungsserie nach Schoppe" (MLS) (Schoppe, 1974), a motor performance test which was developed following factor-analytic evaluation of motor performance (Fleishman, 1962). It enables examination of complex movements, fast oscillating movements, steadiness of the hand and intention movements.

In addition to the original examination we measure regularity of fast tapping movements. As we found in former examinations (Kraus et al., 1987 a, b), this parameter is distinctly disturbed in Parkinson's disease.

Available normal data are not sufficient for a differentiated evaluation of Parkinsonian patients. Therefore we evaluated new normal data from a control group matched for age. We also calculated the age-dependence of the results for all subtests. Interpretation of the MLS results is possible intraindividually as well as by comparison with the age corrected normal values. The simplest way of evaluation which interprets the MLS in its entirety is the calculation of quotients and differences from the original data received from the subtests. Quotients between the results of left and right hand give a measure for side accentuation, the difference between first and second time intervall of the tapping test gives a parameter describing fatigue.

Interpretation of the MLS results for a whole group of Parkinsonian patients by calculation of mean and standard deviation is not straight forward: because of different side accentuation and different constellations of symptoms, especially in cases with moderate symptoms, patients may have normal values in particular subtests.

This all makes it necessary to interprete the MLS as a battery of tests in its entirety. We therefore calculated discriminant analysis between a group of Parkinsonian patients and a group of healthy controls.

Methods

The performance test developed by Schoppe (1974) consists of several subtests: tapping, aiming, tracing, steadiness and plugging (Purdue Pegboard).

By means of this device changes in speed of voluntary movements (plugging, tapping) can be recorded. Furthermore, changes in unsteadiness composed of tremor and various other disturbances (steadiness) can be examined. Tracing and Aiming, two further subtests concerning motor skills, are more appropriate for evaluating atactic disturbances than Parkinsonian symptoms.

All subtests were carried out separately for right and left hand using the original instructions of the motor performance test.

The "Plugging" test is devised for examination of voluntary movements. The subject is required to place 25 pins (diameter 2.5 mm, length 5 cm) from a frame into a series of appropriate holes (hole diameter 2.8 mm) in the contact board, separately as quickly as possible. What is measured in this test is the time interval between plugging the first and the last pin.

"Tapping" is an examination of a fast agonist-antagonist movement. The subject is required to tap on a contact board as fast as possible using a contact pencil. In its original version by Schoppe this test measures only the number of contacts during the two halves of a selected time interval (we used 32 seconds). Thus it provides details about performance speed and, in a simple manner, about alterations of speed (e.g. caused by fatique) by comparing the results of both halves of the whole interval.

In addition to the original MLS we examine regularity of tapping with help of a computer, what is measured here is the range of the log. distribution of tap intervals (Kraus et al., 1987 a, b).

The "Steadiness 1" test was performed in the smallest hole of the series in the board and subjects had no support for their executing arm. They were asked to hold a contact pencil during 32 seconds vertically in a hole with a diameter of 4.5 mm without touching either rim or bottom of the hole. Measured in this test were the number and duration of contacts.

The "Steadiness 4" test was performed in the hole of 8.5 mm diameter in the same manner.

In the "Aiming" test the examine has to hit the contact pencil on 20 contacts; measured in this test are total time required to perform the test, number of hits, number of errors and duration of errors.

In the "Tracing" test the patient has to follow a punched-out path with a stylus as precisely as possible. Measured in this tests are total time required to perform the test, number of errors and duration of errors.

Controls and patients

For the present examination all controlls and patients were righthanded and between 40 and 85 years old.

Out of a total of 97 controls (56 women, 41 men) aged on average 62 ± 12 years, for the different subtests groups of 50 to 59 persons were evaluated. They presented neither neurological nor orthopaedical diseases and their state of health was estimated to be age-appropriate by the examiner. Only right handed controls were chosen for evaluation of normal values.

54 women and 53 men with an average age of 62.5 ± 8.6 years formed the patient group; average duration of disease was 3.6 years, all of them were right-handed. They were assessed by the examiner as follows: 11% severe, 40% moderately severe cases, 49% were slightly diseased. The Hoehn and Yahr stages lay between stage 1 and 4. The major symptoms were tremor in 13% of the patients, motor deficiency was prominent in 72% and in 15% of the cases tremor and motor deficiency were estimated to be equivalent. Apart from a few recently diseased patients, all the patients were under specific medication against Parkinsonism.

Results

Our results of the original MLS test for the control group are shown in Table 1.

Most of the subtests give data which are not normally distributed. Because of the unknown distribution we describe our data material by median (50% limit) and the 95% limit estimated directly from the age corrected data.

For the present age interval (40—85 years) the real age-dependance in most of the subtests was exponential, but linear correction was estimated to be sufficient. In some of the subtests there was no

Table 1. Median, 95% limits (age corrected data) and dependence on age for controls

		Right			Left		
		50%	95%	dependence from age	50%	95%	dependence from age
Plugg.	TD	443.05	567.41	3.169	473.34	618.94	3.165
Tap.	NC1	90.39	69.83	−0.299	80.82	65.02	−0.252
	NC2	84.12	65.54	−0.311	76.37	60.30	−0.214
	NC1−NC2	5.21	10.41	0	4.06	13.42	0
Stead. 4	NE	1.17	6.00	0.127	2.08	8.53	0.197
	DE	0.90	5.41	0.135	1.49	15.73	0.226
Stead. 1	NE	13.38	28.11	0.595	17.79	33.16	0.684
	DE	14.81	51.03	0.987	20.86	74.84	1.254
Trac.	NE	20.92	30.20	0	25.33	34.17	0
	DE	32.41	63.52	0.555	38.23	65.59	0.592
	TD	104.25	220.95	0	107.63	234.65	0
Aim.	NE	0.21	2.09	0	0.67	3.35	0
	NH	20.03	20.84	0.007	20.00	20.95	0
	DE	0.20	1.27	0.034	0.23	2.77	0
	TD	106.49	151.78	1.062	118.71	164.78	1.116
Stead. 4	DE/NE	0.21	1.57	0.022	0.34	1.32	0.019
Stead. 1	DE/NE	0.95	2.20	0.021	0.95	2.14	0.031
Trac.	DE/NE	1.50	5.69	0.028	1.63	4.45	0.030
Aim.	DE/NE	0.12	1.24	0.019	0.15	1.21	0

Plugg. Plugging, *Tap.* Tapping, *Stead. 4* Steadiness 8.5 mm, *Stead. 1* Steadiness 4.8 mm, *Trac.* Tracing, *Aim.* Aiming, *NC1, NC2* number of contacts/interval 1, interval 2, *TD* total duration, *NE* number of errors, *DE* duration of errors, *NH* number of hits, the dependence from age is the gradient.

significant correlation at all (Table 1). The range between the best and the worst result increased with increasing age for all subtests.

Because of different constellations of symptoms data of the whole patients' group are very heterogeneous. We therefore calculated a cluster analysis of the patients' group to differentiate subgroups. We chose a state with three clusters for further evaluation; the clusters showed the most differences in the state. We processed each cluster in a different discriminant analysis against the control data.

The patients' cluster with the best test results consists of 44 persons, the remaining clusters consist of 38 and 21 patients. Choosing the separation with no false positiv classification we get all together 2 false negative out of 103 patients, if we choose a separation with no false negativ classification we get 13 false positive. The data of the different subtest are not normally distributed; therefore the discriminant function can not be used for the evaluation in single persons.

Discussion

We expanded the "Motorische Leistungsserie nach Schoppe" (MLS) concerning the regularity of tapping movements and some new calculated parameters. This motor performance test consists of serveral subtests which give a simple description of different aspects of motor dexterity. In earlier examinations we found significant differences between patients and controls for most of the subtests (Kraus et al., 1987 a, b); furthermore we found a very good correlation between the parameters of the Webster rating scale and the results of the MLS for more than 400 Parkinsonian patients (submitted for publication).

The presented age-corrected normal values are helpful for estimation of patients' data in the monitoring and staging of treatment.

We use the presented quotients of duration and number of errors for estimation of the subtest itself: Most of the conspicuous quotients depend on different effects which do not correspond to the motor disturbance (for example long duration of errors because of impaired visual efficiency). If the quotient exceeds the 95% normal value, the subtest cannot be evaluated.

The difference in the side accentuation of motor performance estimated by means of the apparatus is sometimes the only conspicuous parameter in some very early cases. Here we use a "normogram" considering correlation between the results of left and right hand. Normal values, calculated as a resultant using a trigonometrical formula are also prepared.

In earlier examinations we found significant differences for most of the subtests, although there was a clear overlap between the two distributions.

The discriminant analysis allows interpretation of the MLS battery in its entirety. The MLS records the symptomatics of Parkinsonism very much better than all single subtests.

With regard to findings of different tests using technical equipment and the method of clinical diagnostics, we started out with the hypothesis that only a battery of tests enables one to make an appropriate evaluation of motor disturbances. Our results confirm this hypothesis.

We have already been using the MLS for a long time as a standardized multidimensional test in routine examination for following the course of the disease and for monitoring treatment. With the age-corrected normal data as a baseline, the method can also contribute to early detection of motor disturbances.

We are now examining a more extended version (including for example apparative assessment of diadochokinesia) (Kraus et al., 1987 c) to improve the specifity.

We found the examination of the regularity of fast tapping movements to be a highly significant separating tool as well between patients and controls as well as within the patients group (Kraus et al., 1987 a).

Our apparative examination of diadochokinetic movements (Kraus et al., 1987 c) showed conspicuousness in some cases in which the clinical examination did not lead to any result: For example the coordination between left and right hand was found to be disturbed.

Already the MLS in its simpliest form can catch motor disturbances on very early stages. Our hitherto results show that with help of modification and expansion of the test battery a quite better separation and even a contribution to early differential diagnosis can be expected.

References

Barbeau A (1966) The problem of measurement of akinesia. J Neurosurg (Chicago) 24 [Suppl] I, part III: 331–334

Bowen FP, Brady E, Yahr M (1973) Sensorimotor coordination in Parkinson's disease before and after levodopa therapy. Neurology 23: 1101–1106

Duvoisin RC (1971) The evaluation of extrapyramidal disease. In: Ajuriaguerra J (ed) Monoamines, noyaux gris centraux et syndrome de Parkinson. Symposion Geneve 1970. Masson, Paris, 313–325

Evarts EV, Teräväinen H, Calne DB (1981) Reaction time in Parkinson's disease. Brain 104: 167–186

Fleishman EA (1962) A factor analysis of fine manipulative tests. J Appl Psychol 46/2: 96105

Flowers KA (1975) Ballistic and corrective movements on an aiming task: intention tremor and Parkinsonian movement disorders compared. Neurology 25: 413–421

Hallett M, Khoshbin S (1980) A physiological mechanism of bradykinesia. Brain 103: 301–314

Kraus PH, Klotz P, Fischer A, Przuntek H (1987 a) Assessment of Symptoms of Parkinson's Disease by Apparative Methods. In: Riederer P, Przuntek H (eds), MAO-B-inhibitor selegiline (R-(-)-deprenyl), a new therapeutic concept in the treatment of Parkinson's disease. J Neural Transm [Suppl] 25: 89–96

Kraus PH, Klotz P, Fischer A, Przuntek H (1987 b) Beurteilung der Symptome des Morbus Parkinson mit apparativen Methoden. In: Riederer P, Przuntek H (Hrsg), Morbus Parkinson; Selegilin (R-(-)-Deprenyl), Movergan®, Ein neues Therapiekonzept. Springer, Wien New York

Kraus PH, Keck B, Klotz P, Przuntek H (1987 c) Computer aided analysis of diadochokinesia, Electroenceph Clin Neurophysiol 66/5: 57

Marsden CD, Schachter M (1981) Assessment of extrapyramidal disorders. Br J Clin Pharmacol 11: 129–151

Schoppe KJ (1974) Das MLS-Gerät: Ein neuer Testapparat zur Messung feinmotorischer Leistungen. Diagnostica 20: 43–46

Schwab RS, England AC, Peterson E (1959) Akinesia in Parkinson's disease. Neurology 9: 65–72

Schwab RS, Prichard JS (1951) An assessment of therapy in Parkinson's disease. Arch Neurol Psychiat (Chicago) 65: 489–501

Teräväinen H, Calne DB (1980) Quantitative assessment of parkinsonian deficits. In: Rinne UK, Klingler M, Stamm G (eds) Parkinson's disease—current progress, problems and management. Elsevier, North-Holland, Biomedical Press Amsterdam, New York: 145–164

Webster DD (1968) Clinical analysis of the disability in Parkinson's disease. Modern Treatment 5: 257–282

Correspondence: Dr. P. H. Kraus, St. Josef University-Hospital, Gudrunstrasse 56, D-4630 Bochum 1, Federal Republic of Germany.

Measuring body movements in neurological disease, with special reference to Parkinson's disease

G. Steg[1], B. Johnels[1], P. E. Ingvarsson[1], and M. Thorselius[2]
Departments of [1]Neurology, Sahlgren Hospital, and [2]Applied Electronics, Chalmers University of Technology, Gothenburg, Sweden

Summary

Computer assisted optoelectronic movement analysis has important applications in the quantitative study of human movements and motor disorders. In combination with specific test tasks it has proved useful in the quantitative analysis of pathophysiological mechanisms and effects of therapy in Parkinson's disease. A compound movement task, the PLM-test, was designed to test the integration of postural uprising, locomotion and a goaldirected manual movement in an everyday lifting task. Parkinson patients show a desintegration of this compound movement and it is of pathophysiological and nosological importance that symptom profiles related to basal motor control mechanisms can be differentiated and quantified. Pharmacotherapeutical effects are quantified in single dose effect curves of dopaminergic drugs.

Introduction

In daily life, most movements that cause the motor disability in Parkinson patients are compound in nature. Since objective instrumental methods hitherto have been applicable only to simple movements, under mechanical restrictions, motor disability usually is assessed using subjective rating scales. We have utilized an opto-electronic measurement technique to inflict as little restraints as possible and a motor test—the Posturo-Locomotion-Manual (PLM) test—which is relatively easy to execute, but requires a certain degree of postural control of the body, walking capacity and some manual dexterity. A computer program ensures objective and automatic

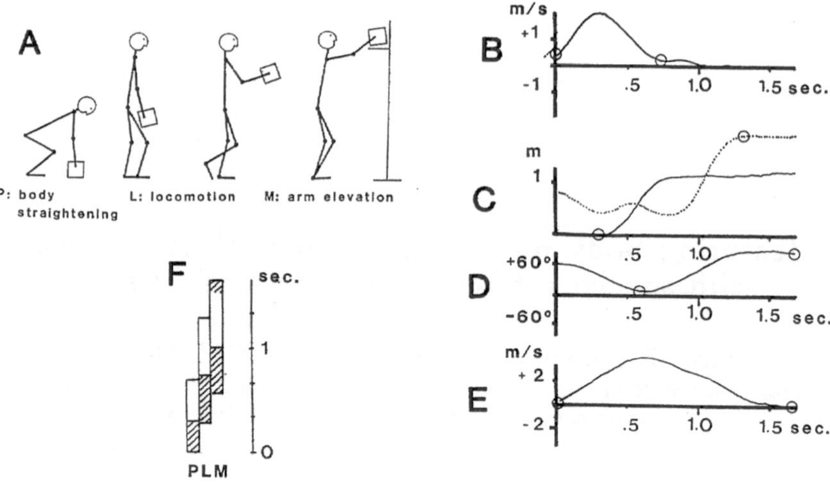

Fig. 1. PLM test procedure and data printout. *A* The subjects are asked to lift a light object from the floor to a shelf 1,5 meters away. The compound body movement is composed of three more or less overlapping phases: The P-phase (straightening up of the body), the L-phase (locomotion) and the M-phase (arm elevation). *B* Data printout of one test movement performed by a normal subject aged 75 years. Vertical speed of the head is shown, with circles marking start and stop of the postural *(P)* phase. *C* Circles marking start and stop of horizontal movement of feet define the locomotion *(L)* phase. *D* Start of the manual *(M)* phase is defined as start of rise in the calculated angle between arm and trunk, i.e. start of arm elevation. *E* Start and stop of the test movement, i.e. the movement time (MT) is derived from the graph showing calculated difference between the vertical and horizontal speeds of the test object. *F* The PLM coordination is shown by degree of overlapping between the bars which plot the P, L and M phase durations. Hatched area in bar graphs show the active (acceleration) part of the movement

evaluation of the motor performance, by measuring the overall motor capacity (movement time for the test) as well as subdivisions of the test movement into a postural (P—rising from floor), a locomotor (L—walking forward a few steps) and a manual (M—placing the object on a shelf placed at the height of the subjects chin) phase (see Fig. 1). These three movement phases represent motor functions that may be selectively disturbed in Parkinson's disease, thus creating a "symptom profile" for each patient (Johnels et al., 1987). The method is applicable for quantification of anti-parkinsonian drug effects. The work on patients with Parkinson's disease demonstrated here is more fuly reported elsewhere (Johnels et al., 1987, 1988; Ingvarsson et al., 1987).

Methods

Subjects: 40 normal subjects aged 21—85 years and 10 patients with Parkinson's disease of different severity performed the PLM-test, some of them on several occasions. Motor test: In the PLM-test subjects were required to bend down to catch a small lightweight object formed as a metal handle, lift it up, walk forward 1.5 meters and place the object on a shelf at the height of their chin. The object should then be returned to the start position and the test movement was then repeated immediately up to eight times (see Fig. 1 A). Equipment: The subject was equipped with light-emitting diodes (LEDs) placed over their temple, shoulder, elbow, hip and both feet. The LEDs were fastened to a hair ribbon, a zip-suite overall sewn from stretch cloth and a pair of elastic diving-boots. LEDs were also placed on the handle of the object. The spatial location of each diode within the measurement field was determined by a 3-D opto-electronic camera system (IROS 3-D, Remplir AB, Göteborg) with a repetition rate of 50—100 Hz. Space coordinate data were recorded on the hard disk of a personal computer (IBM AT 3) and later analyzed in a mini-computer (VAX-station II). Static space resolution was about 1 millimeter.

Results

Normal subjects

Preparation for the test was easy and completed within a few minutes. The normal subjects performed well without hesitation or problems. They were allowed to train the test movement one or two times in ad lib speed and were thereafter requested to perform as fast as possible during all further recordings. Movement Time (MT) ranged between 0.8 and 2.1 seconds, increasing with age. Variation during one recording was 3—5 percent and between different days less than 10 percent. Normal values for MT, P-, L- and M-phases were calculated by linear regression from these data and used for reference in the calculation of the relative performance and determination of the motor deficit in the patient group.

Parkinsonian patients

Baseline recordings

In the treated Parkinson patient with advanced disease the symptoms often vary strongly and sometimes unexpectedly during the day—the "on-off" phenomenon (Nutt, 1987). To assess the degree of disability in each patient, "baseline" recordings were performed in the morning, 12 hours after the last dose of anti-parkinson drugs. 3—4

Table 1. Relative PLM test results in PD patients before treatment

Patient Number	Age Years	P.D. Years	MT	P	L	M
				(times normal values)		
1	38	7	2.70	1.46	2.52	2.40
2	43	6	1.56	1.38	1.52	1.70
3	48	9	3.16	1.73	1.84	3.45
4	51	9	2.01	1.29	1.70	1.91
5	60	7	1.78	1.30	1.68	1.78
6	60	15	2.74	1.12	2.21	1.47
7	60	16	2.14	1.58	1.62	1.94
8	60	17	2.11	1.30	1.32	1.99
9	66	8	2.46	1.45	1.72	2.15
10	75	2	1.58	1.13	1.70	1.82

Average values of Movement time (MT), Postural (P), Locomotor (L) and Manual (M) phase durations, multiplied by the age adjusted values from normal subjects

recordings were made with 15 minutes interval before any treatment was given. In five of the eight patients which were tested at at least three occasions, baseline MT recordings were quite reproducible, usually giving less than 10 percent (and never more than 35 percent) difference in performance on different test days. The remaining three repeatedly tested patients, however, had a more variable performance on different recording dates.

Comparison to normal values

Compared to performance of normal controls, corrected for age. The MT and PLM phase times were increased in all patients tested after 12 hours without medication. The duration of the different P-, L- and M-phases of the test movement were increased in parallel with MT but to a varying degree in the individual patient, thus indication dominating dysfunctions in postural (P), locomotive (L) or manual (M) motor systems respectively in different patients (see Table 1).

Disability profiles

To make clear to what extent specific dysfunction in one of the P, L and M movement phases contributed to the total disability, i.e. the prolongation of MT, the phases were plotted against MT for every measurement done in the "off-state" before treatment (see Fig. 2).

Fig. 2 A demonstrates a case where there was a prolongation of the M-phase and to some degree also of the P-phase with increasing MT. The L-phase values were statistically unrelated to MT. Thus the time for the arm-hand movement increasing relatively more than the

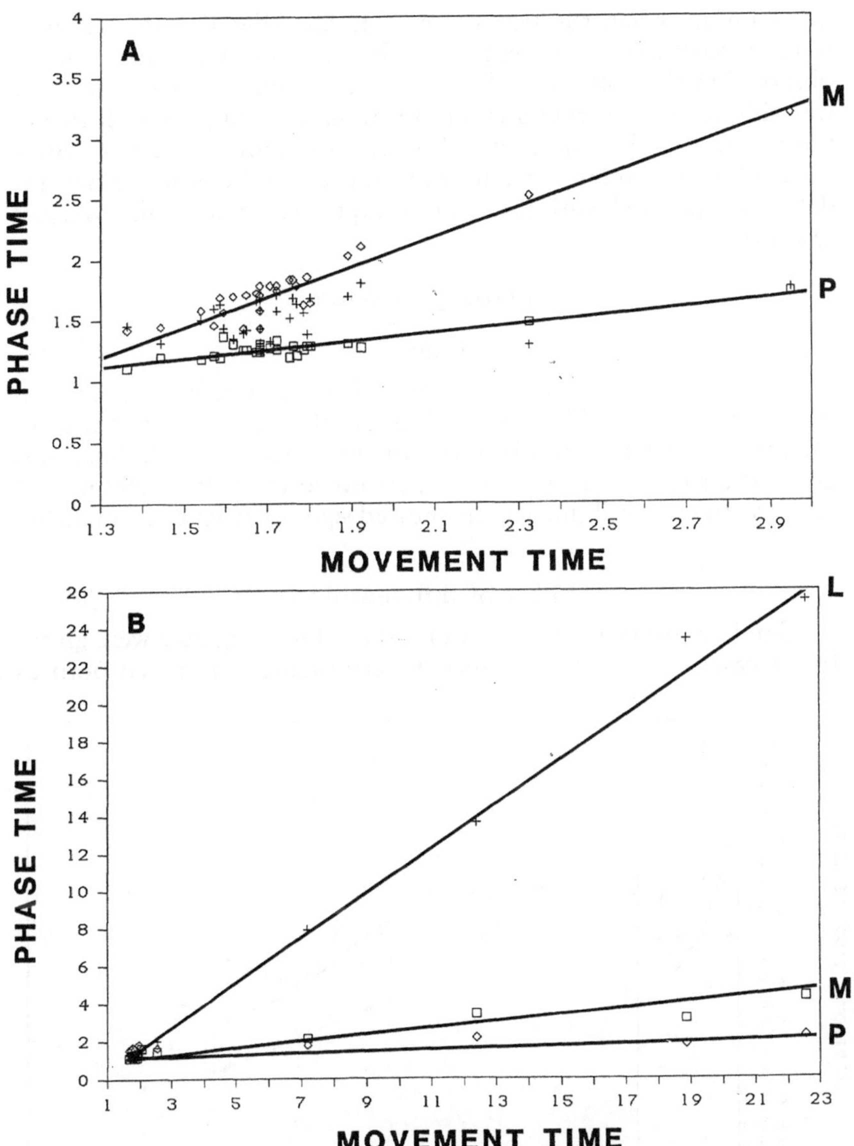

Fig. 2. Clinical disability profiles of two PD patients. Regression curves are achieved by plotting all data achieved from the "off-state" recordings of individual phase time against the movement time (MT). Thus, the slope of the regression curve indicate the degree of prolongation of each of the partial movements of the PLM test, i.e. the relative disability profile. *A* Disability profile of patient with dominating dysfunction of hand dexterity and coordination. *B* Pattern of predominant locomotion difficulties

other phases when the patients performance deteriorated, giving a steep slope of the regression line for the M-phase. This patient, when observed in the "off"-state, had a severe arm-hand dysfunction. He was able to rise up from a chair and walk without greater difficulty but not to use his hands. Fig. 2 B shows an example of a case where the L-phase was increased out of proportion to the other phases, i.e. data that agree well with the clinical impression of a dominating gait disorder.

Effects of treatment

L-dopa

After baseline recordings in the "off"-state L-dopa in combination with a decarboxylase inhibitor (benserazide 28.5 mg in Madopar® or carbidopa 10 mg in Sinemet® per 100 mg L-dopa) was given as a single dose. Tests were performed every 15 to 20 minutes after treatment until the patient showed signs of fatigue, normally up to 2 hours.

Effect of different doses

Single doses of levodopa from 100 to 300 milligrams were given. In all cases except one the motor performance improved both as

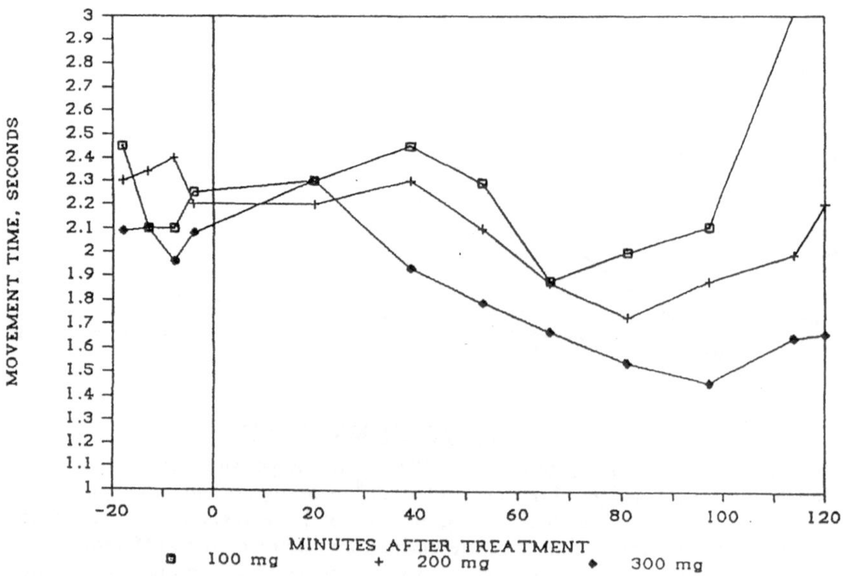

Fig. 3. Effect on movement time by different doses of L-dopa. 100–300 mg Madopar® was given at time 0. Times values found in age matched normal subjects

judged by inspection and in the MT data. With few exceptions there was a clearcut dose-response relationship, as can be seen from Fig. 3. In the case demonstrated, motor performance was barely brought into the normal range for a short time after a dose of 300 mg L-dopa.

Effect on movement phases

L-dopa generally had effect on all the phase durations of the PLM test movement. The effect was most marked and consistent on the L-phase while it was least prominent on the P-phase.

Bromocriptine

Five patients were given single doses of bromocriptine (Parlodel®), ranging from 2.5 mg to 20 mg. No consistent improvement was seen during up to 3 hours observation. In one patient the P-phase was shortened without shortening of MT, L and M-time indicating a more selective action on postural control systems in the nervous system for bromocriptine than for L-dopa.

Discussion

The normal way to quantify the clinical symptoms and disability has been to use verbal so called "rating scales", listing the symptoms or functions and giving each a score following an nonparametric, arbitrary scale. The merit of these tests over simple physiological measurements has been that they permit observation of motor function in activities of daily life (ADL). These scales, however, are subjective, unprecise and unreliable (Diamond and Markham, 1983). They are little suited for pathophysiological research, i.e. elucidation of the mechanisms that cause the symptoms. As they are subjective, it is difficult to compare studies performed at different centra, or even by different observers within the same clinic. Their aid in early diagnosis of PD, or evaluation of treatment in early stages of disease is, at best, quite limited. Thus, it takes long time and studies on numerous patients to reach a general agreement on the effects of a certain treatment. Drug trials of this type necessitate costly toxicity tests. The net result is an impediment to development of new treatments.

The opto-electronic technique provides objective and precise measurements of human movements. It is thus suited for pathophysiological research. Utilizing simple tests with relevance for ADL functions it can identify deviations from normal movement patterns.

Compound movements can be divided into their component movement phases, which permits an evaluation of specific motor functions relating to physiology. When this new technique was used to evaluate patients with Parkinson's disease in the induced "off-state" after 12 hours without medication, it was found that the individual patients vary profoundly regarding which motor function that was most severely disabled, i.e. disability (symptom) profiles have been demonstrated. The two most common drugs in the treatment of Parkinson's disease have been given in acute single-dose experiments. The clinical effect of these are quite different as demonstrated by the measurements. L-dopa showed a shortlasting but marked effect towards normalization of movement time and the P, L and M movement phases, in a dose-related way, while bromocriptine showed a less marked effect most clearly demonstrated on the P-phase.

Thus, in the present application, Parkinson's disease, the opto-electronic technique presented here has helped us gain insights into pathophysiological mechanisms underlying the movement disorder, and shown to be of benefit in diagnosis of disease as well as in evaluation of different drug treatments. It has even been used in ongoing therapeutic trials with transplantation of human fetal substantia nigra tissue to PD patients. Although experience with detection and diagnosis of PD in early stages of disease is limited, our first results have been quite promising, giving us hope that we have a sensitive, reproducible method of detecting even borderline or slightly disabled patients. In one such patient repeated PLM tests—before and after instigation of L-dopa treatment—confirmed an uncertain diagnosis, which had direct beneficial treatment implications.

There seems to be a great development potential for clinical applications of the opto-electronic movement measurement techniques in different movement disorders.

References

Diamond SG, Markham CH (1983) Evaluating the evaluations: Or how to weigh the scales of parkinsonian disability. Neurology (Cleveland) 33: 1098–1099

Ingvarsson PE, Johnels B, Steg G (1987) Dissolution of motor program coordination in Parkinson's disease. In: Grillner S, Stein PSG, Forssberg H, Stuart DG, Herman RM (eds) Neurobiology of vertebrate locomotion. Macmillan, London, pp 105–120

Johnels B, Ingvarsson PE, Rydgren U, Thorselius M, Steg G (1987) Measuring motor function in Parkinson's disease. In: Marsden CD,

Conrad B, Benecke R (eds) Motor disturbances I. Academic Press, London, pp 131–144

Johnels B, Ingvarsson PE, Thorselius M, Valls M, and Steg G (1988) Parkinsonian disability profiles quantified by optoelectronic movement analysis (submitted)

Nutt JG (1987) On-off phenomenon: Relation to levodopa pharmacokinetics and pharmacodynamics. Ann Neurol 22: 535–540

Correspondence: Prof. Dr. G. Steg, Department of Neurology, Sahlgren Hospital, S-413 45 Göteborg, Sweden.

Long-term measurement of tremor: early diagnostic possibilities

E. Scholz, M. Bacher, A. Bellenberg, S. Hart, H. C. Diener, and **J. Dichgans**

Department of Neurology, University of Tübingen, Federal Republic of Germany

Summary

Long-term recording is able to detect small amounts of low intensity tremor, which due to its intermittent manifestation may be missed during the routine clinical examination. Examples are given from patients with an established diagnosis of Parkinson's disease in whom tremor could be observed only under special conditions, such as mental or emotional stress, or was just reported by the patient but could not be observed by the examiner. For reevaluation, data sample epochs classified by the automatic analysis as exhibiting tremor can be stored and visually inspected. Problems may arise with the diagnostic classification of tremor based on its frequency, since a considerable variation in frequency is observed in long-term recordings from parkinsonian patients. This may partially originate from the otherwise very advantageous fact that subjects are not restricted to a fixed position of the arms but are recorded under every day conditions.

Introduction

Our method of 24-hour-tremor-recording was designed for automatic tremor quantification in patients with an established diagnosis of Parkinson's disease (Scholz et al., 1988). One of its possible borderline indications might be the detection of minimal tremor in early cases. In the case of parkinsonian rest tremor this means detecting rare and short episods of the typical 4—5 Hz tremor in incipient manifestation of the tremor type of the disease. It is common clinical knowledge that tremor may be missing during the clinical examination at one time but may be visible at another, such that the diagnostic classification may be delayed. In these early cases tremor may only be detectable in certain situations, such as during excitement, mental drive or fatigue. In a pilot study, we therefore combined the long-term recording with a "stress-experiment" using the "Stroop"-task, which has earlier been used to enhance tremor (Gresty et al., 1984).

Technical equipment of 24-hour-tremor-recording

EMG tremor activity was recorded with a 4-channel Medilog recorder. A pair of surface electrodes (silver/silver chloride pre-gelled electrodes, 1 cm in diameter, commercially available as 24 hour ECG electrodes, Kendall Ltd.) was placed 10 cm apart over the extensor carpi radialis and flexor carpi ulnaris muscles of each forearm.

24 hours of EMG data were stored on analog tape. Data were processed using a PDP 11/44 computer and specially designed software (Bacher et al., 1988). Low frequency artefacts were eliminated by means of a digital FIR-high-pass filter with a cut-off frequency of 10 Hz. The signal was squared, low-pass filtered and linearized again in order to obtain an estimate of the activity level. The cut-off frequency of the digital FIR-low-pass filter was 20 Hz. Power spectra were calculated for periods of 5.12 seconds (real time). Spectra from 3 sequences (15.36 seconds real time) were averaged. Various parameters were calculated including tremor frequency, tremor intensity, and tremor occurrence. Tremor intensity was estimated by means of the signal-to-noise ratio (SNR) of the dominant peak between 3 and 15 Hz. Tremor frequency was measured with an accuracy of 0,195 Hz due to the frequency resolution of the spectrum. The programme is able to distinguish between EMG-activity with a rhythmic modulation of amplitude (= tremor) and non-modulated EMG-activity (tonic background activity, voluntary activity). In order to differentiate tremor from unmodulated or absent EMG-activity, two criteria were used. First, the signal-to-noise ratio had to be greater than 4. Second, the frequency of the dominant peak had to be between 3.7 and 10 Hz. For control of the automatic analysis, all or part of the original data can be stored in a file.

"Stroop" task

Both a colour naming task and the stroop task were presented to patients. In the colour naming task, dots in different colours had to

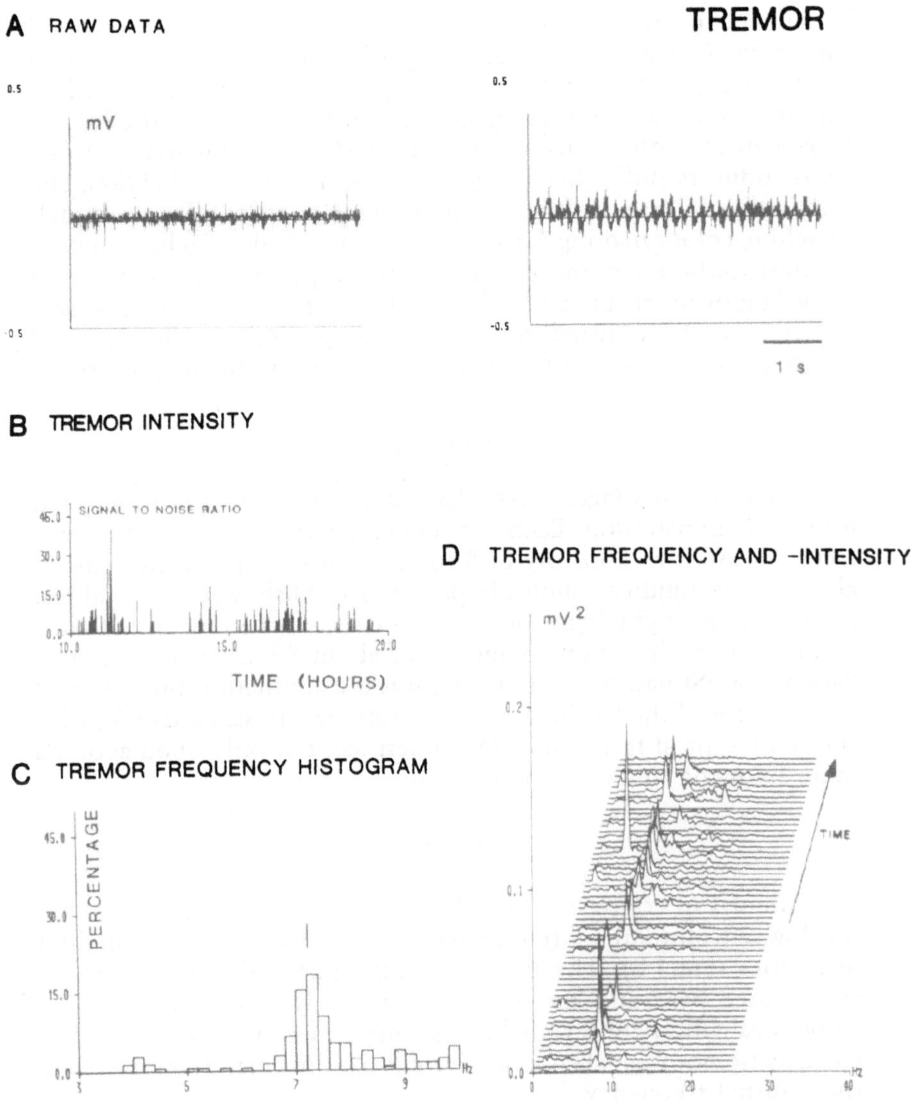

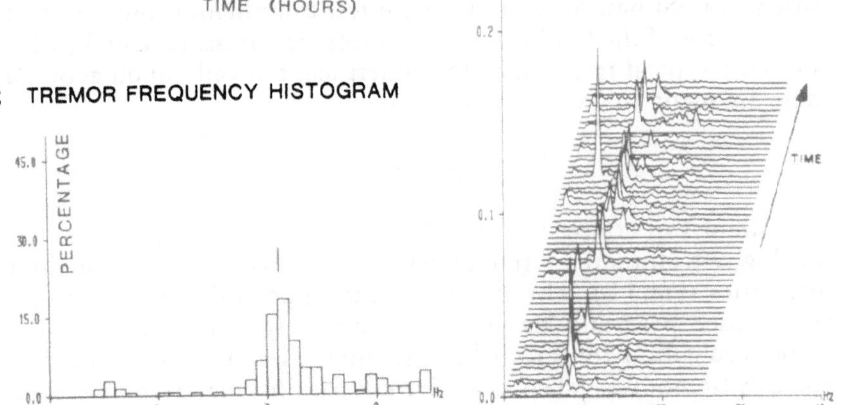

Fig. 1. 1 A reveals 5 to 6 Hz tremor activity, although the mean peak lies at 7.2 Hz (1 C) with a clearcut peak as in all our parkinsonian patients. 1 D shows a considerable stable tremor frequency over time with the usual gross variation in tremor intensity

be named. In the "stroop" task, however, the patient was presented with a large matrix of words comprising the names of common colours (red, yellow, blue and green). Each word was printed in coloured ink, but the printed names and the colour of the ink diverged in an unpredictable way. The tasks are comparably strong stress-inducers and yield reproducible results (Gresty et al., 1984). The standardized time schedule comprised 5 segments of 60 seconds each (baseline, colour naming, baseline, stroop test and again baseline). In contrast to the longterm recording, stress experiments also involved recording of heart rate and galvanic skin response so as to compare different vegetative reactions. All data were processed with a personal computer and stored on floppy disks for further off-line processing.

Patients

Three patients (aged 34/49/53 years respectively) were selected for this demonstration. Each of them suffered from an early Parkinson's disease with a history of 8 to 18 months and discrete signs of akinesia and rigidity. Clinically patient 1 (GS) showed a typical rest tremor of the right hand only with increased mental drive (arithmetic). The high tremor frequency of about 8 Hz was exceptional. Patient 2 (DS) had no tremor on clinical examination but reported gross tremor of the left hand under emotional stress. Patient 3 (MM) showed a typical rest tremor in the left leg, but only upon action a more frequent tremor of both hands.

Results

Patient 1 in contrast to the clinical examination (see above) has the highest amount of tremor with 7.14 percent and the highest intensities (Fig. 1 B). The frequency histogram exhibits a clear-cut small based peak at 7 Hz. Tremor is diffusely distributed over the whole evaluation time with highest intensities around 11 a. m. The patients tremor was taken as a clear-cut parkinsonian tremor despite the unusual frequency.

→

Fig. 2. *A* shows the raw data obtained in the Stroop test during baseline (left) and colour naming (right). Note the increase in tremor amplitude with unchanged frequency. *B* short low intensity tremor bursts are distributed over the whole 10 hour evaluation period. The frequency histogram (*C*) shows a main peak at 9–10 Hz, caused by the evaluation procedure (see text for discussion). Each line in *D* shows an average plot of 1 minute with the highest amount of tremor out of 10 minutes ("worst case plot")

TREMOR

A RAW DATA

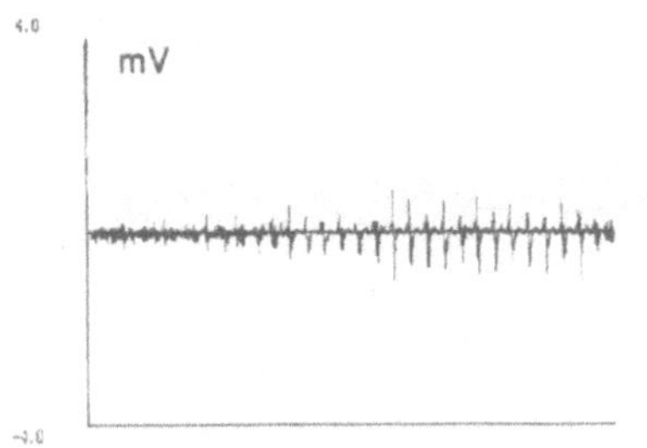

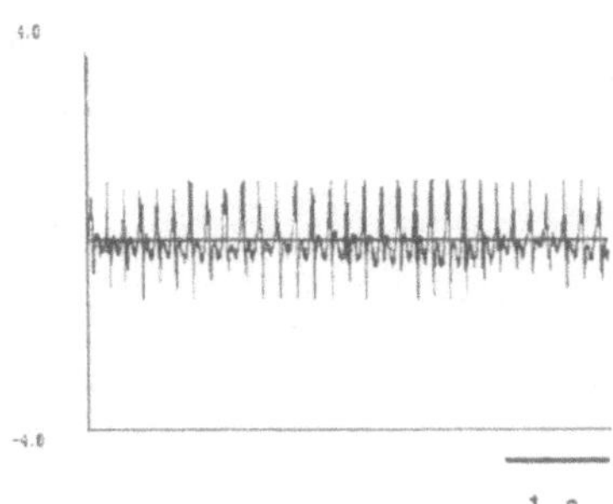

B TREMOR INTENSITY

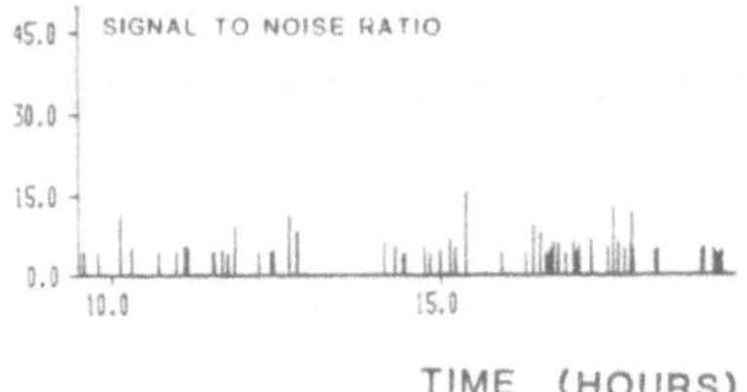

C TREMOR FREQUENCY HISTOGRAM

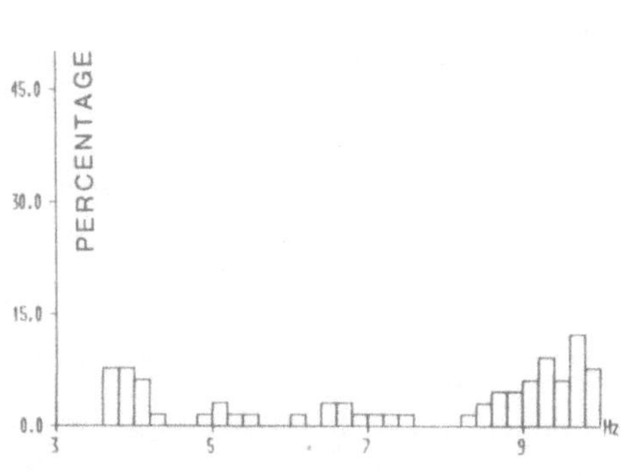

D TREMOR FREQUENCY AND -INTENSITY

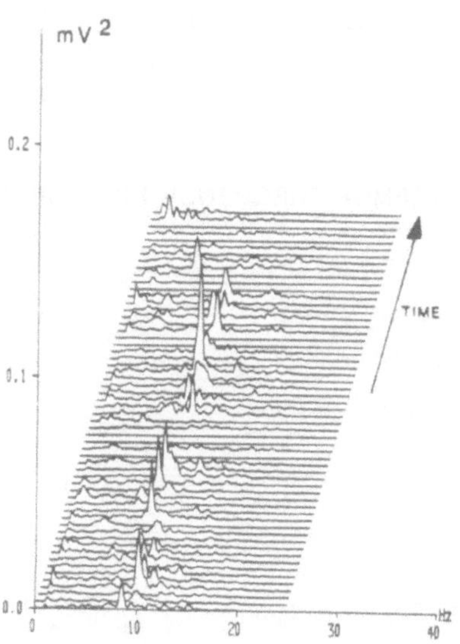

Fig. 2

TREMOR

A RAW DATA

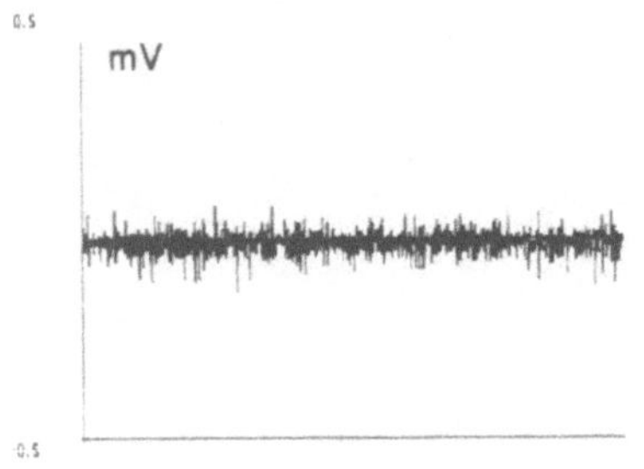

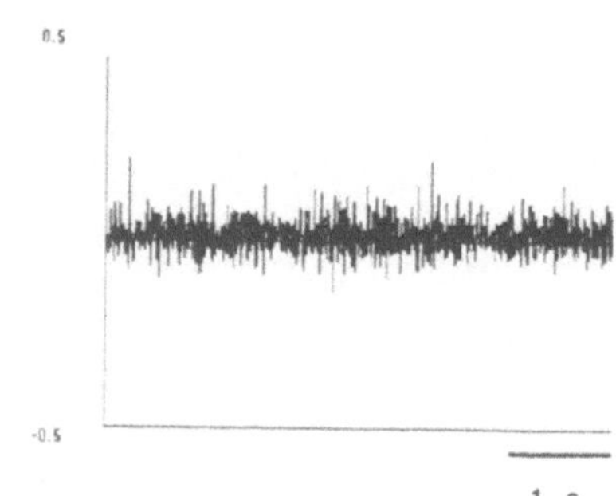

B TREMOR INTENSITY

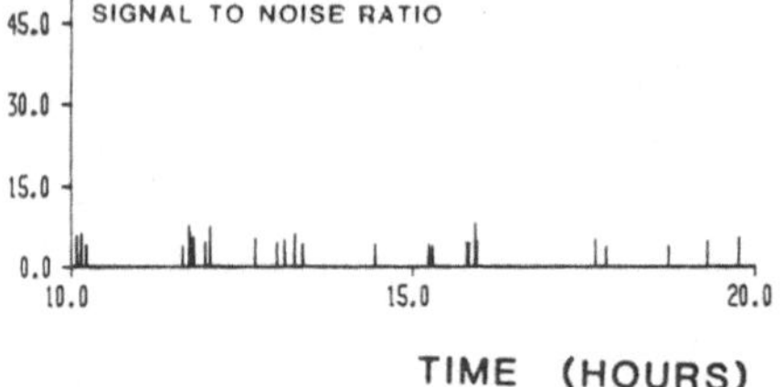

C TREMOR FREQUENCY HISTOGRAM

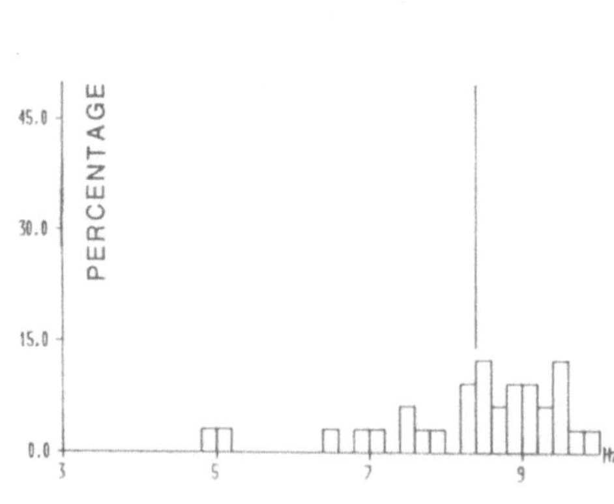

D TREMOR FREQUENCY AND -INTENSITY

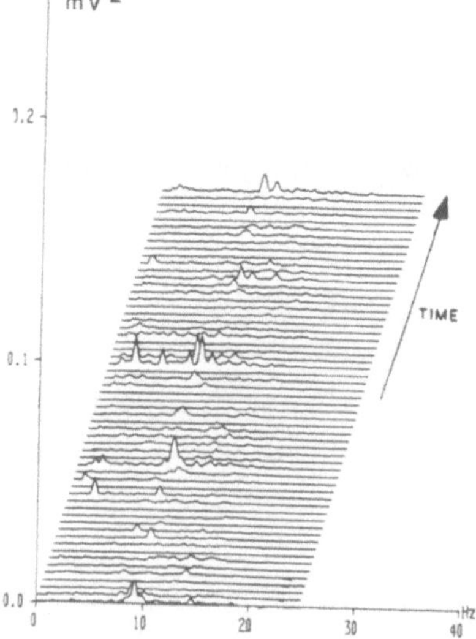

Fig. 3

The records of patient 2 (Fig. 2) show rhythmic activity in 3.52 percent of the 10 hour evaluation time (65 periods out of 2304, each 15.36 s long). The frequency histogram reveals several peaks, the largest around 10 Hz, with only a minor portion around 4 and 5 Hz (see Fig. 2 C). The EMG-raw data gained in the stroop experiment (Fig. 2 A) show a tremor with a frequency of about 5—6 Hz. Tremor amplitude increases from baseline (on the left) to colour naming in the stroop task (Fig. 2 A on the right). Visual evaluation of the long-term recording reveals only very short epochs of a low-frequency tremor (5—6 Hz), but long trains of EMG-activity modulated around 9—10 Hz, which is plotted in Fig. 2 D. We therefore concluded, that patient 2 had only short periods of parkinsonian rest tremor (as when under provocation) and predominantly an enhanced physiological tremor.

Patient 3 shows tremor in 1.45 percent of the evaluation periods. The frequency histogram shows a broad peak around 9 Hz and nearly no activity around 5 Hz (see Fig. 3 C). This frequency distribution was often observed in our patients with essential tremor, but at lower frequencies around 7 Hz. In agreement with the clinical diagnosis the patient most probably suffered from rest tremor in the leg but from an enhanced physiological tremor in the hands. The "Stroop"-experiment shows an increase in EMG-activity without overt tremor modulation. This observation agrees with the assumption of an enhanced physiological tremor (Fig. 3 A).

All patients showed a considerable increase in EMG-activity during the colour naming and the stroop tasks respectively. Tremor amplitude shows an increase with a constant frequency of tremor bursts. Tremor EMG-plots in selected periods during the day and in the stress-experiment are virtually identical.

Discussion

Long-term EMG tremor recordings are a valuable means of recording tremor and treatment effects. Recordings on three consecutive days with our evaluation procedure show reproducible data for tremor occurence and median tremor intensity (Fig. 4). The amount of tremor (1.4 to 7.1 percent) in the incipient cases presented above

Fig. 3. *A* shows non-modulated EMG activity as in voluntary activity. Although the plot in *B* reveals bursts of "tremor", *C* shows no clearcut peak, but rather frequencies diffusely distributed around 9 Hz as in *D*. Tremor is assumed to be enhanced physiological tremor

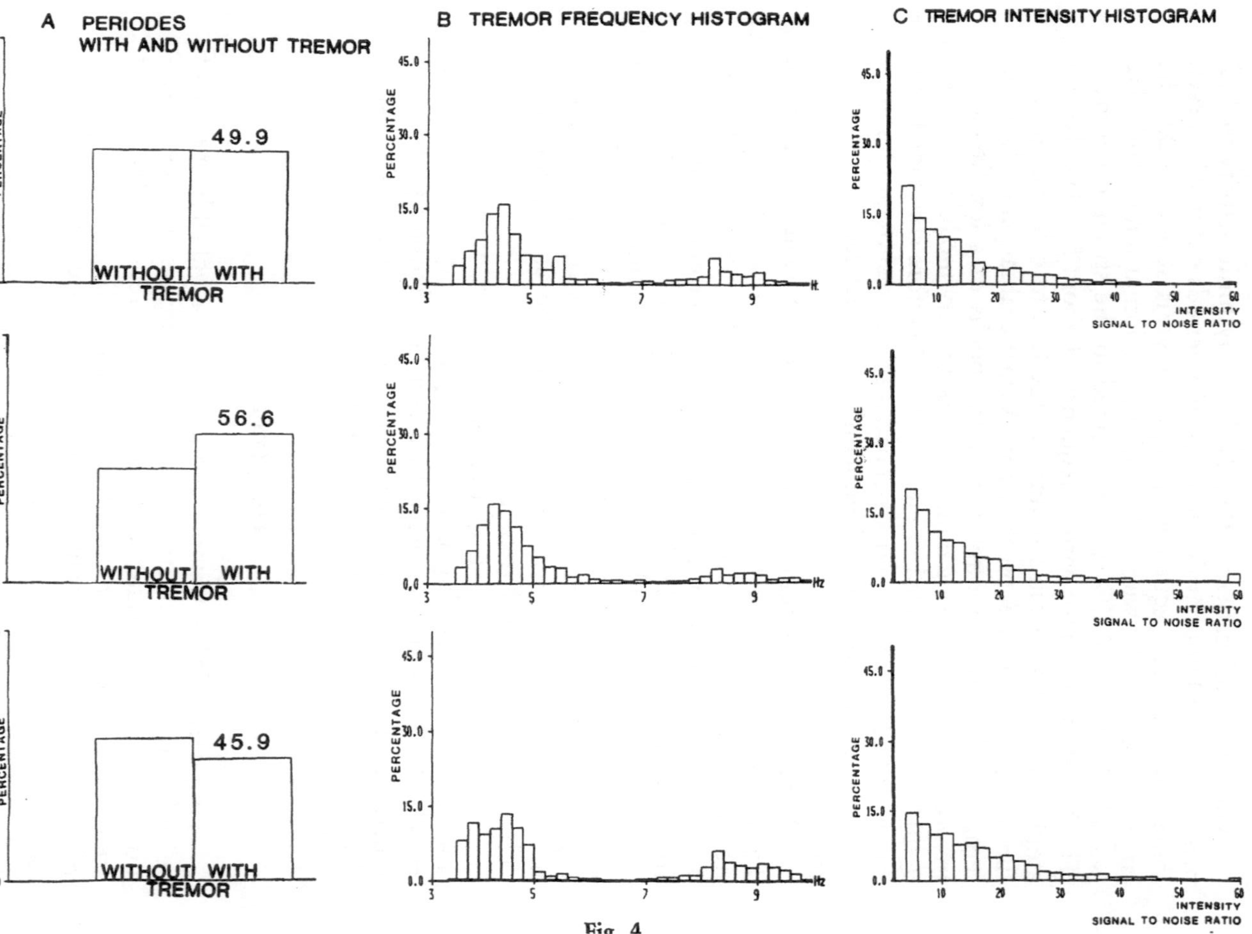
A PERIODES
WITH AND WITHOUT TREMOR
B TREMOR FREQUENCY HISTOGRAM
C TREMOR INTENSITY HISTOGRAM
PERCENTAGE
49.9
56.6
45.9
WITHOUT WITH
TREMOR
Hz
INTENSITY
SIGNAL TO NOISE RATIO

Fig. 4

was much less than in our group of patients with the full-blown disease, who usually display tremor between 20 and 50 percent of time. Nevertheless, the method is able to detect small amounts of tremor and to help classification by an analysis of tremor frequencies. Likewise, tremor intensity was considerably lower in this minimal tremor group, so that the method is sensitive enough for low intensity tremor.

Neither tremor intensity nor occurrence depended on the time of day in these incipient cases, and thus fatigue had no overt influence on tremor.

One problem with the method concerns tremor frequency. Short-term recordings of tremor report a variance in frequency of 0.2 to 0.3 Hz (Findley et al., 1981). Increasing loads have been reported to decrease the frequency of both physiological and enhanced physiological tremor while the frequency of parkinsonian and essential tremor reportedly remain unchanged (Hömberg et al., 1987). This latter statement is contradicted by the observation of a decrease in tremor frequency by up to 1 Hz with increasing load in patients with Parkinson's disease (Rack and Ross, 1986). In long-term recordings of patients with advanced Parkinson's disease we observed a mean variation in tremor frequency of 2.2 Hz with a main frequency peak at 4.2 to 5.8 Hz and with considerable stability of this main peak and its variance over 3 days. Patients with essential tremor had an even broader peak (mean 3.4 Hz) typically around 5 to 7 Hz, such that there was great overlap. The mean frequency alone therefore is not a reliable factor for distinguishing these different forms of central tremor. The same holds true for differentiation of enhanced physiological tremor. In the early diagnosis of Parkinsonian rest tremor, one would like to have a fixed definition of what is a typical parkinsonian tremor frequency in order to distinguish it from other tremors. Patient 1 with the most intense clinical symptoms and the highest amount of tremor showed a clearcut narrow tremor peak, as is usually observed with Parkinson's rest tremor (see Fig. 4 B) but with an atypically high frequency peak at 7 Hz. In the other two patients – although Parkinson's disease was clinically confirmed – the relatively

◄

Fig. 4. Gives a representative example of the reproducibility of long-term tremor recordings in a patient with Parkinson's disease on three consecutive days. Data of day one are represented on the first line, day 2 on line 2, and day 3 on the lower line. The amount of tremor (*A*) shows minor variations. The frequency reveals a stable main peak at 4.5 Hz with the typical frequency distribution, showing a peak width of about 2 Hz (*B*). *C* gives the intensity distribution, exhibiting a minor shift to larger intensities on day 3

low number of periods with tremor manifestation and the variation in frequency exhibited a large scatter of measured frequencies without a clearcut peak. In patient Nr. 2 the predominant high mean frequency tremor of 9–10 Hz during long-term recording speaks in favor of an enhanced physiological tremor, whereas the data obtained with the "Stroop" test reveal high amplitude tremor with a frequency of about 6 Hz compatible with the diagnosis of parkinsonian tremor. The third patient clearly suffered from an enhanced physiological tremor of the hands, although she had a typical rest tremor of the leg. Artificially induced stress clearly enhances EMG-tremor-activity. The "Stroop" test is a reliable and easily applicable method of provocation (Gresty et al., 1984). In this context, we use this procedure as a control for the long-term data. During the test we observed tremor of considerable amplitude not visible during the usual clinical evaluation, nor during long-term recording in one patient. For early diagnosis, the "Stroop" test gives a more potent means for provoking and consequently classifying tremor while long-term recordings provide information about the clinical relevance of tremor in the daily life situation.

References

Bacher M, Scholz E, Diener HC (1988) 24 hours continuous tremor quantification based on EMG-recordings. Electroenceph Clin Neurophysiol (in press)

Findley LJ, Gresty MA, Halmagyi GM (1981) Tremor, the cogwheel phenomenon and clonus in Parkinson's disease. J Neurol Neurosurg Psychiatr 44: 534–546

Gresty MA, McCarthy R, Findley LJ (1984) Assessment of resting tremor in Parkinson's disease. In: Findley LJ, Capildeo R (eds) Movement disorders: tremor. Macmillan, London, pp 321–329

Hömberg V, Hefter H, Reiners K, Freund H-J (1987) Differential effects of changes in mechanical limb properties on physiological and pathological tremor. J Neural Neurosurg Psychiatr 50: 568–579

Rack PMH, Ross HF (1986) The role of reflexes in the resting tremor of Parkinson's disease. Brain 109: 115–141

Scholz E, Bacher M, Diener HC, Dichgans J (1988) The evaluation of treatment by 24-hour tremor recordings in patients with Parkinson's disease. Neurology 235 (in press)

Correspondence: Dr. E. Scholz, Department of Neurology, University of Tübingen, Liebermeisterstrasse 18–20, D-7400 Tübingen, Federal Republic of Germany.

Tremor and electrically elicited long-latency reflexes in early stages of Parkinson's disease*

G. Deuschl and **C. H. Lücking**
Department of Neurology, University of Freiburg, Federal Republic of Germany

Summary

Clinical and electrophysiological characteristics of tremor and long-latency reflexes in hand muscles have been investigated in patients with Parkinson's disease (PD) at early stages of the disease with mostly hemiparkinsonian symptoms. The tremors in the early stage of PD are often untypical in two respects: firstly the resting tremor frequencies are often higher than in case of advanced disease which seems to be a specific characteristic of early PD; secondly the combination of resting and postural tremor is rather common in early PD. Enhancement of long-latency reflexes at the clinically unaffected side may be found suggesting that this can be an early electrophysiological sign of the disease.

Introduction

The classical attribute of tremor in Parkinson's disease is its predominant manifestation during resting conditions. Nevertheless, it is frequently found that the tremors vary considerably according to the frequency and activation conditions among different parkinsonian patients. Different types of tremor have been subdivided mainly according to the differences concerning postural and action tremors (Schwab and Young, 1971; Findley et al., 1981; Lücking et al., 1986). Besides clinical investigation the criteria for these different distinctions are myographic recordings of the EMG and accelerometric recordings including frequency analysis. Recently the analysis of hand muscle reflexes was included as well. By means of clinical

* Supported by the Deutsche Forschungsgemeinschaft (SFB 325).

assessment, myographic recordings and long-latency reflexes the following types of tremor in moderate and advanced PD have been subdivided (Lücking et al., 1986):

Type I: Tremor at rest with cessation during voluntary movements. Tremor frequency mostly between 4 and 5.5 Hz. Reciprocally alternating tremor bursts. Normal long-latency reflexes.

Type II: Resting tremor and postural and action tremor of about the same frequency. Tremor frequencies mostly between 5 and 6.5 Hz. Reciprocally alternating tremor bursts. Abnormal long-latency reflexes.

Type III: Resting and postural tremor of different frequencies (mostly about 5 Hz for the resting and non-harmonic higher frequencies for the postural tremor). Reciprocally alternating tremor bursts for resting and synchroneous bursts for postural tremor. Mostly normal long-latency reflexes.

Type IV: Postural tremor with higher frequencies. Comparable to enhanced physiological tremor.

Meanwhile further techniques have been developed by measuring the power spectrum of the accelerogram while weighting the hands with additional loads (Elble, 1986; Hömberg et al., 1987). This may help in separating the tremors which are dependent on the mechanical conditions from those depending on central oscillators or enhanced reflex circuits. Especially the separation of enhanced physiological tremor from central tremors of various origins is possible.

The above mentioned methods have been used to investigate patients with newly diagnosed Parkinson's disease in order to clarify whether those investigations can help for the early diagnosis of PD.

Methods and patients

Twenty-four patients (15 men, 9 women) with PD at a very early stage have been investigated. The mean age of the population was 58.7 years (SD: 7.4 years, range 49–78 years).

Two or three pairs of antagonistic muscles were recorded with surface electrodes and were displayed on a commercial ink writer. Resting and postural tremor activity was examined. Tremor frequency was determined from an accelerometric signal with a commercial Fast-Fourier analyser (Nicolet Med 999) during resting and postural conditions with and without a weight of 500 grams fixed on the dorsum of the hand. Peak frequencies have been determined. Long-latency reflexes in thenar muscles have been elicited by electrical stimulation of the median nerve. The normal pattern of thenar reflexes following stimulation of the median nerve consists of an Hoffmann-reflex (HR, mean latency: 29 ms) and a long-latency reflex at 50 ms (called LLR II). The latter can be separated from an earlier LLR I at 42 ms (mean) which is frequently enhanced in extrapyramidal disorders and especially in PD. Details of the technique have been described elsewhere (Deuschl et al., 1985; Deuschl et al., 1987).

Results

The first clinical diagnosis of Parkinson's disease in our patients was mostly made at the time of the investigations reported here. The diagnosis was accepted if at least two of the following symptoms were present: Resting tremor, bradykinesia, rigidity or typical postural abnormalities. The first symptoms of the disease were evaluated retrospectively and classified into symptoms of bradykinesia, tremor and pain (as shoulder-arm-syndrome) but not depression or personality changes. The duration of these early and frequently unspecific symptoms was between 3 months and 2 years (mean: 1.3 years). At the time of the investigation the Webster-Score was below 12 points in all the patients. All of them were still fully engaged in their jobs. Of the 24 patients 20 (84%) had a unilateral disease with none of the above mentioned symptoms on the unaffected side. 12 had a hemiparkinsonian syndrome on the right side (50%), 8 on the left side (34%). 4 (16%) had the disease bilaterally. 6 of the 24 patients (25%) were already treated with dopaminergic drugs.

Different types of tremor were found in these patients. 17 of them had resting tremor. 13 of them had resting tremor as the first symptom of their disease. Three patients (12%) had no clinically visible tremor. The tremor frequencies of resting tremors lay between 4 and 9 Hz mostly at lower frequencies between 4 and 5.9 Hz (Fig. 1). The mean tremor frequency was 5.7 (SD: 1.2 Hz). In 5 patients the frequencies lay above 6 Hz.

16 patients had postural tremor. Their frequencies lay between 5.5 and 8.5 Hz (Fig. 1) with a mean of 6.7 Hz (SD: 1.0 Hz). These frequencies fit well with those found in moderate or advanced PD.

Separating the observed tremors in the 21 patients according to the above mentioned classification 6 patients had pure resting tremor (type I), two patients had resting and postural tremor of similar frequency with enhanced LLR I (type II), 9 patients had the combination of resting tremor with low frequency and postural tremor of high frequency (type III) and 4 patients had enhanced physiological tremor (type IV).

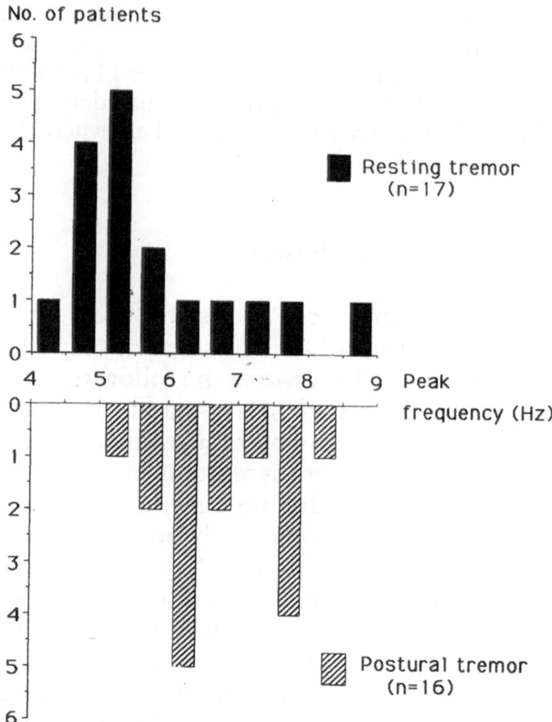

Fig. 1. Peak frequencies of resting and postural tremor in PD. Resting tremor frequencies lay over 6 Hz in 5 patients

Concerning the analysis of frequency shifts during weighting of the limb the patients without clinically visible tremor and those with the enhanced physiological tremor had pronounced shifts of their peak frequencies (Fig. 2 C). All the other patients did not exhibit these frequency shifts (Fig. 2 A, B). The remaining patients had stable frequencies of their dominant tremor peak with and without additional loads (Fig. 2 A, B).

Fig. 2. Power spectra during resting state and unloaded or loaded postural conditions. *A* corresponds to tremor-type II with the same tremor frequency during these three conditions. *B* corresponds to tremor-type III with resting and postural tremor of different frequencies. Both spectra during posture show a stable peak around 8 Hz and an additional peak at 4.5 Hz is discernible during the loaded condition. This constellation is typical for a central postural tremor. *C* corresponds to tremor-type IV with a shift of the peak frequency during loaded conditions indicative of enhanced physiological tremor

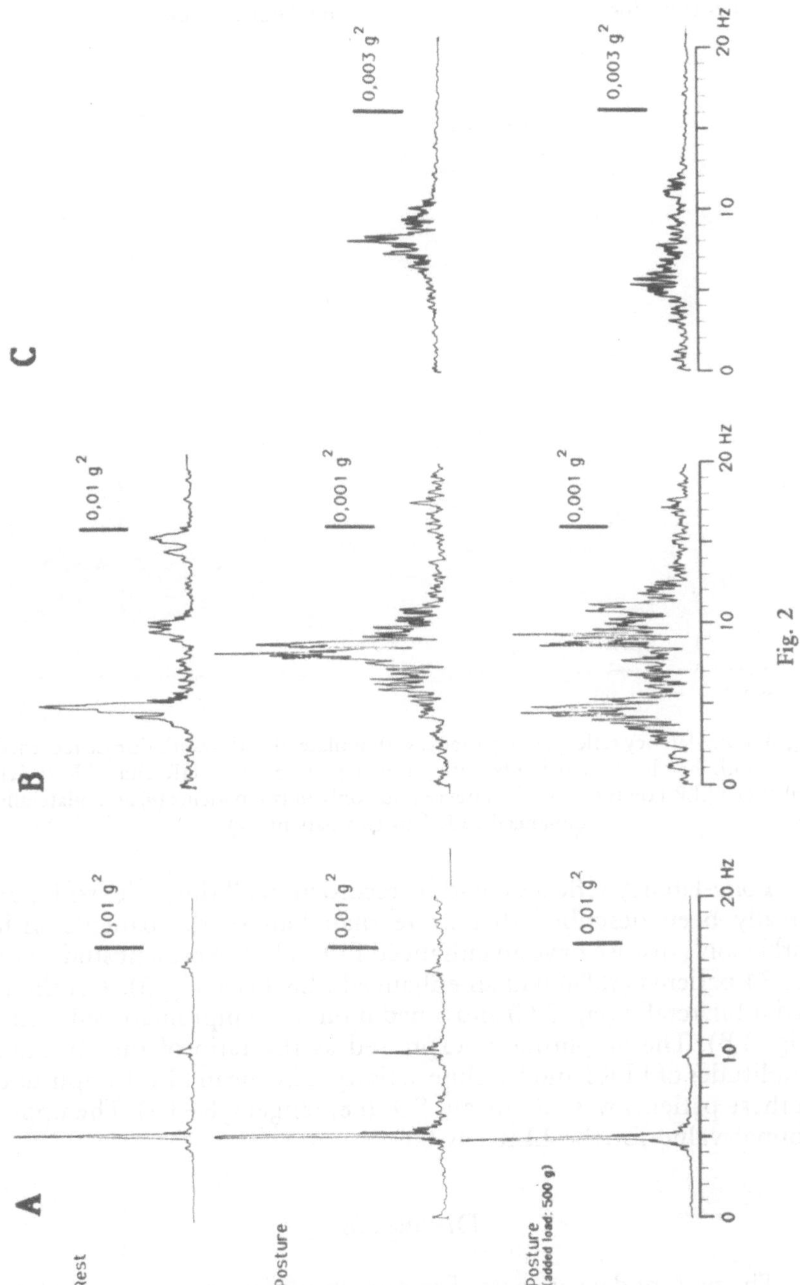
A
B
C
Rest
Posture
Posture
(added load: 500 g)
0,01 g2
0,001 g2
0,003 g2
0
10
20 Hz

Fig. 2

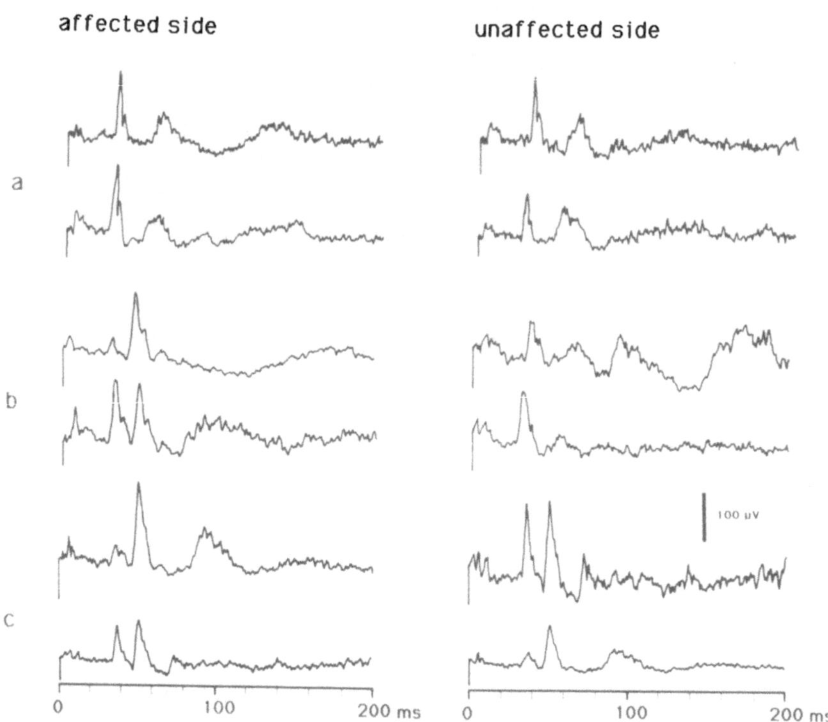

Fig. 3. Long-latency reflexes of 6 patients with unilateral Parkinson's disease recorded from both hands. Two patients with normal patterns with HR and LLR II (*a*). Enhanced LLR I on the clinically affected side only in two patients (*b*) and bilaterally enhanced LLR I in two patients (*c*)

Long-latency reflexes could be recorded in all the patients. It has already been described that more than half of the patients with Parkinson's disease have an enhanced LLR I. In the present study 9 of the 24 patients (38%) had an enhanced LLR I (see Fig. 3). 4 of them had it bilaterally (Fig. 3 C) and 5 had it on the symptomatic side only (Fig. 3 B). The amplitude is expressed as the ratio of the absolute amplitudes of LLR I and baseline activity. The mean LLR I amplitude in these patients was 2.0 (mean, SD: 0.8, range: 0.8—3.7). The upper normal value for the LLR I is 0.8.

Discussion

The present data are part of an ongoing study of tremor in Parkinson's disease. Here we report on the results of patients at a very early stage of their disease. In most of the patients the first diagnosis of PD was made during the consultation. Most of them had strictly unilateral disease. The results of this highly selected group of patients will be discussed in relation to the significance for early diagnosis of PD. The differences to the findings in moderate or advanced Parkinson's disease shall be outlined.

The resting tremor occurring in Parkinson's disease is known to be frequently complicated by postural and action tremors (Schwab and Young, 1971; Findley et al., 1981; Lücking et al., 1986). All these studies investigated patients who suffered from the full-blown disease and to our knowledge we report for the first time about a larger group of patients at a very early stage of their disease.

Three findings of the present study seem to be interesting for early diagnosis of PD. Firstly, the most striking feature of the resting tremors was the rather high mean frequency of 5.7 Hz. This is due to extraordinary high frequencies between 6 and 9 Hz in 6 of the 17 patients with resting tremor. Resting tremor frequencies over 6.5 Hz were never observed in our earlier study of advanced PD. They seem to be a special condition of early PD. It should be emphasized that such high frequencies are obvious for clinical investigation and even neurologists who are familiar with the early diagnosis of PD can be bothered by observation of such tremors.

Secondly, it seems interesting that the quantitative distribution of the tremor types according to Lücking et al. (1986) is different in early PD and in advanced PD. Only 70% of the present patients had resting tremor compared to 90% of the patients with the full-blown disease. The patients with early PD only rarely had resting and postural tremor of the same frequency (type II) compared with 50% of the patients with advanced disease. On the contrary the patients with early PD more often had the combination of a resting tremor with a disturbing postural tremor of different frequencies (type III) which was found in 38% of the present population compared with 15% of the patients with advanced PD. The percentage of patients with enhanced physiological tremor was similar in the present and the former study. The distinction between enhanced physiological (type IV) and those postural tremors which are comparable to essential tremor (type III) was made by weightening the hand with a load of 500 gr. Under this condition the former lower their frequency whereas the latter do not (Elble, 1986; Hömberg et al., 1987). It is concluded from these data that postural tremors are common in early Parkinson's disease and they occur on their own or in combination with the resting tremors. Especially in cases with isolated postural tremor this finding does not contradict the diagnosis of PD. However, we would like to emphasize that these postural tremors are clinically and electrophysiologically indistinguishable from tremors of various other origins. Therefore, in contrary to the resting tremors, these postural tremors are not specific.

The third interesting finding is the demonstration of abnormal hand muscle reflexes in PD. An enhanced LLR I was demonstrated in 38% of the patients. Five of them with a clinically unilateral disease had enhanced reflexes on the symptomatic side only. The others had bilateral enhancement despite strictly unilateral clinical symptoms. Up to now no data are available about the development of enhanced LLR prior to the onset of clinical symptoms in PD. This could only be clarified in a prospective study of normal elderly subjects. However, the present finding of enhanced LLR I at the clinically unaffected side of hemiparkinsonian patients already demonstrates, that the LLR I-enhancement can precede the onset of clinical symptoms of PD. The LLR-enhancement is not specific for PD as it can be seen in other extrapyramidal diseases (Deuschl et al., 1987). However, in connection with the clinical investigation this finding can help for the differential diagnosis.

References

Deuschl G, Lücking CH, Schenck E (1987) Essential tremor: electrophysiological and pharmacological evidence for a subdivision. J Neurol Neurosurg Psychiatr 50: 1435–1441

Deuschl G, Schenck E, Lücking CH (1985) Long-latency responses in human thenar muscles mediated by fast conducting muscle and cutaneous afferents. Neurosci Lett 55: 361–366

Elble J (1986) Physiologic and essential tremor. Neurology 36: 225–231

Findley LJ, Gresty MA, Hamalgyi GM (1981) Tremor, the cogwheel phenomenon and clonus in Parkinson's disease. J Neurol Neurosurg Psychiatr 44: 535–546

Hömberg V, Hefter H, Reiners K, Freund HJ (1987) Differential effects of changes in mechanical limb properties on physiological and pathological tremor. J Neurol Neurosurg Psychiatr 50: 568–579

Lücking CH, Deuschl G, Strahl K, Schenck E (1986) Tremor im Früh- und Spätstadium der Parkinson-Krankheit. In: Fischer PA (Hrsg) Spätsyndrome der Parkinson-Krankheit. Editiones Roche, Basel, S 99–113

Schwab RS, Young RR (1971) Non-resting tremor in Parkinson's disease. Trans Am Neurol Assoc 96: 305–307

Correspondence: Dr. G. Deuschl, Department of Neurology, University of Freiburg, Hansastrasse 9, D-7800 Freiburg i. Br., Federal Republic of Germany.

Evoked potentials in Parkinson's disease

J. Jörg
Department of Neurology, University of Wuppertal,
Wuppertal, Federal Republic of Germany

Summary

Diagnosis of Parkinson's disease ist not yet possible with neurophysiological methods. Therefore we investigated the median somatosensory and visually evoked potentials (SEP and VEP) to single and double stimulation, VEP using a special stimulation technique. Also we investigated the peripheral autonomic surface potentials (PASP) and motor evoked potentials (MEP) employing spinal and transcranial stimulation.

In Parkinson's disease the absolute and relative refractory periods of cortical N 1 peak of SEP did not differ from that of normal subjects. After single stimulation the cortical N 20 latencies were only delayed in 2 of 17 cases. In 8 of 10 cases central conduction time was normal. The SEP peaks of brain stem were also normal and there was no correlation between symptomatology and SEP results.

The data of motor evoked potentials (MEP) demonstrated in all patients a normal central motor conduction time of 5.0 ms. In 3 of 10 patients PASP's were delayed but not reduced. In 30% of patients the VEP data showed a significant delay of P 2 latency, often independently of clinical status or age. But there were no more P 2 alterations after special visual stimulation techniques.

Introduction

The investigation of evoked potentials and refractory period employing paired stimuli is a useful method for detectiving segmental demyelination of the peripheral or central nervous system. In Parkinson's disease normal median somatosensory evoked potentials (SEP) and delayed visual evoked potentials have been described, but to date an early diagnosis of "Parkinson's disease" is not yet possible with neurophysiological methods. Therefore we investigated the median somatosensory and special visual evoked potentials (SEP and

VEP) to single and double stimulation. In a few cases it was also possible to examine the motor evoked potentials (MEP) after scalp and spinal stimulation (Gerhard and Jörg, 1986).

Thermoregulatory dysfunctions with an absence of sweating on the trunk and limbs and compensatory hyperhidrosis on the face and neck are common features of Parkinsonism (Aminoff and Wilcox, 1971). The PASP may prove a useful quantitative test of sympathetic function (Knezevic and Bajada, 1985). Therefore we investigated the peripheral autonomic surface potentials (PASP).

This study furthermore investigated whether there is a correlation between neurophysiological results and the clinical syndromes.

Material and methods

SEP

The median nerve was stimulated at the wrist using surface electrodes; recording regions for SEP were over the hand area contralaterally to the side of stimulation ($C/P_{3 \text{ or } 4}$—$F_{3 \text{ or } 4}$), at the contralateral mastoid process (proc. mastoideus), at the spine of the second vertebra and at Erb's point. After single stimulation the double stimuli were applied with varying interstimuli intervals between 50 and 1 msec. The peaks of the SEP were named N 1, P 1, N 2, P 2. Refractory period of the first negative peak N 1 (so-called N 9, N 12, N 14 or N 20) were recorded at Erb's point, cervical vertebra 2, brain stem (mastoid) and contralateral cortex; a subtraction technique was used.

VEP

The VEPs were recorded from scalp electrodes placed on positions O_z and F_z. As stimuli we used a checkerboard stimulation with a whole field and small foveal field; in a few cases we applied a grid with horizontal and vertical lines and we investigated the visual refractory period.

MEP

The motor evoked potentials (MEP) were recorded from thenar muscles, following transcranial and spinal stimulation with a digitimer. These stimulation points were on contralateral scalp with an anode over the motor hand area and spinal at cervical vertebral 6/7.

PASP

The peripheral autonomic surface potential (PASP) was recorded from the palmar surface of the hand and the volar surface of the middle finger. It was elicited by randomly timed electrical stimuli over the median nerve at the contralateral wrist.

Subjects

In a first series SEPs of 42 normal subjects aged 15 to 72 years were examined contralaterally and cortically (27 women, 15 men). On 17 occasions the medianus was stimulated on both sides with single and paired stimuli. In a second series median nerves of 16 volunteers (aged 15 to 50 years) were stimulated once or twice at the right wrist and the recordings were placed at Erb's point, at the spine of the second cervical vertebra, proc. mastoideus (A_1) and C/P_3. In 24 patients with Parkinson's disease (17 women, 7 men) aged 55 to 89 years, we investigated checkerboard VEPs and median SEPs from the upper cervical spinal cord, mastoideus and contralateral scalp. There the central conduction time was measured. In 7 cases we were able to determine the cortical refractory period, in the other cases this procedure was too stressfull and took too long. In 5 normal subjects and 5 patients we investigated the special visual stimulation technique described earlier and the MEP examination. In 10 normal subjects and 10 patients we investigated the palmar PASP's.

Results

In normal subjects median SEPs were obtained from all sites after single stimulation with typical latencies and amplitudes defined for every localization; the N_1 latencies were cortically at 20.0 ± 1.3 msec, mastoideus at 15.3 ± 1.3 msec, cervical spine at 12.8 ± 1.3 msec and Erb's point at 10.9 ± 0.9 msec. There was no significant correlation between latency and age. The absolute refractory time of N_1 latency was 1.25 msec at Erb's point, 2.13 msec at second vertebra, 11.5 msec at the brain stem and 13.6 msec at the cortical hand area. All absolute RT of the 4 points differed significantly ($p < 0.05$).

In Parkinsons' diseases the absolute and relative RT of cortical N_1 peak did not differ from that of normal individuals. After single stimulation the cortical N 20 peak latencies were delayed only in 2 of 17 cases (25.4 respectively 23.3 msec), in one case there was the very infrequent combination of Parkinson's syndrome and MS. On 10 occasions we were able to measure central conduction time and 8 times we found normal values. Also, there were normal results in most cases for latencies and amplitudes of second vertebra, brain stem, interpeak differences between vertebra to mastoid and mastoid to cortex and right/left differences. There was no correlation between the symptomatology exhibited in "hemiparkinson" cases ($n = 2$) and SEP results. Nor were any disturbed peaks or amplitudes after the first SEP complex, if the analysis time was applied to 200 msec. In examination of VEP employing checkerboard whole field stimulation, the latencies of P_2 and N_3 were considerably longer in Parkinson's disease than in control subjects (P_2 [both sides] of normal subjects: 98.4 ± 6.3 msec, P_2 of Parkinson's disease right 110.4 ± 9.8 msec, left 110.8 ± 10.1). For both eyes in 5 of 15 cases the P_2 latency was longer than 112 msec, i.e. it was pathological. There were no significant P_2 latency differences between checkerboard stimulation and grid with horizontal and vertical lines. Only the P_2 latencies following foveal stimulation are longer than those following whole field stimulation, but without differences between healthy subjects and Parkinson's patients.

Application of motor evoked potentials (MEP) resulted in a central motor conduction time of 5.0 ± 0.4 msec and no differences between normal subjects and 5 patients with Parkinson's disease.

The PASP consisted of a biphasic potential with an initial negative and a later positive peak. The mean palmar PASP latency was 1.45 ± 0.15 ms. In all patients PASP's could be obtained; in 3 patients the onset of PASP was delayed. The amplitudes were not reduced compared to normal.

To sum up the neurophysiological results: sometimes the PASP's and frequently P_2 latencies of VEPs are pathological in Parkinson's disease, but not to a large extent and not in all cases. So the methods we have used do not exhibit a diagnostic value for parkinsonism.

Discussion

Unlike in MS or spinal tumours the SEP and motor evoked potentials (MEP) have no diagnostic value in Parkinson's disease. Nor could we find any alterations of the cerebral refractory period in 7 cases examined for cortical N 20 peak.

In contrast to the normal SEPs and central conduction times, the VEPs of Parkinson patients were often delayed; this is independent of clinical status and age and has been described in the literature (Sollazo, 1985; Enzensberger and Fischer, 1986; Nightingale et al., 1986). Nightingale et al. (1986) suggest that the abnormal VEPs in Parkinson patients may be, at least in part, due to a biochemical and electrophysiological disorder in the retina. Our VEP data show a significant increase in the mean P 100 latency not only following checkerboard whole field stimulation, but also after foveal stimulation or streak-pattern either horizontal or vertical. But special visual stimulation did not yield more information than the checkerboard stimulation.

Our data confirm previous reports concerning VEP-delays in Parkinson's disease (Sollazo, 1985). But these results are without diagnostic value because in only 30–40% patients is the P_2 delay outside the normal range.

Sollazo (1985) found an improvement of P 100 delay after acute and after chronic dopa therapy in primary parkinsonism only; other authors (Enzensberger and Fischer, 1986) did not find VEP alterations before and during dopa therapy. We studied patients with dopa therapy before and during chronic deprenyl-therapy; we found a clinical improvement but no improvement of P 100 delay.

Both sympathetic skin activity and the sympathetic skin response have been used in the past to evaluate autonomic function in patients with suspected dysautonomia. The physiological mechanisms underlying the production of the PASP are still poorly understood (Shahani et al., 1984). The PASP was present in all hands of all patients and in most cases PASP's were normally configured.

References

Aminoff MJ, Wilcox CS (1971) Assessment of autonomic function in patients with a Parkinsonian syndrome. Br Med J 4: 80–84

Enzensberger W, Fischer PA (1986) Hirnelektrische Befunde bei Parkinson-Patienten im Langzeitverlauf. In: Fischer PA (Hrsg) Spätsyndrome der Parkinson-Krankheit. Editiones Roche, Basel, pp 159–171

Gerhard H, Jörg J (1986) Motor conduction time of tractus corticospinalis by spinal stimulation. Z EEG-EMG 17: 197–200

Jörg J, Gerhard H (1987) Somatosensory motor and special visual evoked potentials to single and double stimulation in "Parkinson's disease"—an early diagnostic test? J Neural Transm [Suppl] 25: 81–88

Jörg J, Gerhard H (1988) Median somatosensory evoked potentials to single and double stimulation in normal subjects and in Parkinson's disease. In: Barber C, Blum Th (eds) Evoked potentials III. Butterworth (in print)

Knezevic W, Bajada S (1985) Peripheral autonomic surface potential. A quantitative technique for recording sympathetic conduction in man. J Neurol Sci 67: 239–251

Nightingale S, Mitchell KW, Howe JW (1986) Visual evoked cortical potentials and pattern electroretinograms in Parkinson's disease and control subject. J Neurol Neurosurg Psychiatr 49: 1280–1287

Shahani BT, Halperin JJ, Boulu Ph, Cohen J (1984) Sympathetic skin response—a method of assessing unmyelinated axon dysfunction in peripheral neuropathies. J Neurol Neurosurg Psychiatr 47: 536–542

Sollazo D (1985) Influence of L-Dopa/carbidopa on pattern reversal VEP: behavioural differences in primary and secondary parkinsonism. Electroencephalogr Clin Neurophysiol 61: 236–242

Correspondence: Dr. J. Jörg, Department of Neurology, University of Wuppertal, D-5600 Wuppertal, Federal Republic of Germany.

diagnostic value because in only [illegible] 40% patients is the P100 delay outside the normal range.

Schiavo (1985) found an improvement of P100 delay after chronic and after [illegible] l-dopa therapy in [illegible] parkinsonian patients. [illegible] and [illegible] did not find VEP alterations before and during dopa therapy. We studied patient, [illegible] before and during chronic [illegible] therapy, we found a clinical improvement but no improvement of P100 delay.

Both [illegible] skin [illegible] and the sympathetic skin responses have been used in the test to evaluate autonomic function in patient with [illegible]. The physiological [illegible] of the PASR are still poorly understood [illegible]. [illegible] present in all patients and in most cases [illegible] was normally configurated.

References

[illegible] M, [illegible] (197?) [illegible]
[illegible]
[illegible]
[illegible]
[illegible] (1987?) [illegible] double stimulation [illegible]
[illegible] (1984) [illegible] quantitative technique for recording sympathetic [illegible] Neurol [illegible]
Nightingale S, Mitchell KW, Howe JW (1986) Visual evoked cortical potentials and pattern electroretinograms in Parkinson's disease and control subjects. J Neurol Neurosurg Psychiatry 49: 1280–1287
[illegible] (1984) [illegible] peripheral neuropathies [illegible]
[illegible] (1985) [illegible] behavioural differences [illegible] Electroencephalog. Clin. Neurophysiol [illegible]

Correspondence: Dr. [illegible], Department of Neurology, University [illegible], Federal Republic of Germany.

Brain mapping of EEG and evoked potentials during physiological aging and in Parkinson's disease, dementia and depression

K. Maurer, R. Ihl, W. Kuhn, and **T. Dierks**
Department of Psychiatry, University of Würzburg, Federal Republic of Germany

Summary

Electrical brain activity measured by topographical EEG and evoked potential (EP) recording was analyzed in 33 patients (suffering from dementia of Alzheimer type (10 DAT), parkinsons disease (13 PD) and depression (10 MDD) and 20 controls. P 300 in EP and four frequency bands in EEG (delta 0—3.5 Hz, theta 4—7.5 Hz, alpha 8—11.5 Hz, beta 12—15.5 Hz) were evaluated with statistical probability mapping.

In DAT delta and theta were increased, alpha was decelerated and beta diminished. The P 300 developed its maximum in frontal regions. In PD delta activity increased, theta activity decreased, alpha and beta showed no alteration. P 300 was not altered. In 5 patients with PD a dementive syndrom was observable. These patients showed the same alterations as the DAT group. However, beta did not change. No P 300 could be recorded. In MDD only alpha activity was decelerated. P 300 was not altered.

The results indicate, that topographical EEG and EP analysis is capable to distinguish between all groups investigated. Differences in P 300 topography may reflect the pathoanatomical involvement of hippocampal circuits in dementia of Alzheimer type.

Introduction

It is well known that the electrical activity of the brain changes during physiological aging showing a) a decrease in the mean alpha rhythm frequency by 0.5 to 1.0 Hz (Obrist, 1954), b) an increase in the amount of beta waves (Gibbs and Gibbs, 1951) and c) an increase in slow activity (Ihl et al., 1988). In the field of evoked potentials the age-dependent alterations are slight regarding exogenous components with short latency but become clinically significant when endogenous waves such as P 300 are applied (Hegerl et al., 1985; Maurer et al., 1988).

In Parkinson's disease (PD) there is general agreement that EEG and EP exhibit mostly normal features (Christian, 1987; Maurer et al., 1988). In dementia of Alzheimer's type (DAT) the frequency of the alpha rhythm decreases (Ihl et al., 1988) combined with an excess of slow activity and nonreactivity of the background pattern. Concerning EP, Maurer found delayed early auditory evoked potentials (EAEP) in DAT-patients (Maurer, 1988). Best documented are results gained with the P 300 paradigm (Pfefferbaum et al., 1984; Maurer et al., 1988). Both authors described an amplitude decrease and a latency increase in DAT. In depression neither a reproduceable pathological EEG nor a consistens abnormal EP-pattern concerning exogenous waves is known. The P 300, however, was found to be influenced in depression (Roth et al., 1981; Maurer and Dierks, 1988).

In the following study the newly developed topographical display of EEG and EP (Brain Mapping) was applied to investigate the age factor and disease dependent alterations of EEG and P 300 in Parkinson's Disease (PD), dementia of Alzheimer's type (DAT), Parkinson's disease plus dementive syndrome (PD + DS) and major depressive disorder (MDD). The main aim was to find out whether imaging of EEG and P 300 provides new information concerning early detection and differential diagnosis in PD, DAT, PD + DS and MDD.

Methods

EEG data were obtained from 19 electrodes placed upon the scalp according to the 10—20 system. Spontaneous EEG was recorded in the resting condition with eyes closed. For data collection the Brain Atlas III (Bio-logic Systems Corp.) was used. From the EEG-recording stored onto magnetic disks 12 artifact-free two-second epochs were chosen randomly and transformed through Fourier transformation into frequency spectra. For the purpose of imaging the scalp areas around the original 19 electrodes were filled in by linear interpolation. Maps of activity were constructed for 4 frequency bands: delta (0—3.5 Hz), theta (4—7.5 Hz), alpha (8—11.5 Hz) and beta (12—15.5 Hz). For eliciting the auditory P 300 we recorded AEP after frequent low-pitched (1000 Hz, 80 dB) and infrequent high-pitched (2000 Hz, 80 dB) tones. The tones were presented in a random sequence with a 20% probability of hearing the infrequent high-pitched tone. The subjects were asked to pay attention and to count the rare tones (target stimuli). After separate averaging of the responses due to target and non-target stimulation a positive wave could be recorded with a latency of about 300 ms and a maximal parietal positive field. In the case of P300 we recorded the electrical activity from 20 electrodes with mastoid as reference.

Patients

Ten adult control subjects (mean age 43.8 years) and 10 elderly controls (mean age 74.2 years) were investigated.

Patients were

a) 10 persons diagnosed as having DAT (DSM-III, mean age 70.1 years),

b) 13 patients with Parkinson's disease (mean age 64.1 years). 5 of the 13 patients with Parkinson's disease additionally had a dementive syndrome (mean age 65 years) and

d) 10 patients suffering from major depressive disorder (ICD-9: 296.1, 296.3, 296.4 and 300.4; mean age 45.9 years).

The degree of dementia was evaluated according to BCRS-criteria (Brief Cognitive Rating Scale) in DAT and PD plus DS. DAT and PD + DS patients had a moderate degree of dementia (BCRS: range 3.9 to 5.1). Patients with PD were evaluated with Columbia Score ($\bar{x} = 32.15 \pm 10.7$) and clinical staging after Hoehn and Yahr ($\bar{x} = 2.76 \pm 1.36$).

Patients with PD and PD plus DS were under medication of L-dopa combined with benserazide or carbidopa. 6 patients received bromocriptine, 6 lisuride, 2 terguride, 3 selegilin and 1 amantadine. Patients with DAT and MDD were medication free.

The results will be presented in the form of grand means of EEG and P300 activity. To explore areas with significant changes t-value maps were calculated. The Wilcoxon-test was used to find out whether the alpha slowing in DAT turned out to be significant.

Results

EEG

In the case of EEG in each patient group (DAT, PD, PD plus DS and MDD) 5 patients were considered. After calculation of group means patients' data were compared with control data by means of t-test procedures (Fig. 1).

Concerning AEP 300 each group contained 10 or 11 persons except the PD plus DS collective, with 5 patients. The group means of young controls, elderly controls, DAT and MDD will be shown by using 3-D-head formats (Fig. 2 a–d). In the case of PD, results will be presented in the form of a range of amplitudes and by using a

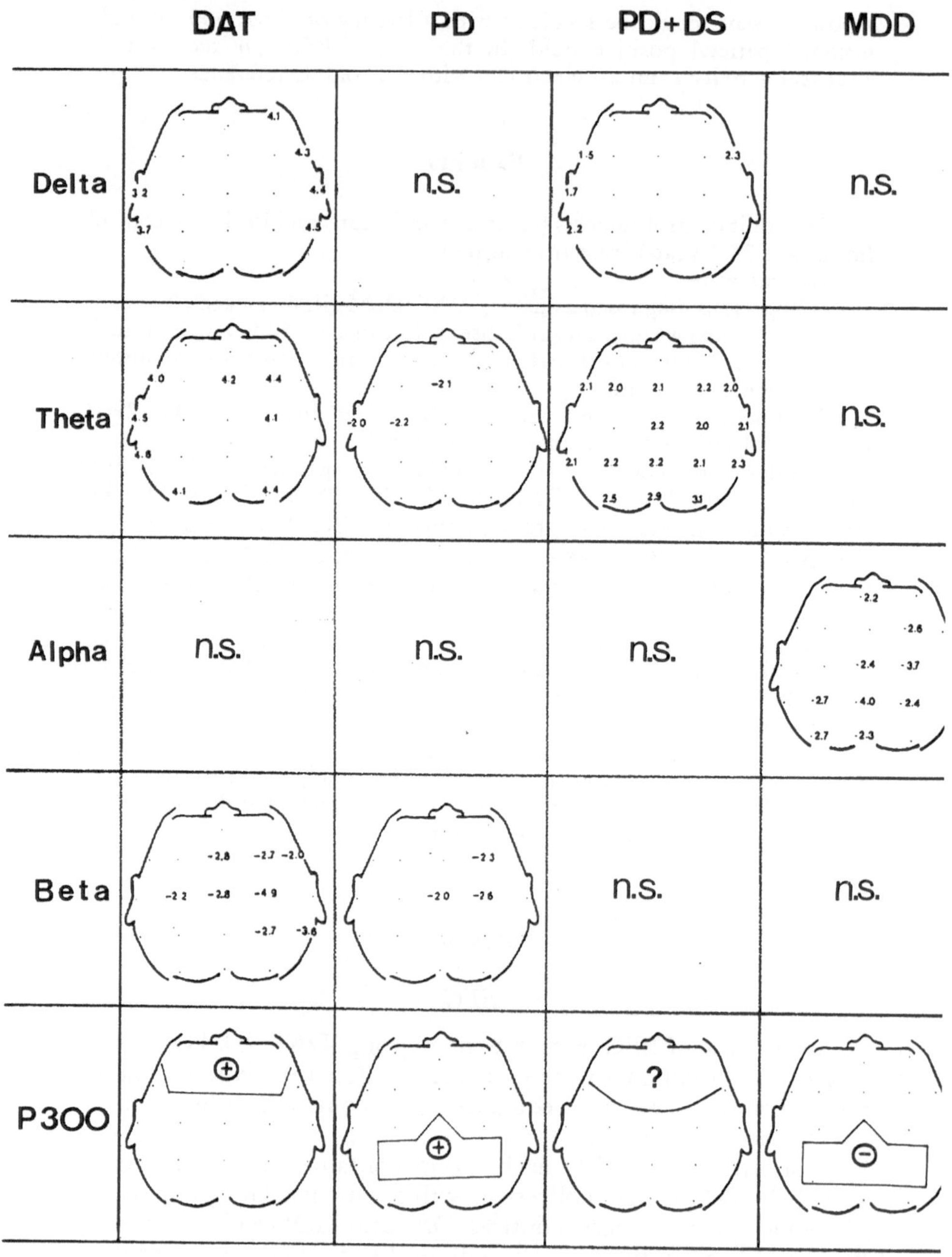

	DAT	PD	PD+DS	MDD
Delta	4.1 4.3 3.2 4.4 3.7 4.5	n.s.	1.5 2.3 1.7 2.2	n.s.
Theta	4.0 4.2 4.4 4.5 4.1 4.6 4.1 4.4	-2.1 -2.0 -2.2	2.1 2.0 2.1 2.2 2.0 2.2 2.0 2.1 2.1 2.2 2.2 2.1 2.3 2.5 2.9 3.1	n.s.
Alpha	n.s.	n.s.	n.s.	-2.2 -2.6 -2.4 -3.7 -2.7 -4.0 -2.4 -2.7 -2.3
Beta	-2.8 -2.7 -2.0 -2.2 -2.8 -4.9 -2.7 -3.6	-2.3 -2.0 -2.6	n.s.	n.s.
P300	⊕	⊕	?	⊖

Fig. 1

Minimum-Maximum display. In 5 cases with PD plus DS it was not possible to record a P 300 at all.

DAT-patients exhibited a considerable increase of delta and theta activity (Fig. 1). The alpha activity did not change to any large extent, whereas slow beta underwent a reduction (Fig. 1). The alpha frequency underwent a remarquable slowing in the order of 2 to 3 Hz and the alpha field moved from its usual parieto-occipital area to more anterior structures.

In PD no significant slowing in the delta range could be observed whereas theta exhibited a tendency to decrease. The alpha and slow beta activity did not alter significantly.

In contrary to DAT no slowing of the alpha frequency and no change in topography took place.

In patients suffering from PD plus DS a similar pattern was observed as in DAT (Fig. 1). Slow beta, however, did not change.

In MDD the only alteration occurred in the alpha frequency band where a decrease of activity took place.

P 300

During physiological aging P 300 amplitudes became lower but maintained their topography with a parietal maximum of positivity (Fig. 2 a and b). In DAT considerably increased amplitudes could be observed bifrontally (Fig. 2 c). In PD the P 300 was normal concerning amplitude and topographical features whereas the 5 cases with PD plus DS no P 300 could be elicited at all. In patients with MDD a slight parietal amplitude decrement was obvious (Fig. 2 d).

Discussion

The main findings in our study were changes in EEG and P 300 topography which allowed a discrimination between DAT, PD, PD plus DS and MDD. Concerning electrophysiology DAT and PD came out to be different in the way that delta and theta increased in DAT whereas in PD no changes in slow activity took place. The

◄

Fig. 1. EEG activity and P 300 results in dementia of Alzheimer's type (*DAT*), Parkinson's disease (*PD*), Parkinson's disease plus dementive syndrom (*PD + DS*) and in major depressive disorder (*MDD*), compared to controls.
Numbers in the head formats represent t-values. Negative t-values indicate a decrease and positive t-values an increase in activity (n.s. = not significant). t-values of the four main frequencies are indicated. In the case of P 300 the corresponding topographical field are shown (⊕ = positive P 300 field; ⊖ = decrement of P 300 field)

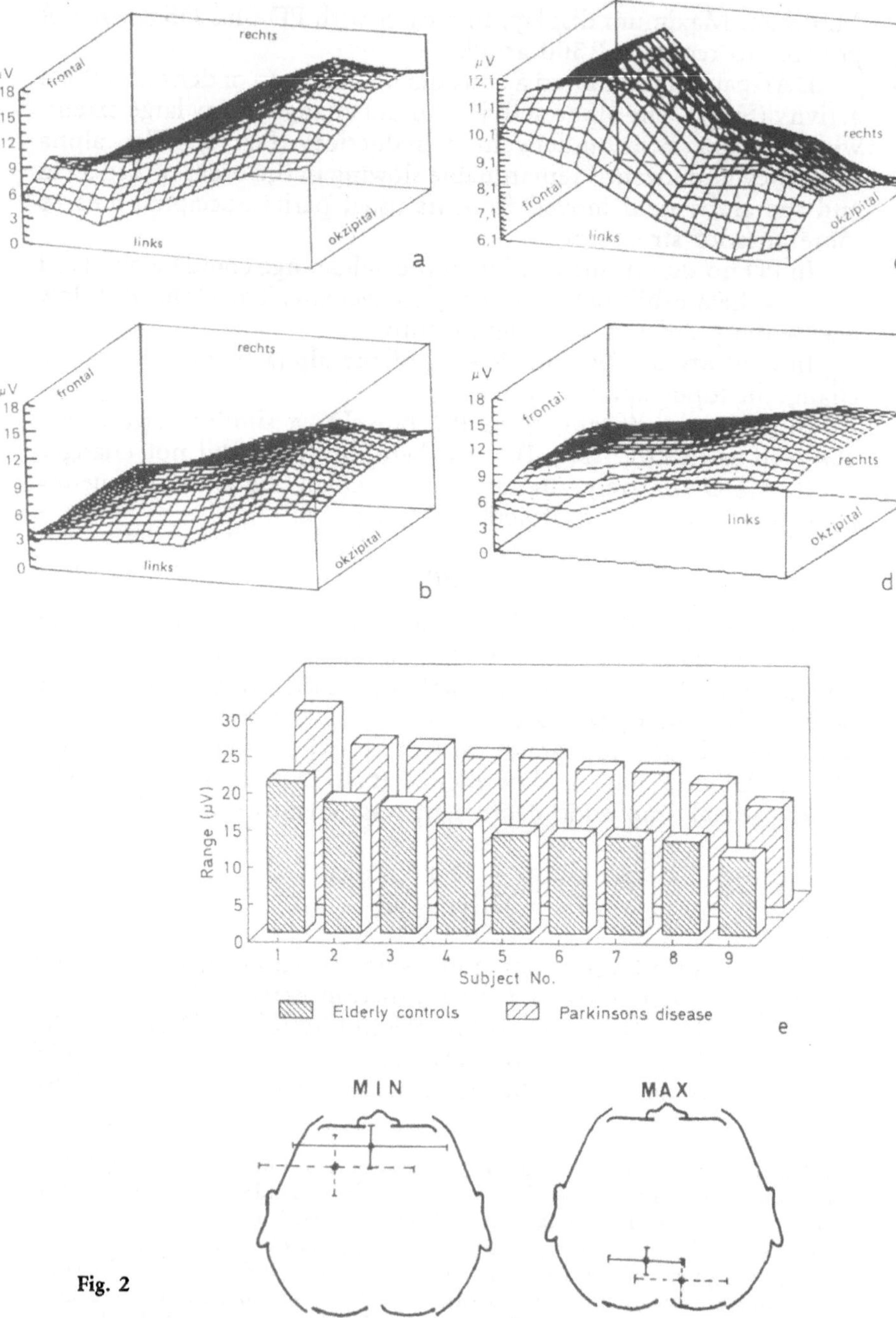
a
b
c
d
e
f
µV
frontal
rechts
links
okzipital
Range (µV)
Subject No.
Elderly controls
Parkinsons disease
MIN
MAX

Fig. 2

parieto-temporo-occipital lesion pattern in DAT is in accordance to a post mortem study (Brun and Gustafson, 1976) with findings of a neural degeneration affecting the temporo-parieto-occipital areas most severely. The fact that slow activity was not changed in PD (in our collection even a "hypernormal" behaviour was observed with the tendency of a theta decrement) is a point also apparent from the electrophysiological standpoint that cortical structures responsible for the EEG surface activity are not affected in the early stage of the disease. Whereas the alpha-activity was not different between both disease the altered topography with an anterior shift of the alpha field in DAT also allowed differentiation between DAT and PD. This shift could not be observed in PD where a normal topography of alpha activity was maintained. The beta activity was not lowered in PD in the same way and intensity as in DAT, which can be regarded as another sign that the typical lesion pattern of DAT does not occur in PD (Ihl et al., 1988).

There is good evidence that the P 300 is generated in limbic and paralimbic structures such as the hippocampus and amygdala (Halgreen et al., 1982; Okada et al., 1983). This explains the dramatic change in P 300 morphology and topography in our DAT group where the P 300 field moved toward frontal structures. This "hypofrontal" P 300 behaviour was not observed in patients with PD where a normal P 300 with even elevated P 300 amplitudes could be observed (Fig. 2 e and f) as further indication that the hippocampal formation is not involved in PD.

In patients with PD plus DS a similar pattern as observed in DAT occurred. However beta activity was not as much diminished as in DAT. If we regard the role of beta activity in cognitive processing it can be assumed that another factor besides solely mental impairment occurs, for example bradyphrenia imitating a dementive syndrome.

The fact that in our 5 cases with PD plus DS no P 300 could be elicited may indicate a slightly stronger impairment of limbic structures when both diseases occur in the same patient. This hypothesis is underlined by the fact that in patients with PD plus DS the disease lasted as long as maximally 10—12 years.

Patients with MDD exhibited a decrease of alpha activity and P 300 in a way which was not observed in DAT, PD and PD plus DS. Since both conditions came out to be reversible under antidepressant

◀———

Fig. 2. Cognitive P 300 fields. *a* control group (mean age: 43.8 years); *b* elderly control group (mean age: 74.2 years); *c* DAT patients (mean age: 70.1 years); *d* patients with major depressive disorder (mean age: 45.9 years); *e* range of amplitudes in patient with Parkinson's disease; *f* Minimum-Maximum display of P 300 in Parkinson's disease

therapy this alteration may be due to more functional alterations linked to degree of depression.

Summarizing, quantification and topography of EEG and P 300 activity turned out to be valuable in the differential diagnosis of DAT, PD, PD plus DS and MDD. The functional EEG imaging test results in DAT, PD and PD plus DS are in accordance with findings obtained by other imaging procedures such as CBF and PET and fit into our understanding of a more parieto-temporo-occipital and limbic lesion pattern in DAT and a subcortical extrapyramidal lesion in PD.

References

Brun A, Gustafson L (1976) Distribution of cerebral degeneration in Alzheimer's disease. Arch Psychiatr Neurol 223: 15–33

Gibbs FA, Gibbs EL (1951) Change with age, awake. In: Gibbs FA, Gibbs EL (eds) Atlas of electroencephalography, vol I: Methodology and methods. Addison-Wesley, Reading, Mass, pp 82–88

Halgren E, Squires NK, Wilson CL, Crandell PH (1982) Brain generators of evoked potentials: The late (endogenous) components. Bull Los Ang Neurol Soc 47: 108–123

Hegerl U, Klotz S, Ulrich G (1985) Späte akustisch evozierte Potentiale – Einfluß von Alter, Geschlecht und unterschiedlichen Untersuchungsbedingungen. Z EEG-EMG 16: 171

Ihl R, Dierks T, Maurer K, Frölich L (1988) Lokalisation kognitiver Störungen bei Demenz vom Alzheimer-Typ. Psycho 14: 381–382

Maurer K (1988) Evozierte Potentiale (AEP-VEP-SEP). Enke, Stuttgart, S 1–66

Maurer K, Dierks T (1988) Topographie der P 300 in der Psychiatrie – I. Kognitive Felder bei Psychosen. Z EEG-EMG 18: 21–25

Maurer K, Ihl R, Dierks T (1988) Topographie der P 300 in der Psychiatrie – II. Kognitive Felder bei Demenz. Z EEG-EMG 19: 26–29

Obrist WD (1954) The electroencephalogram of normal aged adults. Electroenceph Clin Neurophysiol 6: 235–244

Okada YC, Kaufmann L, Williamson SJ (1983) The hippocampal formation as a source of slow endogenous potentials. Electroenceph Clin Neurophysiol 55: 417–426

Pfefferbaum ABG, Wenegret JM, Roth WT, Kopell BS (1984) Clinical application of the P 3 component of event-related potentials. II. Dementia, depression and schizophrenia. Electroenceph Clin Neurophysiol 59: 104–124

Roth WT, Pfefferbaum A, Kelly AF, Berger PA, Kopell BS (1981) Auditory-event related potentials in schizophrenia and depression. Psychiatr Res 4: 199–212

Correspondence: Dr. K. Maurer, Department of Psychiatry, University of Würzburg, Füchsleinstrasse 15, D-8700 Würzburg, Federal Republic of Germany.

Positron emission tomography in Parkinson's disease glucose metabolism

M. Huber, K. Herholz, V. Holthoff, and W.-D. Heiss

Department of Neurology, University of Cologne,
and Max-Planck-Institut für neurologische Forschung, Cologne
(Director: Prof. Dr. W.-D. Heiss), Federal Republic of Germany

Summary

^{18}FDG-PET results in 11 patients with Parkinson's disease are presented. Reductions in total glucose utilization were noted in three non-demented and two demented patients, in the latter there was also a significant cortical metabolic reduction. In two patients with lateralized tremor increased metabolic rates in the cerebellum and contralateral sensorimotor region were noted. There was no asymmetry of basal ganglia glucose metabolism in patients with lateralized Parkinson's syndrome. Decline in dopaminergic inhibitory innervation of the striatal nuclei due to neuronal degeneration in some patients leads to increased metabolism in the basal ganglia.

Introduction

Selective degeneration of dopamine-containing neurons in the compact zone of the substantia nigra with resultant reduction in the dopaminergic innervation of the striatal nuclei represents the major pathological process in Parkinson's disease (PD). Positron emission tomography (PET) provides a noninvasive imaging technique to measure the three-dimensional distribution and magnitude of local abnormalities in cerebral blood flow and metabolism. Energy requirements of the brain are almost exclusively supplied by glucose and its metabolism, as visualized by PET, reflects mainly local neuronal activity. It is the topic of this paper whether the loss of dopaminergic innervation leads to alterations of metabolic activity in basal ganglia or cerebral cortex. It may also be of interest if cerebral metabolic changes which cannot be explained by the dopaminergic deficit alone occur in PD.

Methods

All PET studies were performed under resting conditions with a four-ring positron camera (Scanditronix PC-384) (Eriksson et al., 1982). Brain glucose metabolism was measured using the ^{18}FDG technique (Reivich et al., 1979) and a dynamic scanning and blood-sampling technique previously described by Heiss et al. (1984). Relative regional metabolic rates were calculated by dividing regional values through the individual global metabolic rate.

Patients and results

We studied 11 patients with Parkinson's syndrome (19—74 years of age). Mean duration of disease was 9.8 years (3—23 years), all patients were receiving levodopa, bromocriptine or amantadin medication. In 5 patients PD symptoms were clearly lateralized. One patient was also suffering from chronic heart failure, two had moderate to severe dementia according to clinical criteria. In the latter, CT scans showed evidence of cortical cerebral atrophy. In the two demented patients a significant reduction both in total and cortical glucose utilization was noted while three patients showed moderately reduced values. On the average, there were no gross changes in global cerebral glucose metabolism in patients with Parkinson's syndrome when compared to age-matched controls (Table 1). Mean regional metabolic rates for frontal, sensorimotor, temporal, parietal, occipital, and visual cortex, when compared with control values, were found normal. In two patients with lateralized tremor a slightly increased metabolism was noted in the contralateral sensorimotor cortical region and the cerebellum. There was no asymmetry of basal ganglia metabolism in

Table 1. Relative metabolic rates of different cerebral regions and absolute values of glucose metabolism in Parkinson's disease compared with controls

	rCMR (Glu)			
	increased	decreased	absolute values	normal values (sd)
	(no. of patients)		μmol/100 g/min	μmol/100 g/min
Total	1	5	31.35	34.60 (3.82)
Sensorimotor Cortex*	2	—	36.55	39.70 (4.74)
Caudate nucleus*	3	—	38.20	42.85 (4.53)
Lentiform nucleus*	3	—	43.60	45.10 (6.25)
Cerebellum*	3	—	31.00	33.15 (4.07)

* Alterations with respect to relative metabolic rates

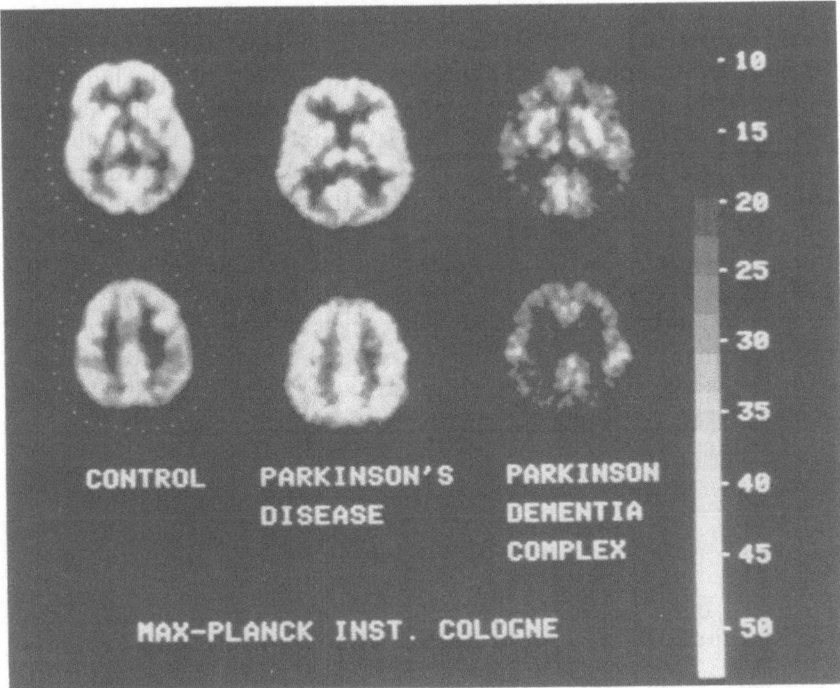

Fig. 1. PET scan of a Parkinson patient showing relatively increased metabolism in caudate and lentiform nucleus as compared with Parkinson dementia complex with reduced cortical metabolism and a control scan

patients with lateralized symptoms. Increased relative metabolic rates in the caudate and, more pronounced, in the lentiform nucleus were found in three patients. In the demented patients, however, there was a marked decline in cortical glucose utilization.

Discussion

Our observations are in accordance with Kuhl et al. (1984), who reported moderate reduction of average cerebral glucose metabolism in patients with L-DOPA treated PD, but differences were not significant. Regional analysis of glucose metabolism did not reveal any association with duration, stage or severity of any symptoms. There was no selective metabolic change in the striatum that paralleled depression of striatal dopamine. It was therefor concluded that striatal pathway degeneration does not have major effects on striatal or cortical glucose utilization. Martin et al. (1984) found increased glucose utilization in the lentiform nucleus contralaterally to the most severly affected limb in some individual patients. In three patients we noted moderately increased metabolism in the lentiform nucleus but there was no asymmetry according to the clinical syndrome. Our findings in the demented patients are also in accordance with a study of 14 PD patients compared to Alzheimer (AD) and multiple infarct dementia (MID) patients by Kuhl (1985). Bilateral cortical hypometabolism which was most pronounced in the posterior parietal cortex clearly differentiated PD and AD patients from the MID group. The patterns of hypometabolism were similar in both diseases which may suggest common pathological mechanisms in both forms of dementia. The results in two patients with lateralized tremor may suggest involvement of cerebellar and contralateral sensorimotor neurons in the pathophysiology of parkinsonian resting tremor of the upper limb.

References

Eriksson L, Bohm C, Kesselberg M (1982) A four ring positron camera system for emission tomography of the brain. IEEE Trans Nucl Sci NS 29: 539–543

Heiss WD, Pawlik G, Herholz K, Wagner R, Göldner H, Wienhard K (1984) Regional kinetic constants and cerebral metabolic rate for glucose in normal human volunteers determined by dynamic positron emission tomography of (^{18}F)-2-fluoro-2-deoxy-D-glucose. J Cereb Blood Flow Metab 4: 212–223

Kuhl DE, Metter EJ, Riege WH, Markham CH (1984) Patterns of cerebral glucose utilization in Parkinson's disease and Huntington's disease. Ann Neurol 15 [Suppl]: S 119–125

Kuhl DE, Metter EJ, Benson DF, Ashford JW, Riege WH, Fujikawa DG, Markham CH, Mazziotta JC, Maltese A, Dorsey DA (1985) Similarities of cerebral glucose metabolism in Alzheimer's and Parkinsonian dementia. J Cereb Blood Flow Metab 5 [Suppl] 1: S 169–170

Martin WRW, Beckman JH, Calne DB, Adam MJ, Harrop R, Rogers JG, Ruth TJ, Sayre CI, Pate BD (1984) Cerebral glucose metabolism in Parkinson's disease. Can J Neurol Sci 11: 169–173

Reivich M, Kuhl D, Wolf A, Greenberg J, Phelps M, Ido P, Casella V, Fowler J, Hoffman E, Alavi A, Som P (^{18}F)-fluorodeoxyglucose method for the measurement of local cerebral glucose utilization in man. Circ Res 44: 127–137

Correspondence: Dr. M. Huber, Department of Neurology, University of Cologne, Joseph-Stelzmann-Strasse 9, D-5000 Köln 41, Federal Republic of Germany.

Parkinson's disease studied using PET

K. L. Leenders

MRC Cyclotron Unit, Hammersmith Hospital, London, U.K.

Summary

Positron emission tomography makes it possible to measure quantitatively certain aspects of regional brain tissue energy metabolism and dopaminergic neurotransmitter activity in vivo in man. It has been shown that this method can be applied in the study of pathophysiology of Parkinson's disease and other conditions. Striatal influx of the radiolabelled tracer L-^{18}F-fluoro-dopa is related to the clinical severity of the disease. A preclinical diagnosis of Parkinson's disease should in principle be possible, although no PET studies with this particular aim have been undertaken yet.

Combination with a tracer for the dopamine D_2 receptors may help in making the diagnosis. In Parkinson's disease dopamine turnover is markedly decreased while the density of dopamine D_2 receptors is essentially unchanged compared to age-matched controls. On the other hand in neurodegenerative conditions accompanied by parkinsonism both "pre-" and "post-synaptic" binding of tracers seem to be impaired.

Introduction

In recent years it has become possible to measure in vivo human striatal dopaminergic function using radio-labelled tracers and positron emission tomography (PET). Further validation and expansion of this method may lead to elucidation of pathophysiology of movement disorders since in many of these conditions a disturbance of one or more neurotransmitter systems has been demonstrated.

Studies relating clinical features with changes in cerebral neurotransmitter function measured in vivo by PET can now be undertaken. Particularly longitudinal studies starting in an early phase of the disease and using various types of tracers seem to be promising (see discussion). However, to my knowledge no PET study diagnosing patients in an early phase of the disease or even at a pre-clinical stage has been performed yet. Cross-sectional cohort studies are less suitable due to the usual large interindividual variance of the results combined with the rather small number of patients which currently can be scanned. This is not just caused by the relatively long duration of the scanning procedures. Also the radiochemistry is often complicated and needs to be performed immediately before a scan because of the short radioactive half-life of the used radionuclides. However, data handling and analysis is the most time consuming aspect of measuring tissue function with PET to date. Reduction of count measurements into manageable units and conversion of time-activity curves into meaningful pharmacological or biochemical entities is a formidable task. It seems that the developments in this field are still in an early stage. The inevitably low patient throughput per scan laboratory, in combination with the small number of PET centres worldwide, will make it understandable why accumulation of biological or clinical results with PET is a slow process.

Positron emission tomography (PET)

The essential measurement of PET consists of the detection of radiolabelled tracers in the body e.g. the brain. A special type of radionuclide needs to be incorporated as a "label" in the tracer molecules. These radionuclides decay by emitting a positron, e.g. oxygen-15, carbon-11 or fluorine-18. A positron is a particle with the same mass as an electron but positively charged. The main three reasons for choosing this type of radionuclide are the following. First, they are nuclides of physiological atoms, which means that their incorporation into the required tracer molecules does not, or only slightly, change the chemical properties of the tracer. Secondly, the short radio-active half-life (minutes to a few hours) allows administration of tracer in a dose sufficient to obtain measurable signals while keeping the radiation dose low enough for human use. Thirdly, the characteristic physical features accompanying positron emission are the basis for quantitation of the regional radio-activity. Namely shortly after emission from a decaying nucleus a positron annihilates with an electron. This results in conversion of the masses of the positron and electron into two high energy gamma rays (511 keV) travelling simultaneously into opposite directions. The construction of most PET tomographs is such, that a ring of detectors surrounds the body. Simultanuous stimulation of two oposite detectors (coincidence event) by the two gamma rays resulting from emission of a positron allows exact determination of the direction from where the event took place. After collection of sufficient coincidence events ("counts") within a certain time frame (seconds to minutes), the distribution of local radioactivity in the scanned cross-section (plane) can be calculated by standard tomographical reconstruction techniques. Thus for each region in the brain a time-activity curve can be determined in absolute units of radio-activity (microcurie per ml tissue). The built up and washout of tracer in a certain brain tissue region become more meaningful when they can be compared with the dose delivered to the brain via the arterial system. To obtain an "arterial input curve" a series of blood samples is usually taken from a small indwelling radial artery cannula after administration of the tracer. Whether the next step, namely calculation of a pharmacological or biochemical entity related to the tracer activity, can be achieved depends on the specific properties of the tracer molecule. The mathematical models which are used for this purpose vary widely in complexity. Assumptions and limitations are discussed in the literature (e.g. Phelps et al., 1986).

Dopaminergic tracers

Pre-synaptic tracers

a) L-^{18}F-fluorodopa (^{18}F-dopa)

This analogue of L-dopa can be used as a tracer for L-dopa transport from blood to brain, dopamine formation and subsequent conversion into metabolites (Garnett et al., 1983; Firnau et al., 1987; Leenders et al., 1986 a, 1986 b). After i.v. administration, only a small fraction is taken up by the brain ($\pm$ 0.1% of the injected tracer dose). L-dopa transport across the blood-brain barrier is an active, energy dependent and strictly stereo-selective process in competition with other large neutral aminoacids (Leenders et al., 1986 c). ^{18}F-dopa is decarboxylated to ^{18}F-dopamine in the brain, particularly in decarboxylase rich regions like striatum. ^{18}F-dopamine is further metabolised into ^{18}F-HVA and ^{18}F-DOPAC, but Firnau and colleagues (1987) showed that in monkey brain the first 1 to 1.5 hour after ^{18}F-dopa administration the radio-activity in striatum was predominantly ^{18}F-dopamine. The sum of ^{18}F-dopamine, ^{18}F-HVA and ^{18}F-DOPAC formation is determined by the regional decarboxylation rate. In cerebral tissues other than the striatum the O-methylated derivative was the principle labelled compound. In arterial plasma of with carbidopa pretreated subjects the main metabolite after ^{18}F-dopa administration was found to be the O-methylated derivative (Boyes et al., 1986).

b) ^{11}C-nomifensine

Nomifensine (NMF) binds specifically to catecholamine uptake sites on nerve terminals (Slater et al., 1984; Scatton et al., 1985). In striatum specific nomifensine binding is virtually only related to binding to dopaminergic nerve terminals. Unilateral lesions of the nigrostriatal dopaminergic pathway in rats produced a marked (± 80%) decrease of specific striatal binding of ^{3}H-nomifensine. A unilateral lesion of the locus coeruleus did not change tracer uptake. Nomifensine labelled with the positron emitting radionuclide carbon-11 (^{11}C-NMF) has been applied in PET studies (Aquilonius et al., 1987; Leenders et al., 1988 a). A similar experiment as reported by Scatton and colleagues in rats using autoradiography and ^{3}H-NMF (1985) was performed on a Rhesus monkey using PET and ^{11}C-NMF (Leenders et al., 1988 a).

Post-synaptic tracers

a) ^{11}C-methyl-spiperone

Spiperone is a dopamine D_2 receptor antagonist and belongs to the butyrophenone group of neuroleptic drugs. Also the radiolabelled analogue ^{11}C-methyl-spiperone (^{11}C-MSP) has been shown to bind predominantly to D_2 receptors in striatum where receptor concentration is highest. However, also binding to serotonin receptors occurs (Frost et al., 1987), which particularly dominates binding in cortical regions. ^{11}C-MSP has been used in man (Wagner et al., 1983) and Eckernäs and colleagues (1987) discuss the mathematical modelling associated with quantification of ^{11}C-MSP uptake.

b) ^{11}C-raclopride

Raclopride is also a neuroleptic drug (substituted benzamide) and can be radiolabelled to visualise dopamine receptor binding in human brain with PET (Farde et al., 1985). This compound is different from ^{11}C-MSP in several respects. ^{11}C-raclopride appears to bind more specifically to dopamine receptors and also the kinetics of binding are different in that a dynamic equilibrium relatively rapidly (after ± 40 minutes) is obtained. Quantitative analysis with this method was described by Farde and colleagues (1986).

Results and discussion

^{18}F-dopa has been used to study the deficient nigrostriatal dopaminergic system in patients with Parkinson's disease (Leenders et al., 1986). Striatal uptake of this tracer was markedly impaired in patients. The ratio of striatal radio-activity over that in the surrounding non-dopaminergic brain tissue can already serve as a crude indicator of specific "pre-synaptic" dopaminergic activity mainly reflecting decarboxylation of dopa to dopamine and subsequent trapping in the tissue. A more severely affected patient group had a lower mean ratio than the group of patients in an earlier phase of the disease. Figure 1 illustrates the radio-activity distribution throughout the brain after ^{18}F-dopa administration in a patient with Parkinson's disease compared to that in an age-matched healthy volunteer.

A more sophisticated approach to quantify ^{18}F-dopa uptake is possible by applying a multiple time graphical plotting technique (Patlak et al., 1983). This method compares tissue radio-activity uptake

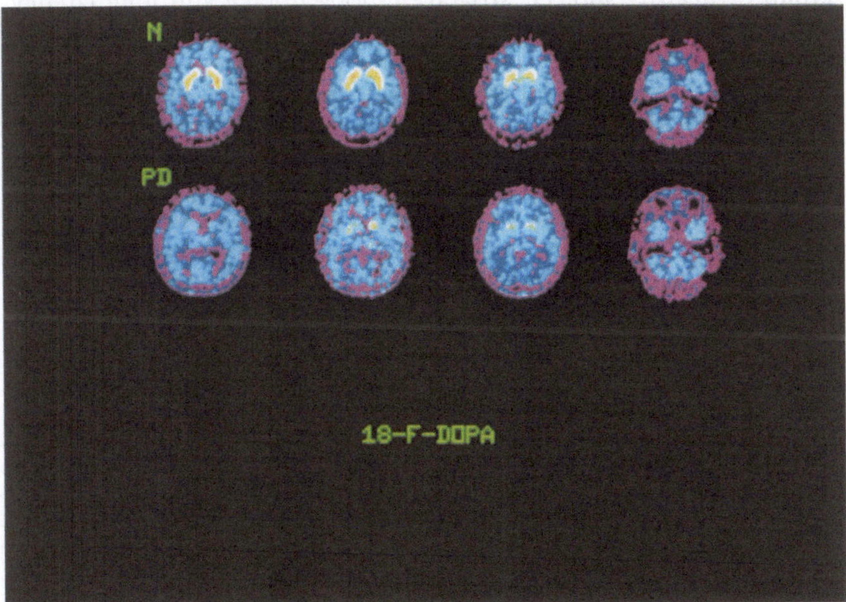

Fig. 1. Cerebral L-^{18}F-6-fluorodopa uptake in a healthy subject (upper row = *N*) and in patient with Parkinson's disease (lower row = *PD*). Three consecutive cross-sections at the level of the striatum are displayed and one cross-section (the very right image) at the level of the cerebellum. Striatal uptake of the tracer in the patient is considerably less compared to the healthy control

with arterial plasma radio-activity and provides an influx constant (K_j) which is a measure for total tracer flux from the blood into a final compartment in the tissue. Such a quantification permits comparison of one individual with another. For instance clinical severity (bradykinesia) in a group of parkinsonian patients correlated well with striatal K_j for ^{18}F-dopa (Leenders et al., 1987). Also comparison of ^{18}F-dopa influx in one subject before or after a certain intervention is possible (Leenders et al., 1987).

PET ^{18}F-dopa studies to assess progression of the disease in the same patients have not been published yet but are currently undertaken. Neither have studies been reported to try and establish early or even pre-clinical diagnosis. The latter should be possible since a large part of the pre-synaptic dopaminergic system needs to be lost (60–80%?) before clinical signs become obvious. This presupposes good knowledge about the normal range of ^{18}F-dopa tracer uptake into striatum in age-matched control groups. Currently at the MRC Cyclotron Unit, Hammersmith Hospital, London, a study is in preparation (Dr. G. Sawle, personal communication) to measure in a group of healthy subjects striatal ^{18}F-dopa uptake and long latency stretch reflexes. The latter is markedly abnormal in patients with Parkinson's disease.

No studies have been published yet using ^{11}C-nomifensine and PET as a possible tracer for striatal dopaminergic nerve endings in patients with Parkinson's disease, although first experimental trials were promising (Aquilonius et al., 1987; Leenders et al., 1988 a).

Combination of ^{18}F-dopa and ^{11}C-nomifensine PET tracer studies suggests in principle the possibility to relate striatal dopamine formation to the pool of striatal nerve endings. It might provide in vivo data about the question whether in an early phase of Parkinson's disease dopamine formation is increased relative to the remaining dopaminergic nerve terminals. Although currently combined tracer studies are carried out, no results are available yet and much work to validate the tracer methods still needs to be done.

Reports concerning ^{11}C-MSP and its application in Parkinson's disease have been scarce so far (Leenders et al., 1985; Hägglund et al., 1986). Untreated parkinsonian patients showed similar striatal ^{11}C-MSP uptake compared to healthy controls. Levodopa drug treatment seemed to reduce ^{11}C-MSP uptake to some extent (Leenders et al., 1985), but the number of patients studied was small. These findings are in agreement with post-mortem results showing virtually no change of dopamine D_2 receptor densities in parkinsonian patients (Bokobza et al., 1984). The in this chronic disease apparently intact post-synaptic dopaminergic system, in the presence of a severe presynaptic lesion, "explains" why dopaminergic drug treatment is effective at all in Parkinson's disease. Patients with other chronic neurodegenerative diseases associated with parkinsonian features like Steele-Richardson-Olszewski syndrome do not respond or only slightly to levodopa therapy. In that condition impaired pre-synaptic dopaminergic function (Leenders et al., 1988 b) is accompanied by striatal dopamine D_2 receptor decreases (Baron et al., 1985), probably due to striatal neuronal cell loss.

^{11}C-raclopride has not been used extensively in patients with Parkinson's disease yet, but a recent report (Lindvall et al., 1987) about 2 patients showed normal striatal values before and after engraftment of homologous adrenal medulla tissue.

The effect of unilaterally administered MPTP (right internal carotid artery MPTP infusion) in a Rhesus monkey has been studied using ^{11}C-raclopride (Leenders et al., 1988 a). Two days after the lesion an increased (± 50%) specific tracer uptake was found in the lesioned striatum in the presence of a clinically impaired pre-synaptic dopaminergic function. At 6 weeks after the lesion increased ^{11}C-raclopride and marked decreased ^{11}C-nomifensine (see above) was found in the lesioned striatum, but normal values on the unlesioned side. Five months and one year after the lesion ^{11}C-raclopride uptake was normal again, while pre-synaptic function was still impaired according to the ^{11}C-nomifensine uptake. This suggests that acute lesions of the nigro-striatal system can provoke a temporary increase of striatal dopamine D_2 receptor density. However, a chronic lesion of this system seems compatible with normal post-synaptic receptor density, at least in the absence of striatal neuronal cell loss (see discussion of ^{11}C-MSP).

On the basis of the available PET scan data it can thus be concluded that characterisation of striatal dopaminergic function in vivo and the relationship of this function to clinical severity is possible. Combination of various tracers, each specific for a certain component of the nigrostriatal dopaminergic pathway, seems to provide a distinction between patients with Parkinson's disease and patients with extrapyramidal signs and symptoms in the context of other neurodegenerative conditions.

References

Aquilonius SM, Bergström K, Eckernäs SA, Hartvig P, Leenders KL, Lundqvist H, Antoni G, Gee A, Rimland A, Uhlin J, Långström B (1987) In vivo evaluation of striatal dopamine reuptake sites using ^{11}C-nomifensine and positron emission tomography. Acta Neurol Scand 76: 283–287

Baron JC, Maziere B, Loc'h C, Sgouropoulos P, Bonnet AM, Agid Y (1985) Progressive supranuclear palsy: loss of striatal dopamine receptors demonstrated in vivo by positron tomography. Lancet ii: 1163–1164

Bokobza B, Ruberg M, Scatton B, Javoy-Agid F, Agid Y (1984) (3 H)-spiperone binding, dopamine and HVA concentrations in Parkinson's disease and supranuclear palsy. Eur J Pharmacol 99: 167–175

Boyes RE, Cumming P, Martin WRW, McGeer EG (1986) Determination of plasma [^{18}F]-6-fluorodopa during positron emission tomography: elimination and metabolism in carbidopa trated subjects. Life Sci 39: 2243–2252

Eckernäs SA, Aquilonius SM, Hartvig P, Hägglund J, Jundqvist H, Någren K, Långström B (1987) Positron emission tomography (PET) in the study of dopamine receptors in the primate brain: evaluation of a kinetic model using ^{11}C-N-methyl-spiperone. Acta Neurol Scand 75: 168–178

Farde L, Ehrin E, Eriksson L, Greitz T, Hall H, Hedstrom CG, Litton JE, Sedvall G (1985) Substituted benzamides as ligands for visualisation of dopamine receptor binding in the human brain by positron emission tomography. Proc Natl Acad Sci USA 82: 3863–3867

Farde L, Hall H, Ehrin E, Sedvall G (1986) Quantitative analysis of D_2 dopamine receptor binding in the living human brain by PET. Science 231: 258–261

Firnau G, Sood S, Chirakal R, Nahmias C, Garnett ES (1987) Cerebral metabolism of 6-[F-18]Fluoro-L-dopa in the primate. J Neurochem 48: 1077–1082

Frost JJ, Smith AC, Kuhar MJ, Dannals RF, Wagner Jr HN (1987) In vivo binding of ^{3}H-N-methylspiperone to dopamine and serotonin receptors. Life Sci 40: 987–995

Garnett ES, Firnau G, Nahmias C (1983) Dopamine visualised in the basal ganglia of living man. Nature: 305: 137–138

Hägglund J, Aquilonius SM, Eckernäs, SA, Hartvig P, Lundquist H, Gullberg P, Långström B (1987) Dopamine receptor properties in Parkinson's disease and Huntington's chorea evaluated by positron emission tomography using ^{11}C-N-methyl-spiperone. Acta Neurol Scand 75: 87–94

Leenders KL, Herold S, Palmer AJ, Turton D, Quinn N, Jones T, Frackowiak RSJ, Marsden CD (1985) Human cerebral dopamine system measured in vivo using PET. J Cereb Blood Flow Metabol 5 [Suppl]: S 517–518

Leenders KL, Frackowiak RJS, Quinn N, Marsden CD (1986 a) Brain energy metabolism and dopaminergic function in Huntington's disease measured in vivo using positron emission tomography. Movement Disorders 1: 69–77

Leenders KL, Palmer AJ, Quinn N, Clark JC, Firnau G, Garnett ES, Nahmias C, Jones T, Marsden CD (1986 b) Brain dopamine metabolism in patients with Parkinson's disease measured with positron emission tomography. J Neurol Neurosurg Psychiatr 49: 853–856

Leenders KL, Poewe WH, Palmer AJ, Brenton DP, Frackowiak RSJ (1986 c) Inhibition of L-[^{18}F]fluorodopa uptake into human brain by amino acids demonstrated by positron emission tomography. Ann Neurol 20: 258–262

Leenders KL (1987) Parkinson's disease. Clinical and experimental advances. J Libbey, London Paris, pp 21–32

Leenders KL, Aquilonius SM, Bergström K, Bjurling P, Crossman AR, Eckernäs SA, Gee AG, Hartvig P, Lundqvist H, Långström B, Rimland A, Tedroff J (1988 a) Unilateral MPTP lesion in a Rhesus monkey: effects on the striatal dopaminergic system measured in vivo with PET using various novel tracers. Brain Research 445: 61–67

Leenders KL, Frackowiak RJS, Lees AJ (1988 b) Steele-Richardson-Olszewski syndrome: brain energy metabolism, blood flow and fluorodopa uptake measured by positron emission tomography. Brain 111: 615–630

Lindvall O, Backlund EO, Farde L, Sedvall G, Freedman R, Hoffer B, Nobin A, Seiger Å, Olson L (1987) Transplantation in Parkinson's disease: two cases of adrenal medullary grafts to the putamen. Ann Neurol 22: 457–468

Patlak CS, Blasberg RG, Fenstermacher JD (1983) Graphical evaluation of Blood-to-brain transfer constants from multiple-time uptake data. J Cereb Blood Flow Metabol 3: 1–7

Phelps ME, Mazziotta JC, Schelbert HR (eds) (1986) Positron emission tomography and autoradiography. Principles and applications for the brain and heart. Raven Press, New York

Scatton B, Dubois A, Dubocovitch ML, Zahniser NR, Fage D (1984) Quantitative autoradiography of 3H-nomifensine binding sites in rat brain. Life Sciences 36: 815–822

Slater P, Crossman AR (1984) Nomifensine. A pharmacological and clinical profile. The Royal Society of Medicine, London, pp 15–19

Wagner HN, Burns HD, Dannals RF, Wong DF, Langstrom B, Duelfer T, Frost JJ, Ravert HT, Links JM, Rosenbloom SB, Lukas SE, Kramer AV, Kuhar MJ (1983) Imaging dopamine receptors in the human brain by positron tomography. Science 221: 1264–1266

Correspondence: Dr. K. L. Leenders, Paul Scherrer Institute, PET Group, CH-5234 Villigen, Switzerland.

[illegible] (198[illegible]) [illegible]

Wong DF, Wagner HN, Dannals RF, Links JM, Frost JJ, Ravert HT, Wilson AA, Rosenbaum AE, Gjedde A, Douglass KH, Petronis JD, Folstein MF, Toung JKT, Burns HD, Kuhar MJ (1984) Effects of age on dopamine and serotonin receptors measured by positron tomography in the living human brain. Science 226: 1393–1396

Pathobiochemistry of the extrapyramidal system: a "short note" review

P. Riederer[1], E. Sofic[1], W. D. Rausch[2], G. Hebenstreit[3], and J. Bruinvels[4]

[1] Clinical Neurochemistry, Department of Psychiatry, University of Würzburg, Federal Republic of Germany
[2] Medical Chemistry, Vet.-Med. University, Wien, Austria
[3] Geriatric Hospital, Mauer-Öhling, Austria
[4] Department of Pharmacology, University of Rotterdam, The Netherlands

Summary

Evidence is presented to show that in Parkinson's disease (PD) there exists a topographical degeneration in the nigrostriatal dopamine system. This finding may account for response variabilities and side effects occuring after antiparkinsonian therapy. Postsynaptic receptors do not show topical differences in both B_{max}- and K_D-values in treated PD. Postsynaptic receptors respond inadequately in about 30% of PD. Denervation supersensitivity may only occur in striatal subareas depleted of dopamine to an extent of >90%. Compensatory mechanisms include presynaptic overactivity and enhanced activity of otherwise reduced tyrosine hydroxylase. Iron (III), total iron and ferritin are significantly increased in the substantia nigra indicating disturbances at the level of redox-equilibrium, respiratory chain activity and energy metabolism. These findings together with accumulation of exogenous or endogenous toxins may be of pathogenic importance in PD. Whether such pathophysiological considerations involve peripheral dopamine or catecholaminergic systems (e.g. the sympatho-adrenal medullary function) is uncertain and requires further experimental studies.

Topographic aspects of denervation

Degeneration of the substantia nigra (SN) does not show an uniform and unspecific distribution, but is featured by unilateral or, more frequently, symmetrical focal neuronal loss mainly involving the central and caudal parts of the zona compacta of SN (Bernheimer et al., 1973; Jellinger, 1986). Therefore, a topographically differentiated denervation of the striatum may occur. This is underlined by the arrangement of the nigrostriatal projection innervating the putamen mainly from the caudo-lateral portion of SN zona compacta, while the caudate nucleus receives its dopamine (DA) containing fibres from the rostro-medial SN zona compacta. Recently it could be shown, that the subregional loss of DA in the putamen and caudate nucleus have opposite patterns. The rostral part of the caudate head is more severely affected than the caudal part (16% and 38% of controls resp.) while the rostral part of the putamen (13% of controls) is less damaged than the caudal part (4% of controls) (Hornykiewicz and Kish, 1986). Our own data are principally in agreement with this finding, showing a decrease of DA in the caudal part of the putamen to 11% of controls, while the rostral part is less affected (39% of controls) (Riederer, 1986). In addition, the more severely denervated caudal putamen shows a more severe reduction of 3,4-dihydroxyphenylacetic acid (DOPAC) (19% of controls) and homovanillic acid (HVA) (41% of controls) while these disturbances occur in a lower percentage in the rostral part (79% of controls for DOPAC and 72% of controls for HVA).

Table 1 also gives evidence for a mainly presynaptic localization of DOPAC. DOPAC is a more sensitive marker for the denervation process compared to HVA. In the more severely affected caudal putamen the reduction of DOPAC is similar to that of DA while the loss of DOPAC is approximately the same as that of HVA in the less affected portion. In agreement with the similar concentration of DA in the rostral and caudal putamen, D 2-receptor densities in both controls and Parkinson's disease (PD) on usual antiparkinson treatment as measured by antagonist binding with 3 H-spiroperidol are not different. The lower B_{max}-values in PD may be caused by down-regulation of receptor density due to drug treatment or by a partial loss of cell density of either cholinergic and/or GABA-ergic cell bodies due to progression of the disease.

It is suggested that the topographic denervation of SN involves various portions of the striatum differentially and it is assumed that supersensitive receptors will only occur in subareas with loss of innervation $> 90\%$.

According to histological examinations (Jellinger, 1986), however, such severe degeneration seems to be present only in small parts of the striatum. In these areas the percentage increase of 3 H-spiroperidol might be much higher than the 29% enhancement described by Seeman et al. (1987). In fact, most of these data overlap

Table 1. Dopamine, DOPAC and HVA in the caudal and rostral putamen of benign Parkinson's disease (n = 3–5) (from Riederer et al., 1986)

Putamen	Dopamine	DOPAC mean as percent of 3–4 controls	HVA
caudal	11	19	41
rostral	39	79	72

	N	age	sex	duration of disease (yrs.)	post mortem time (hrs)	therapy	diagnosis	neuropathology
controls	4	77 (74–83)	1 M/ 3 F		7,5 (4–11)	antibiotics cardiovascular treatment	heart-infarction (2) broncho-pneumonia (2)	age related brain histology
Morbus Parkinson	5	79 (75–82)	2 M/ 3 F	10 (8–13)	6,5 (3–9)	combined l-dopa amantadine	Parkinson's disease (5)	Lewy-bodies; loss of melanin in SN

with the upper range of the control population. But this again may be evidence of a subtle localization of supersensitivity, while most receptors are innervated, or at least, partly innervated.

Assuming, that this hypothesis is relevant, drug therapy would meet normosensitive, activated and supersensitive receptors. As a consequence the physiological and behavioral output of therapeutic strategies would largely depend on the type of topographic denervation.

Therefore, therapeutic efficacy and side effects related to the motor system (hyperkinesias, dystonias) can be differentiated according to the lesions and subareas involved.

Presynaptic chemical pathology (Table 2)

Circadian or long-term associated oscillations of motor performance suggest disturbances in the feedback control of the nigrostriatal-strionigral loop. Normally, DA is synthetized from

Table 2. Presynaptic lesion

	Region	Parkinson's disease % controls
TH	SN	25
DA	SN	13
Fe	SN	177
Fe^{2+}/Fe^{3+}	SN	43
MAO-B	SN	125
GABA-binding	SN	44
Cocain binding	ST	21
TH	ST	7–12
TH	CN	20–50
TH → Fe^{2+}-stimulation	CN	84
DA	PUT	18 C; 38 R benign
DA	PUT	4 malignant
DOPAC	PUT	32 C; 47 R benign
DOPAC	PUT	10 malignant
HVA	PUT	29 C; 47 R benign
HVA	PUT	29 malignant
Fe	CN	92
Fe	PUT	81
Fe^{2+}/Fe^{3+}	PUT	77
D 2-Antigen	PUT	100
D 3-Antigen	PUT	126
Glutathione	PUT	26
Ascorbic acid	PUT	69
MAO-A/MAO-B	ST	100
MAO-A: H_2O_2 stim	ST	80–100
MAO-B: H_2O_2 stim	ST	200–600

TH tyrosine hydroxylase; *DA* dopamine; *DOPAC* 3,4-dihydroxyphenyl acetic acid; *HVA* homovanillic acid; *MAO* monoamine oxidase; *ST* striatum; *SN* substantia nigra; *PUT* putamen; *CN* caudate nucleus; *C* caudal portion; *R* rostral portion

endogenous precursor amino acids including DOPA, is stored most efficiently in vesicles and its release into the synaptic cleft is under nerve impulse control and regulated by pre- and postsynaptic receptor activity. In contrast, denervation of presynaptic neurons in PD progressively reduces the capability to synthetize DA from tyrosine and DOPA. Reduced storage capacity and decline in released DA are expected to be the pathophysiological basis of PD. In fact, supplementation of DA is able to overcome akinesia for some time. There is, however, no correlation between the loss of presynaptic terminals and response to l-DOPA dose. The great individual variability in the sensitivity of behavioral response (akinesia) to l-DOPA supposes disturbances in the nigrostriatal DA-feedback control. As long as the operation of this feedback control can be sufficiently maintained, so long can akinesia be minimized. Endogenous compensatory mechanisms, as indicated by decrease in the DA-homovanillic acid (HVA) ratio, are directed to enhancing the release of DA at a stage at which denervation has already started. If the loss of neurons exceeds about 70%, clinical symptoms are apparent and can be improved by supplementation of missing DA. However, the exogenous precursor substance l-DOPA is generally administered in unphysiological quantities.

Depending on how much of the vesicular storage capacity is left, more or less DA synthesized from a small pool of endogenous and an additional large pool of exogenous l-DOPA will be stored and released by using the nerve impulse traffic. However, as this storage capacity of vesicles declines in PD as denervation proceeds, more and more of DA synthetized from administered l-DOPA will not use the physiological route to stimulate receptors. In contrast, additional uncontrolled spillage into the synapse may occur, and DA synthetized in other compartments may now contribute to synaptic processes. There is good evidence to assume that l-deprenyl's efficacy is in part due to blockade of the deamination of DA in extraneuronal tissue (e.g. glia). Accumulation of non-neuronal DA may contribute to synaptic signal transduction.

This uncontrolled receptor stimulation impairs the endogenous cybernetic capacity to maintain a feed back control. Unphysiological stimulation of presynaptic receptors would be without regulatory consequence, as synthesis of DA via tyrosine hydroxylase is minimized in the case of treated PD, but instead largely derives from exogenous l-DOPA.

Tyrosine hydroxylase, iron and the influence of toxins

Brain tyrosine hydroxylase

Tyrosine hydroxylase (TH), the rate limiting enzyme of the catecholamine pathway, drops with increasing age (McGeer and McGeer, 1976) and shows significantly decreased activity in the nigrostriatal system of PD (Lloyd et al., 1975). More recently it could be shown that the enzyme responds to iron (II) with a stimulation of 1300% in the caudate nucleus of controls and 1100% in PD. Furthermore, TH activity from parkinsonian tissue was increased by a statistically significant 48% in the presence of exogenous protein kinase (Rausch et al., 1988). In addition, calculations of the activity on the basis of TH-protein demonstrated a significant increase of TH activity in the caudate nucleus of PD suggesting a compensatory increase of feedback regulatory enzyme activity due to degeneration of dopaminergic cell bodies, fibres and nerve endings (Mogi et al., 1988).

In SN of PD iron (III) and total iron are significantly increased leading to a change in the iron (II)-iron (III) ratio of 2.44 in controls to 1.06 in PD (Sofic et al., 1988). Furthermore, ferritin, an iron (III) binding protein is significantly increased in this brain region (Riederer et al., 1988). Therefore, TH might be depleted by its cofactor iron (II) and a decreased activity may appear (Table 3). This hypothetical conclusion, however, remains to be verified in future studies. Two arguments seem to support this hypothesis: 1. enhanced oxidation of iron (II) to iron (III) may be triggered by toxins of

Table 3

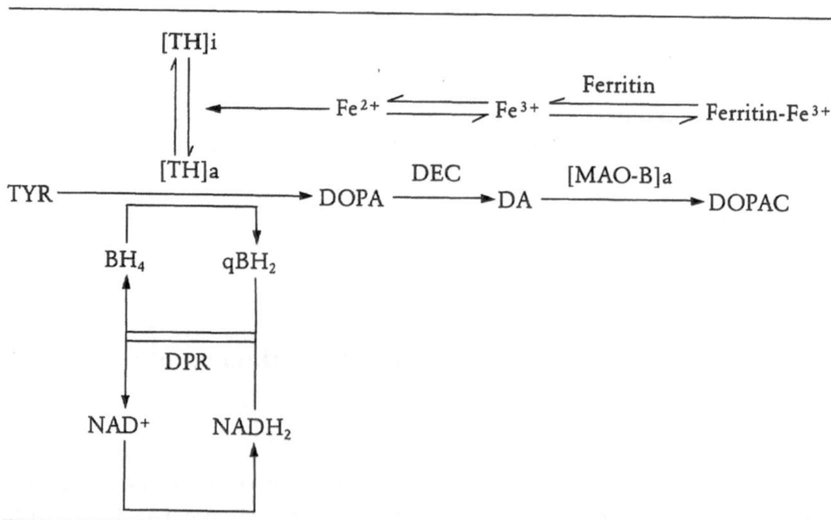

DPR dihydropteridine reductase; *BH_4* (GR)-L-erythro-tetrahydrobiopterin; *qBH_2* quinonoid-7,8-dihydrobiopterin; *TH* tyrosine hydroxylase in activated (a) or inhibited (i) function; *MAO-Ba* monoamine oxidase-B is activated by exogenous or endogenous oxidants

exogenous or endogenous origin (e.g. hydrogen peroxide, radicals derived from it etc.). Indirect evidence for increased lipid peroxidation has been shown recently in SN of PD (Dexter et al., 1986). Secondly, an iron (II) containing compound, oxiferriscorbone, has been shown to be of benefit in the treatment of PD (Birkmayer and Birkmayer, 1986).

The sympatho-adrenal medullary function

Clinical studies show that autonomic dysfunctions occur in PD (Birkmayer and Riederer, 1985). In particular, orthostatic hypotension is frequently observed and the vulnerability of parkinsonian patients to drugs triggering orthostatic hypotension (e.g. l-DOPA, dopaminergic agonists) is remarkable. Although biochemical examinations in peripheral fluids and tissues have been performed only occasionally, some evidence exists to suggest a deficient sympatho-adrenal medullary function. In fact, a dopamine deficiency was not only found in the basal ganglia of parkinsonian brains (Ehringer and Hornykiewicz, 1960) but also in urines of such patients (Barbeau et al., 1961). Although this finding and a loss of HVA may not be specific enough to be correlated to PD, a reduction in adrenal medullary TH activity (Riederer et al., 1978) and a severe loss of DA, NA and adrenaline (Carmichael et al., 1988) has been found to be significant. Furthermore, low plasma renin activity (Barbeau, 1975) has been related to the progression of central or possibly renal dopaminergic receptors in this disorder (Sullivan et al., 1973). Animal studies indicate an association of adrenal TH acitivity with the stimulatory properties of central dopamine receptors (Quik and Sourkes, 1977). As the excitatory role of central DA neurons on adrenal TH activity is lost in PD, a decrease of the latter enzyme activity follows. This cannot be counteracted in full by the DA-inhibiting serotoninergic system which shows reduced functional activity in PD. With regard to the iron (II)-stimulating properties of TH, an induction of adrenal medullary TH activity might in turn influence central TH function (see above).

Recently it has been shown that the adrenal secretion rate of noradrenaline and adrenaline correlated with the respective receptor activity of various peripheral tissues (Ito et al., 1986). Therefore, a reduction in the sympatho-adrenal medullary function causes increased catecholaminergic receptor response to amines and sympathicomimetics including antiparkinson drugs leading to enhanced orthostatic vulnerability.

Postsynaptic signal transduction (Table 4)

Unphysiological stimulation of postsynaptic receptors induces down-regulation of receptor numbers, that might end up in a hyperpolarization block of the receptor by excess DA. Postsynaptic DA-receptors have been divided into two classes, adenylate cyclase dependent D 1-receptors and D 2-receptors that are independent of this enzymatic process. In PD, these receptors show decreased, unchanged or increased receptor density as defined by antagonist binding techniques (Birkmayer and Riederer, 1985). Most of these discrepant data are due to the variety of drug treatment strategies (Guttman et al., 1986). In the largest study Seeman et al. (1987) report a small, but significant increase in D 2-receptor density in untreated patients and no change in a treated group (Guttman et al., 1986; Seeman et al., 1987). These techniques, however, provide us only with the information, that a ligand binds sufficiently to a receptor protein, while there is no evidence from such data for an intact amplification and output system. At least in the long-term treated patient and in the decompensated late phase of PD, dysfunction of postsynaptic receptors may indeed be apparent. This is demonstrated by about 30% of patients with PD that no longer respond to antiparkinson drugs, die in an akinetic crisis but do show comparable D 2 receptor-densities post mortem. In such cases, early work has shown insufficient stimulation properties of adenylate cyclase to DA (Riederer et al., 1978). Furthermore, displacement of radiolabeled spiroperidol in post mortem putamen by classical DA-agonists such as bromocriptine and lisuride shows a tendency towards increased sensitivity (lower IC_{50}-values) while some decrease was noteable for CI 201–678. The mixed agonist-antagonist terguride did not change these properties. Although these preliminary data provide only

Table 4. Postsynaptic disturbances

	Region	Parkinson's disease % controls
Spiroperidol binding (D 2)	PUT	100–129 without DOPA 80–100 with DOPA
Adenylatcyclase Dependent D 1	PUT	73 Responder 10 Nonresponder
D 1-antagonist Binding	PUT	100

PUT Putamen

suggestive evidence, it is mainly the opposite shift induced by the D 1 and D 2 agonist CI 201–678 and the D 2 agonists bromocriptine and lisuride, which should be examined further in denervated tissue. This is best documented by experiments to displace a radioligand (^{3}H-spiperone) by an unlabeled ligand (bromocriptine, lisuride, BHT 920 and others) in the presence or absence of GTP. In such experiments it can be shown that denervated receptors respond differentially compared to innervated ones (Wirth and Riederer, unpublished observation). These data provide evidence for the assumption that postsynaptic receptors change (lose) functional activity in late phases of PD. CI 201–678 showed beneficial effects in PD comparable to those of bromocriptine and lisuride while side effects (e.g. orthostatic hypotension) were much milder. Therefore, the synergistic effect of D 1 and D 2 stimulation seems to be superior to D 2-stimulation alone (Birkmayer et al., 1988).

It is noteworthy in this context, that amantadine, which changes membrane fluidity (Wesemann, 1984) by as yet unknown mechanisms, is a valueable drug in the treatment of akinetic crises (Danielczyk, 1973). Therefore, increased rigidity of neuronal membranes might disturb the receptor integrity by blocking the conformational change between low and high affinity states. Such conformational changes, which are physiologically induced by the endogenous ligand DA, seem to be of importance for the pharmacological efficacy of DA agonists. Goldstein et al. (1985) put forward the hypothesis that DA-agonists with low intrinsic activity need the presence of synaptic DA (therapeutically formed by l-DOPA) to mediate an agonistic effect. However, there is evidence to suggest, that it is the progressive and deleterious lack of DA, which cannot be substituted in sufficient quantities in the decompensated final phase of PD, that limits the therapeutic benefit of receptor stimulation.

References

Barbeau A, Murphy GF, Sourkes TL (1961) Excretion of dopamine in diseases of basal ganglia. Science 133: 1706–1707

Barbeau A (1975) Pathophysiology of the oscillations in performance after long-term therapy with l-DOPA. In: Birkmayer W, Hornykiewicz O (eds) Advances in parkinsonism. Editiones Roche, Basel, pp 424–434

Bernheimer H, Birkmayer W, Hornykiewicz O, Jellinger K, Seitelberger F (1973) Brain dopamine and the syndromes of Parkinson and Huntington. J Neurol Sci 20: 415–455

Birkmayer W, Riederer P (1985) Die Parkinson-Krankheit, 2. Aufl. Wien New York, Springer

Birkmayer W, Birkmayer JGD (1986) Iron, A new aid in the treatment of Parkinson patients. J Neural Transm 67: 287–292

Birkmayer W, Riederer P, Emich C (1988) Treatment of Parkinson's disease with the abeorphine CI 201–678, a rigid analogue of dopamine. Biogenic Amines 5: 269–274

Carmichael SW, Wilson RJ, Brimijoin WS, Melton LJ, Okazaki H, Yaksh TL, Ahlskog JE, Stoddard SL, Tyce GM (1988) Decreased catecholamines in the adrenal medulla of patients with parkinsonism. N Engl J Med 318: 254

Danielczyk W (1973) Die Behandlung von akinetischen Krisen. Med Welt 24: 1278–1282

Dexter DT, Carter C, Agid F, Agid Y, Lees AJ, Jenner P, Marsden CD (1986) Lipid peroxidation as cause of nigral cell death in Parkinson's disease. Lancet ii: 639–640

Ehringer H, Hornykiewicz O (1960) Verteilung von Noradrenalin und Dopamin im Gehirn des Menschen und ihr Verhalten bei Erkrankungen des extrapyramidalen Systems. Klin Wochenschr 38: 1236–1239

Goldstein M, Lieberman A, Meller EA (1985) A possible molecular mechanism for the antiparkinsonian action of bromocriptine in combination with levodopa. TIPS: 436–437

Guttman M, Seeman P, Reynolds GP, Riederer P, Jellinger K, Tourtellotte WW (1986) Dopamine D 2 receptor density remains constant in treated Parkinson's disease. Ann Neurol 19: 487–492

Hornykiewicz O, Kish SJ (1986) Biochemical pathophysiology of Parkinson's disease. In: Yahr MD, Bergmann KJ (eds) Advances in neurology, vol 45. Raven Press, New York, pp 19–34

Ito K, Sato A, Suzuki H (1986) Increases in adrenal catecholamine secretion and adrenal sympathetic nerve unitary activities with aging in rats. Neurosci Lett 69: 263–268

Jellinger K (1986) Overview of morphological changes in Parkinson's disease. In: Yahr MD, Bergmann KJ (eds) Advances in neurology, vol 45. Raven press, New York, pp 19–34

Lloyd KG, Davidson L, Hornykiewicz O (1975) The neurochemistry of Parkinson's disease: effect of l-DOPA therapy. J Pharmacol Exp Ther 195: 453–464

McGeer PL, McGeer EG (1976) Enzymes associated with the metabolism of catecholamines, acetylcholine and GABA in human controls and patients with Parkinson's disease and Huntington's chorea. J Neurochem 26: 65–76

Mogi M, Harada M, Kiuchi K, Kojima K, Kondo T, Narabayashi H, Rausch D, Riederer P, Jellinger K, Nagatsu T (1988) Homospecific activity (activity per enzyme protein) of tyrosine hydroxylase increases in Parkinsonian brain. J Neural Transm 72: 77–81

Quik H, Sourkes TL (1977) Central dopaminergic and serotoninergic systems in the regulation of adrenal tyrosine hydroxylase. J Neurochem 28: 137–147

Rausch WD, Hirata Y, Nagatsu T, Riederer P, Jellinger K (1988) Tyrosine hydroxylase activity in caudate nucleus from Parkinson's disease: effects of iron and phosphorylating agents. J Neurochem 50: 202–208

Riederer P, Sofic E, Konradi C (1986) Neurobiochemische Aspekte zur Progression der Parkinson-Krankheit: Postmortem Befunde und MPTP-Modell. In: Fischer PA (Hrsg) Spätsyndrome der Parkinson-Krankheit. Editiones Roche, Basel, pp 37–48

Riederer P, Rausch WD, Schmidt B, Kruzik P, Konradi C, Sofic E, Danielczyk W (1988) Biochemical fundamentals of Parkinson's disease. Mount Sinai J Med 55 1: 21–28

Riederer P, Sofic E, Rausch WD, Schmidt B, Reynolds GP, Jellinger K, Youdim MBH (1988) Transition metals, ferritin, glutathione and ascorbic acid in parkinsonian brains. J Neurochem (in press)

Seeman P, Bzowej NH, Guan HC, Bergeron C, Reynolds GP, Bird ED, Riederer P, Jellinger K, Tourtellotte WW (1987) Human brain D 1 and D 2 dopamine receptors in schizophrenia, Alzheimer's, Parkinson's and Huntington's diseases. Neuropsychopharmacology 11: 5–15

Sofic E, Riederer P, Heinsen H, Beckmann H, Reynolds GP, Hebenstreit G (1988) Increased iron (III) and total iron content in post mortem substantia nigra of Parkinsonian brain. J Neural Transm 74/3 (in press)

Sullivan JM, Nakano KK, Tyler HR (1973) Plasma renin activity during levodopa therapy. Significance of long and short-term treatment. J Am Med Ass 224: 1726–1727

Wesemann W (1984) Aspekte zum Wirkmechanismus von Amantadine. In: Danielczyk W, Wesemann W (eds) Amantadin workshop. Socio-Medico Verlag, Gräfelfing (Edition Materia Medica, pp 15–23)

Correspondence: Prof. Dr. P. Riederer, Clinical Neurochemistry, Department of Psychiatry, University of Würzburg, Füchsleinstrasse 15, D-8700 Würzburg, Federal Republic of Germany.

Dopaminergic neurotransmission and status of brain iron

M. B. H. Youdim

Department of Pharmacology, Faculty of Medicine, and Rappaport Family Research Institute, Technion, Haifa, Israel

Summary

The highly characteristic uneven distribution of iron in the human and animal brains, with high iron content present in globus pallidus, substantia nigra, ventral pallidum, caudate nucleus, intra penduncular nuclei, red nucleus and dente gyrus, suggests a functional importance for this metal. Our studies have clearly demonstrated that the homeostasis of brain iron is important for normal functioning of this organ. Either increased or decreased availability of iron in the above specific regions significantly alters the GABA and dopamine neurotransmission.

Introduction

The importance of metals in the brain function (Youdim and Ben-Shachar, 1986) and the pathophysiology of neuropsychiatric disorders (Gabay et al., 1985; Prohaska, 1987) have been discussed on numerous occasions. However, until very recently little attention was paid to brain iron, the most abundant metal in the body and earth's crust. Those aspects of iron that were studied were limited to the examination of iron distribution in different brain regions of human and animals (Youdim, 1985). These studies indicated a unique iron distribution and specific regional deposits of the metals. The highest concentrations are found in the globus pallidus and substantia nigra. These regions contain the highest densities of dopaminergic and GABA-ergic neurons. Other brain areas which are rich in iron include ventral pallidum, caudate nucleus, thalamus, inter-penduncular nuclei and dente gyrus of hippocampus-amygdalae complex (Hill and Switzer, 1984). The function of this metal and its possible association with neurological disorders, known to be linked with pathophysiological disturbances in the above brain regions, were hardly examined. What is clear is that the distribution of brain iron has no relationship to its transporter, transferrin or transferrin receptor (Connor et al., 1987; Hill et al., 1985). Thus globus pallidus and substantia nigra which are rich in iron have relatively low concentration of the latter. By contrast, cerebellum, pons and cerebral cortex, which are relatively low in iron content, have the highest transferrin concentration in the brain. This discrepancy does not allow a conclusion to be reached about the transport of iron into the brain and deposition in globus pallidus and substantia nigra.

This brief review will consider the various animal models we used as a procedure for examining the functional aspect of brain iron. However, the study of brain iron is hampered by the inability of the metal to cross the blood brain barrier (BBB). Furthermore, brain iron has an extremely low turnover rate as compared to that in the liver, the major site of iron metabolism and disposition (Ben-Shachar et al., 1986). In our studies we have approached these problems by using animal models in which brain iron metabolism can be altered by iron-deficiency or central iron injection (Youdim and Ben-Shachar, 1987; Youdim and Green, 1977).

Dopamine receptor sensitivity and brain iron

Iron-deficiency induced subsensitivity

Iron deficiency (ID) is the most prevalent nutritional disorder in the world affecting some forty million people. This phenomenon is associated with altered behaviour (reduced learning and cognition, apathy, hypo activity) and brain biochemistry (see Pollitt and Leibel, 1976, 1982; Youdim, 1985; Youdim and Ben-Shachar, 1987; Lozoff, 1988). ID induced in rats by nutritional means results in a highly significant reduction (30—40%) of non-haem iron in the brain without affecting the activities of iron-dependent tyrosine hydroxylase (TH), cytochrome oxidase, tryptophan hydroxylase, succinic dehydrogenase and aldehyde dehydrogenase, or dopamine metabolism in vitro and in vivo (Youdim and Green, 1977; Youdim et al., 1980; Sourkes et al., 1974). It is possible that the remaining iron present in the brain is sufficient to act as a cofactor for these enzymes. Even so, the ID rats to exhibit profound behavioural and other biochemical changes associated with dopaminergic subsensitivity (see Youdim and Ben-Shachar, 1987; Youdim, 1985) (see Table 1 and Table 2). Thus these animals show lower behavioural responses to D-amphetamine, apomorphine of L-dopa-monoamine oxidase inhibitor, suggesting postsynaptic changes at the level of DA D_2 receptor. Indeed, radioligand bindings studies have clearly demonstrated a significant reduction of DA D_2 receptor (^{3}H-spiperone binding) B_{max} in caudate nucleus without a change in α- and β-adreno, serotonin, benzodiazepine and muscrinic-cholinergic receptor (Youdim et al., 1983). The subsensitivity of DA

Table 1. Drug induced behavioural changes in iron deficient (ID) and intraventricular (IV) iron infected rats (Youdim and Ben-Shachar; Ben-Shachar et al., 1987 b)

	ID	IV Fe^{+++} (5 μl/100 mM)
1. Pargyline or tranylcypromine plus tryptophan	Diminished	–
2. Pargyline or tranylcypromine plus 5-hydroxytryptophan	Diminished	–
3. 5-Methoxy, N,N-dimethyl-tryptamine	Diminished	–
4. D-Amphetamine	Diminished	Increased
5. Pargyline or tranylcypromine plus L-Dopa	Diminished	Increased
6. Apomorphine	Diminished	Increased
7. Bromocryptine	Diminished	–
8. Thyrotropin-releasing hormone or analogue CG 3073	Unchanged	–

– Not done

Table 2. Biochemical alteration resulting from iron deficiency induced D_2 receptor subsensitivity (Youdim and Ben-Shachar, 1986)

1. Scatchard analysis of [^{3}H]spiperone binding by caudate nucleus shows a significant reduction in receptor number (measured by B_{max}) but not in affinity (measured by K_D)
2. Behavioral response to apomorphine and amphetamine is reduced
3. Phenobarbital sleeping time is increased
4. There is no change in loss of rightening or in convulsion time in response to leptazol
5. Amphetamine and L-Dopa-induced hypothermia is blocked
6. Serum prolactin is increased
7. Serum testosterone is increased
8. There is an increase in liver and reduction in prostate prolactin receptors ([^{125}I] ovine prolactin binding)
9. Opiate response to β-endophine, leu- and metenkephalins is increased

D_2 receptor and lower receptor sites resulting from ID is not limited to the caudate nucleus, since similar changes may also occur in dopamine neurons of nucleus, accumbens and the pituitary and frontal cortex. The consequences of this are reduced thermoregulation (Youdim et al., 1981) and increased prolactin release (Barkey et al., 1985) and reduced learning in ID rats (Yehuda et al., 1986) respectively. The mechanism by which ID reduces the B_{max} of DA D_2 receptor is not known. However, serum iron is thought to be important for maintaining brain iron and the DA D_2 receptor. The possibility that iron is involved in DA D_2 receptor cannot be excluded since iron can induce synthesis of iron storage protein ferritin by activating the translation of inactive cytoplasmic ferritin mRNAs (Aziz and Munro, 1987). The anaemia per se is not thought to be involved in DA receptor subsensitivity, since phenylhydrazine induced haemolytic anaemia (reduced haemoglobin) in rats has no such effects on dopaminergic neurotransmission or DA D_2 receptor B_{max} (Ben-Shachar et al., 1985). Furthermore, supplementation of ID rats with iron salts (ferrous sulfate) in the diet or i.p. injection of desferal restores the DA D_2 receptor B_{max} as well as the behavioural responses (Youdim et al., 1980; Ashkenazi et al., 1982; Ben-Shachar et al., 1986) as outlined in Table 1 within 7–14 days.

Intraventricular iron injection and dopamine receptor

Supersensitivity

The restoration of the caudate nucleus DA D_2 receptor B_{max} in ID rats after treatment with ferrous salts is associated with normalisation of brain iron and DA D_2 receptor. However, neither brain iron nor DA D_2 receptor can be elevated beyond the control values by continued iron-supplementation of rats for up to six months. This paradox may be explained by our recent observation that during ID there is a significant opening of BBB at the level of capillary endothelial cells, which can be reversed with correction of ID (Ben-Shachar et al., 1988). It could be for this reason that during ID iron can be taken up by the brain and not when BBB is fully reformed again. These findings suggest a significant regulatory mechanism for iron transport into the brain, presumably by transferrin, as initiated by its receptor.

Central iron salt injection has been used as a procedure to produce epileptogenic foci (Wilmore et al., 1978). This procedure is associated with damaging effects of the lipid (free cytotoxic hydroxy radical formation) peroxidation in inhibitory GABA neurons. Czernansky et al. (1983), while studying the epileptic foci after intra amygadaloid iron injection, according to the procedure of Wilmore et al. (1978) noted a time-dependent biochemical and behavioural dopamine receptor supersensitivity in the caudate nucleus. In our own studies ferrous or ferric sulphate (5 μl/100 mM) injected intraventricularly was associated with a time-dependent increased apomorphine induced stereotypy and increased ^{3}H-spiperone binding in the first 3 weeks, which were maintained for up to 17 weeks. These animals also exhibited excessive gnawing in response to apomorphine, a finding which has been attributed by us to the degeneration of striatal GABA neurons, since a reduction of glutamate decarboxylase activity in caudate nucleus was noted (Ben-Shachar et al., 1987 a and b; Zhang et al., 1986). The epileptogenic foci induced by iron salt certainly confirm those originally reported by Wilmore et al. (1978). Iron may thus have a dual effect, namely being stimulatory to nigra-striatal dopamine neurons and inhibitory to stria-pallidal GABA neurons (Yehuda and Youdim, 1988). Indeed, we and others have suggested that free iron is inhibitory to GABA neuron function since central injection of iron reduces glutamic acid decarboxylase activity in the striatum (Ben-Shachar et al., 1987 b; Zang et al., 1986). If iron is inhibitory to GABA neuron it would be expected that in iron-deficiency there will be increased GABAergic activity. ID in rats is thought to affect GABA metabolism (Taneja et al., 1986) by altering its metabolic enzymes.

Whether the high concentrations of iron present in globus pallidus and substantia nigra is a need to maintain the balance between inhibitory action of GABAergic and "excitatory" dopaminergic neurons remains for future investigations. What is clear is that the central iron metabolism is dependent on serum iron and its steady state maintenance is an important factor for normal brain function.

Parkinson's disease and brain iron: treatment with iron-complex

Neurotoxicity

Participation of metals, particularly iron, in generating highly cytotoxic hydroxyl radical (OH) from its interaction with H_2O_2 leading to lipid peroxidation membrane fluidity and cell degeneration in systemic organs is well known (Halliwell and Gutteridge, 1984; 1986). Such a process has also been suggested to participate in the cases of degenerative neurological disorders such as Parkinson's and Alzheimer's diseases (Youdim, 1988; Riederer et al., 1988 a). An abnormality of iron metabolism in the Parkinsonian brains was sought and found. Indeed a number of studies indicate not only increased presence of iron and lipid peroxidation (Riederer et al., 1988 a; Youdim, 1988) but also decreased availability of antioxidants, glutathione and ascorbate in the substantia nigra (Riederer et al., 1985; 1988 b). If an increase of brain iron is associated with the degenerative process, its transport and accumulation in this tissue requires further detailed study. It is not known whether the increase of iron and ferritin noted in the Parkinsonian substantia nigra is due to a reduction in brain iron turnover or excess uptake (transport) into the brain. At least it is known that the BBB is not damaged in Parkinson's disease and is not different from that of animals. However, our animal studies show that iron does not readily cross the BBB and intraventricular iron injection does not necessarily induce dopaminergic neurotoxicity but may affect GABAergic neurons. On the contrary, iron causes dopaminergic supersensitivity and GABAergic degeneration, via increased ^{3}H-spiperone binding and decreased glutamic acid decarboxylase respectively (Ben-Shachar et al., 1987 a and b; Yehuda and Youdim, 1988).

Iron therapy of Parkinson's disease

The increase of iron and ferritin in substantia nigra of Parkinsonian, the inability of iron to cross the BBB and unaltered dopamine D_2 receptor in Parkinson brains must be reconciled with the recent clinical studies of Birkmayer and Birkmayer (1986), who reported significant decrease of disability score of Parkinsonism in response to iron-complex therapy. There is an urgent need to substantiate the latter studies since it may represent a new therapeutic direction. If iron-complex, plus L-dopa therapy does indeed have beneficial effect, the question of its mechanism of transport and action remains unanswered. Can the iron-complex be transported as a complex with L-dopa into the brains? The hydroxyl groups are known for their ability to chelate metals. The two hydroxyl groups on phenyl ring of L-dopa can and may provide the sites for such chelation. Once such a complex has entered the brain a number of processes could take place. Riederer et al. (1988 a) have provided evidence that iron, a cofactor for tyrosine hydroxylase (Weiner, 1988) can substantially activate the enzyme from caudate nucleus of control and Parkinsonian brain in vitro. Availability of adequate brain tyrosine uptake may provide the basis of the formation of dopamine from the activated enzyme. There is no evidence that in vivo iron can stimulate tyrosine hydroxylase activity since neither iron-deficiency nor iron-overload alters tyrosine hydroxylase activity in vivo (Youdim and Green, 1977; Youdim et al., 1980; Sourkes et al., 1974). The stability of iron in Parkinsonian brains would indicate that an adequate supply of iron is present (Riederer et al., 1988 a; 1988 b). If iron-L-dopa combination does have therapeutic action, what would be the functional action of iron on L-dopa decarboxylation? Dopa decarboxylase is not a rate limiting enzyme, even though its activity is reduced in Parkinsonism, iron is not involved in its reaction mechanism. The possibility that the clinical beneficial effect of iron, if any, could be related to the activation of dopamine D_2 receptor cannot be discounted, since we have provided evidence for this phenomenon in intraventricularly iron injected rats. However, a word of caution is required here. The iron induced dopamine receptor supersensitivity is a long term process requiring at least 1–2 weeks post treatment. This contrasts with the rapidity (hours rather than days) with which iron is reported to produce its anti-Parkinson effect (Birkmayer and Birkmayer, 1986). In the final analysis any of the above proposed mechanisms might facilitate the action of iron in Parkinsonian disease. But it must not be forgotten that iron has important functions in myoglobin and oxygen carrying capacity of red blood cells. The vast literature on the role of iron in increasing muscle mobility and working capacity and their alterations due to iron status will need to be carefully considered (Edgerton et al., 1982) in relation to its central usefulness in Parkinsonism. There has been a growing body of evidence accumulating that iron-status can influence maximal physical performances, submaximal endurance and spontaneous activity in adults (Lozoff, 1988). The importance of these findings in relation to central and peripheral effect of iron should be investigated.

References

Ashkenazi R, Ben-Shachar D, Youdim MBH (1982) Nutritional iron and dopamine binding sites in rat brain. Pharmacol Biochem Behav 17 [Suppl] 1: 43–47

Aziz N, Munro HN (1987) Iron regulates ferritin mRNA translation through a segment of its 5' untranslated region. Proc Natl Acad Sci USA 84: 8478–8482

Barkey R, Ben-Shachar D, Amit T, Youdim MBH (1985) Increased hepatic and reduced prostatic prolactin binding in iron-deficient and neuroleptic treatment: correlation with changes in serum prolactin and testosterone. Eur J Pharmacol 109: 193–200

Ben-Shachar D, Finberg JPM, Youdim MBH (1985) The effects of chelating agents on dopamine D_2 receptor. J Neurochem 45: 999–1005

Ben-Shachar D, Ashkenazi R, Youdim MBH (1986) Long term consequences of early iron-deficiency on dopaminergic neurotransmission. Int J Devel Neurosci 4: 81–88

Ben-Shachar D, Jacobowitz D, Youdim MBH (1987 a) Dopamine D_2 receptor modulation by brain non-haem iron. Br J Pharmacol 90: 2 P

Ben-Shachar D, Youdim MBH (1987 b) Neuroleptic induced dopamine receptor supersensitivity and tardive dyskinesia may involve altered brain iron metabolism. Br J Pharmacol 90: 95 P

Ben-Shachar D, Yehuda S, Spanier I, Finberg JPM, Youdim MBH (1988) Selective alterations in blood brain barrier and insulin transport in iron-deficient rats. J Neurochem 50: 1434–1438

Birkmayer W, Birkmayer JGD (1986) Iron a new aid in the treatment of Parkinson's patients. J Neural Transm 67: 287–292

Connor JR, Phillips TM, Lakshman MR, Barron KD, Fine RE, Csiza CK (1987) Regional variation in the levels of transferrin the CNS of normal and myelin-deficient rats. J Neurochem 49: 1523–1529

Czernansky JG, Holman CA, Bonnet KA, Grabowsky K, King R, Hollister LE (1983) Dopaminergic supersensitivity at distant sites following iron induced epileptic foci. Life Sci 32: 385–391

Edgerton VR, Ohira Y, Gardner GW (1982) Effect of iron-deficiency anaemia on voluntary activities of rats and human. In: Politt E, Leibel RL (eds) Iron deficiency, brain biochemistry and behavior. Raven Press, New York, pp 141–160

Gabay S, Harris J, Ho BT (eds) Metals in neurology and neurobiology, vol 15. Alan R Liss, New York

Halliwell B, Gutteridge JM (1984) Oxygen toxicity, oxygen radicals, transition metals and disease. Biochem J 219: 1–17

Halliwell B, Gutteridge JM (1985) Oxygen radicals and the nervous system. TINS 1: 22–26

Hill JM, Switzer III RC (1984) The regional distribution and cellular localization of iron in the rat brain. Neuroscience 11: 595–602

Hill JM, Ruff MR, Weber RJ, Pert CB (1985) Transferrin receptors in rat brain: neuropeptide-like pattern and relationship to iron distribution. Proc Natl Acad Sci USA 82: 4553–4557

Lozoff B (1988) Behavioural alterations in iron-deficiency. Adv Pediat (in press)

Pollitt E, Leibel RL (1976) Iron-deficiency and behavior. J Pediatr 88: 372–381

Pollitt E, Leibel RL (eds) (1982) Iron deficiency, brain biochemistry and behaviour. Raven Press, New York

Prohaska JR (1987) Functions of trace elements in brain metabolism. Physiol Rev 67: 858–901

Riederer P, Sofic E, Rausch W, Kruzik P, Youdim MBH (1985) Dopaminforschung heute und morgen. In: Riederer P, Umek H (Hrsg) L-Dopa-Substitution der Parkinson-Krankheit. Springer, Wien New York, S 127–144

Riederer P, Rausch WD, Schmidt B, Kruzik P, Konradi C, Sofic E, Danielczyk W, Fisher M, Ogris E (1988 a) Biochemical fundamentals of Parkinson's disease. Mount Sinai J Med 55: 21–28

Riederer P, Sofic E, Rausch WD, Schmidt B, Youdim MBH (1988 b) Transition metals, ferritin, glutathione and ascorbic acid in Parkinsonian brains. J Neurochem (in press)

Sourkes TL, Quik M, Falardeau M (1974) Effects of iron and copper deficiences on monoamine metabolism. Adv Neurol 5: 253–258

Taneja V, Mishra K, Agarwal KN (1986) Effect of early iron-deficiency in rat on the γ-aminobutyric acid shunt in brain. J Neurochem 46: 1670–1674

Weiner N (1979) Tyrosine hydroxylase. In: Youdim MBH (ed) Aromatic amino acid hydroxylases and metal disease. Wiley, Chichester, pp 141–190

Wilmore LJ, Sypert GW, Manson JV, Hurd RW (1978) Chronic focal epileptiform discharges induced by injection of iron into rat and cat cortex. Science 200: 1501–1503

Yehuda S, Youdim MBH, Mostofsky DI (1986) Brain iron deficiency causes reduced learning capacity in rats. Pharmacol Biochem Behav 25: 141

Yehuda S, Youdim MBH (1988) Brain iron-deficiency: biochemical and behavioural aspects. In: Youdim MBH (ed) Brain iron metabolism. Taylor and Francis, London (in press)

Youdim MBH, Green AR (1977) Biogenic monoamine metabolism and function in iron-deficient rats: behavioural correlates. In: Porter RR (ed) Iron metabolism. Elsevier, Amsterdam, pp 201–225

Youdim MBH, Green AR, Bloomfield MR, Michel BD, Heal D, Grahame-Smith DG (1980) The effects of iron-deficiency on brain biogenic monoamine biochemistry and function. Neuropharmacology 19: 259–267

Youdim MBH, Yehuda S, Ben Uriah Y (1981) Reversed circadian rhythm of dopaminergic-mediated behaviors and thermoregulation in rats produced by iron-deficiency. Eur J Pharmacol 84: 295–301

Youdim MBH, Ben-Shachar D, Yehuda S, Ashkenazi R (1983) Brain iron and dopamine receptor function. In: Mandel P, De Feudis FV (eds) CNS receptors from molecular pharmacology to behavior. Raven Press, New York, pp 309–322

Youdim MBH (1985) Brain iron metabolism: biochemical and behavioural aspects in relation to dopamine neurotransmission. In: Lajtha A (ed) Handbook of neurochemistry, vol 10. Plenum, New York, pp 731–756

Youdim MBH, Ben-Shachar D (1987) Minimal brain damage induced by early iron deficiency: modified dopaminergic neurotransmission. Israel J Med Sci 23: 19–25

Youdim MBH (1988) Iron in the brain: implications for Parkinson's and Alzheimer's diseases. Mount Sinai J Med 55: 97–101

Zang ZH, Qi Y, Wu XR, Zuo OH (1986) Ferrous induced seizures and changes in lipid peroxidation. GAD activity and GABA uptake in the epilepsy-prone versus epilepsy-resistant rats. Neuroscience (abstract)

Correspondence: Prof. Dr. M. B. H. Youdim, Department of Pharmacology, Faculty of Medicine, Technion, P.O.B. 9649, Haifa, Israel.

Dopaminergic modulation of neuropeptide gene expression in the rat striatum

V. Höllt[1] and B. Morris[2]

[1] Physiologisches Institut der Universität München
[2] Max-Planck-Institut for Psychiatry, Martinsried, Federal Republic of Germany

Summary

The advent of measuring specific mRNAs by molecular hybridization techniques has facilitated the study of neuropeptide gene regulation by pharmacological agents. In this report this approach was used to measure the levels or proenkephalin mRNA in the striatum of rats chronically treated with dopamine antagonists by in-situ hybridization. Chronic administration of haloperidol (2.4 mg/kg/day) for 7 days increased the levels of proenkephalin mRNA, whereas the specific D 1 antagonist SCH 23390 (2.4 mg/kg/day) decreased the proenkephalin mRNA in the striatum. These result suggest that there is a tonic suppression, via D 2 receptors, and a tonic enhancement, via D 1 receptors, of proenkephalin synthesis in the striatum.

In addition, recent data will be summarized showing that the mesostriatal dopamine system has no influence on the expression of striatal prodynorphin neurones, but causes a tonic enhancement on the neurones expressing protachykinins (the precursors for substance P and substance K).

In Parkinson's disease the levels of proenkephalin-peptides in the striatopallidal pathway are reduced. In view of the enhancing influence of D 1 receptors on the expression of proenkephalin neurones the use of D 1 agonists in Parkinson's disease might be considered.

Introduction

The mesostriatal dopaminergic system has a major influence on the function of the striatum. Dopamine has been shown to modulate the electrical activity of striatal neurones and to affect the release of a variety of striatal neurotransmitter and neuropeptides. The actions of dopamine in the striatum appear to be mediated via two distinct receptors, D 1 and D 2. The D 1 receptor induced effects have often been shown to be different from that caused by D 2 receptor activation. Thus, D 1 agonists stimulated striatal adenylate cyclase activity and GABA release, while D 2 agonists had the opposite effect (Girault et al., 1986; Onali et al., 1984). The major efferent pathways from the striatum contain GABA and various neuropeptides (Graybiel, 1986). In the rat, cells synthesizing protachykinin and prodynorphin project to the substantia nigra, especially to the pars reticulata. The precursor molecules are cleaved into the respective neuropeptides. α- and β-protachykinins give rise to substance P and/or substance K and prodynorphin is cleaved into several opioid peptides called dynorphins, neoendorphins and leu-enkephalins. Proenkephalin is another opioid peptide precursor which is synthesized by striatal cells with major projections into the globus pallidum. The precursor molecule gives rise to leu-enkephalin and several copies of met-enkephalin as well as of met-enkephalin containing peptides (Höllt, 1986).

The cleaved neuropeptides are released from the nerve terminals upon stimulation of the neurones. The precursors are translated from their mRNAs in the rouph endoplasmic reticulum within the perikarya of the neurones. Through the use of nucleic acid hybridization techniques it is now possible to quantify the levels of RNA coding for neuropeptides. In particular, by using in-situ hybridization, it has proved possible to gain information on both the anatomical localization and on the relative concentrations of neuropeptide mRNAs in the rat brain. The mRNA levels may be regarded as a dynamic index of neuropeptide turnover, whereas the peptide levels are more difficult to interpret, since they reflect the balance between release and biosynthesis of the peptides. The present paper summarizes recent data demonstrating the great potential of in-situ hybridization histochemistry for the study of the dopaminergic regulation of neuropeptide genes within in the striatum of rats.

Methods and materials

Haloperidol and SCH 23390 were administered to male Sprague-Dawley rats via s.c. implanted osmotic minipumps at a flow rate of 1 μl/rat/hour (= 2.4 mg/kg/day). After 7 days the rats were killed, their brains frozen and 10 μm section prepared for in-situ hybridization, as previously described (Morris et al., 1986; Morris et al., 1988 a). Briefly, sections were mounted on slides, dried and fixed with paraformaldehyde. Following treatment with acids and proteases the sections were prehybridized in high salt buffers containing DNA and formamide. Thereafter, the sections were hybridized with the radiolabelled probe in the same buffer with added tRNA at 45° C for 24 hours. The probe used was a restriction fragment containing DNA of about 1000 bases complementary to rat proenkephalin mRNA (a generous gift from Dr. Sabol, NIH, Bethesda, U.S.A.; Yoshikawa et al., 1983) which was labelled with α-32 P—dATP by nick—translation. After washing and dehydration the section were apposed to Cronex MRF film. Densitometry was performed using a Leitz Texture Analysis System with microscope-linked densitometer.

Results

Fig. 1 shows the effect of chronic treatment of rats with haloperidol and SCH 23390 on the levels of proenkephalin in the

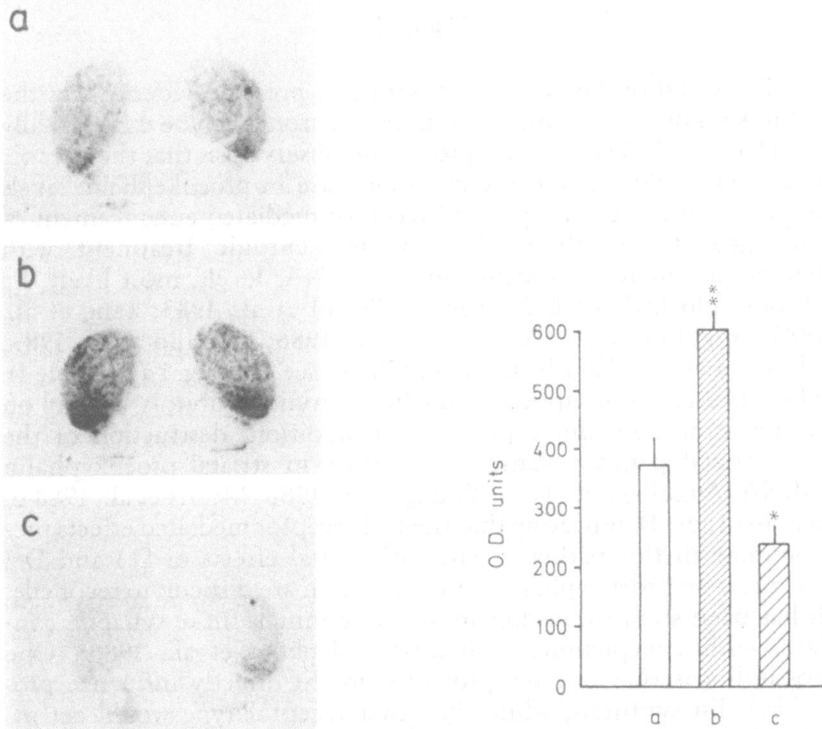

Fig. 1. In-situ hybridization of proenkephalin mRNA in the striatum in control *a*, haloperidol-treated *b*, SCH 23390-treated rat *c*. The right panel shows quantitation by measurement of the optical densities. Results are mean ± S.E.M. of 6 experiments; * $p \leq 0.05$; ** $p \leq 0.01$ (one way analysis of variance and Neuman-Keul's test)

striatum of rats. The hybridization signals on the autoradiograms illustrate the uniform distribution of proenkephalin synthesizing neurones in the striatum. The grain densities were markedly increased following chronic haloperidol treatment reflecting the increase in the biosynthesis of proenkephalin. In contrast, chronic administration of the specific D 1 antagonist SCH 23390 resulted in a decrease in the intensity of the hybridization signal relativ to control rats. Quantitativ densitometry of several in-situ hybridization experiments showed a significant increase in the grain density after chronic haloperidol and a significant decrease after SCH 23390. If the cells were coated with liquid emulsion to provide greater anatomical resolution, it appeared that the number of silver grains per cell was increased by haloperidol and decreased by SCH 23390 (data not shown).

Discussion

The result of this series of experiments provide evidence that the expression of proenkephalin in striatal neurones can be differentially affected via D 1 and D 2 receptors. The observation that the specific D 1 SCH 23390 blocker induces a decrease in proenkephalin levels implies that there is a tonic D 1 receptor mediated enhancement of proenkephalin synthesis. In contrast, chronic treatment with haloperidol increases proenkephalin mRNAs levels, most likely via chronic blockade of D 2 receptors (Sabol et al., 1983; Tang et al., 1983; Angulo et al., 1986; Sivam et al., 1986; Romano et al., 1987; Morris et al., 1988 a; Morris et al., 1988 b; see also Fig. 1 and Table 1). Thus, D 2 receptors appear to mediate a tonic inhibitory control on striatal proenkephalin expression. In addition, destruction of the mesostriatal pathway causes an increase in striatal proenkephalin mRNA (Angulo et al., 1986; Young et al., 1986; Morris et al., 1988 b; see also Table 1) indicating that the D 2 receptor mediated effects predominate in this pathway. The differential effects of D 1 and D 2 receptors on proenkephalin gene expression are difficult to reconcile. It has been shown that dopaminergic terminals make synaptic contacts with enkephalinergic neurones (Kubota et al., 1986). One possibility is that one receptor type might directly influence proenkephalin synthesis, while the other receptor type would act indirectly via interneurones or via a presynaptic action on an excitatory neurone. Cyclic AMP has been demonstrated as a second messenger involved in the induction of proenkephalin mRNA in adrenal medullary cells (Kley et al., 1987). It is noteworthy to mention that

Table 1. Dopaminergic modulation of neuropeptides in striatal projection neurones

Peptide family	Localization	Substantia nigra lesion		Modulation in rats D_2 antagonism		D_1 antagonism		Status in Parkinson's disease
		mRNA	Peptides	mRNA	Peptides	mRNA	Peptides	Peptides
Proenkephalin								
met-enkephalin leu-enkephalin met-enkephalin-arg-phe met-enkephalin-arg-gly-leu	striatopallidal (external segment)	↑	↑	↑	↑	↓	nd	↓
Prodynorphin								
dynorphins neoendorphins leu-enkephalin	striatopallidal (internal segment) striatonigral	=	nd	=	(↑↓=)	=	nd	=
Protachykinins								
substance P substance K	striatopallidal striatonigral	↓	↓	↓	↓	nd	nd	↓

nd not determined

D 1 receptors activate whereas D 2 receptors inhibit adenylate cyclase activity in the striatum (Onali et al., 1984). In Parkinson's disease levels of proenkephalin-derived peptides are decreased in the pallidum (Agid and Javoy-Agid, 1985). On the other hand, destruction of the mesostriatal system or chronic D 2 blockade in rats causes an increase in the enkephalin peptide levels in the pallidum (Hong et al., 1978). The reason of this difference is unknown and in-situ hybridization data of humans are needed. In view of the enhancing influence of D 1 receptors on expression of enkephalinergic neurones the use of D 1 agonists in the therapeutic regime for parkinsonism might be considered. Dopamine effects on the striatal prodynorphin system are less clear. Thus prodynorphin-derived peptides in the striatonigral pathway have been reported to be increased (Quirion et al., 1985), decreased (Tandon et al., 1986) or unchanged (Nylander and Terenius, 1986; Li et al., 1986) after chronic haloperidol treatment. The reasons for these equivocal results are unknown. Recent data indicate that the destruction of the mesostriatal pathway or chronic treatment with haloperidol or SCH 23390 does not affect prodynorphin mRNA levels in the rat striatum (Young et al., 1986; Morris et al., 1988 b). It thus appears that the expression of prodynorphin in the striatonigral pathway is not regulated by dopaminergic receptors. Recently, we obtained evidence that the biosynthetic activity of prodynorphin containing neurones in the caudomedial striatum is under positive control of a serotoninergic raphé-striatal pathway (Morris et al., 1988 c).

Levels of dynorphin A were unchanged in the globus pallidus and the substantia nigra of post-mortem parkinsonian brains (Taquet et al., 1985). This might indicate the integrity of the striatopallidal and striatonigral dynorphinergic pathways described in animals, if existing in man.

The peptides of the protachykinin family (substance P, substance K) show a similar striatal distribution as the prodynorphin peptides. However, the protachykinin producing neurones are under positive control of the mesostriatal dopaminergic system (Bannon et al., 1986; Young et al., 1986; Table 1). In fact, chronic treatment with haloperidol or destruction of the mesostriatal pathway causes a decrease in the mRNA levels coding for protachykinins. This decrease in the biosynthesis is associated with a decrease in the peptide levels in the substantia nigra. Since dopamine is thought to exert a tonic inhibitory influence in the striatum, its positive effect might be indirectly caused by disinhibition of inhibitory striatal neurones. It is likely that these effects are mediated via D 2 receptors, although the effect of D 1 compounds has not studied yet.

In Parkinson's disease the levels of substance P in the substantia nigra and pallidum are decreased. It is likely that the decrease in peptide content may indicate a decrease in protachykinin biosynthesis induced by the deficiency of the mesostriatal dopaminergic system. In situ hybridization experiments with postmortem human brains should answer this question.

In conclusion, we have summarized recent data showing that the mesostriatal dopaminergic system controls the expression of neuropeptides in striatal projection neurones in a different manner. In addition, dopamine can exert opposite effects on the expression of a single neuropeptide system via D 1 and D 2 receptors.

References

Agid Y, Javoy-Agid F (1985) Peptides and Parkinson's disease. Trends Neurosci 9: 30–35

Angulo JS, Davis LG, Burkhart BA, Christoph GR (1986) Reduction of striatal dopaminergic neurotransmission elevates striatal proenkephalin mRNA. Eur J Pharmacol 130: 341–343

Bannon MJ, Lee JM, Giraud P, Young A, Affolter HU, Bonner T (1986) Dopamine antagonist haloperidol decreases substance P, substance K and preprotachykinin mRNAs in rat strionigral neurones. J Biol Chem 261: 6640–6642

Girault JA, Spampinato U, Glowinski J, Besson MJ (1986) In vivo release of GABA in the rat neostriatum II oposing effects of D 1 and D 2 receptor stimulation in the dorsal caudate putamen. Neuroscience 19: 1109–1118

Graybiel AM (1986) Neuropeptides in the basal ganglia. In: Martin JB, Barchas JD (eds) Neuropeptides in neurologic and psychiatric disease. Raven Press, New York, pp 135–161

Höllt V (1986) Opioid peptide processing and receptor selectivity. Ann Rev Pharmacol Toxicol 26: 59–77

Hong JS, Yang HYT, Fratta W, Costa E (1978) Rat striatal metenkephalin content after chronic treatment with cataleptogenic and noncataleptogenic drugs. J Pharmacol Exp Ther 205: 141–147

Iorio LC, Barnett A, Leitz FH, Houser VP, Korduba CA (1983) SCH 23390, a potential benzazepine antipsychotic with unique interactions on dopaminergic systems. J Pharmacol Exp Ther 226: 462–468

Kley N, Loeffler JP, Pittius CW, Höllt V (1987) Involvement of ion channels in the induction of proenkephalin A gene expression by nicotine and cAMP in bovine chromaffin cells. J Biol Chem 262: 4083–4089

Kubota Y, Inagaki S, Kito S, Tagaki H, Smith AD (1986) Ultrastructural evidence of dopaminergic input to enkephalinergic neurones in rat neostriatum. Brain Res 367: 374–378

Li S, Siwam SP, Hong JS (1986) Regulation of concentration of dynorphin A (1–8) in the striatonigral pathway by the dopaminergic system. Brain Res 398: 390–392

Morris BJ, Haarmann I, Kempter B, Höllt V, Herz A (1986) Localization of prodynorphin messenger RNA in rat brain by in-situ hybridization using a synthetic oligonucleotide probe. Neurosci Lett 69: 104–198

Morris BJ, Höllt V, Herz A (1988a) Dopaminergic regulation of striatal proenkephalin mRNA and prodynorphin mRNA: contrasting effects of D 1 and D 2 antagonists. Neurosci 25: 525–532

Morris BJ, Herz A, Höllt V (1988b) Localization of striatal opioid gene expression and its modulation by the mesostriatal dopamine pathway: an insitu hybridization study. J Mol Neurosci (in press)

Morris BJ, Reimer S, Höllt V, Herz A (1988c) Regulation of striatal prodynorphin mRNA levels by the raphé-striatal pathway. Mol Brain Res (in press)

Nylander I, Terenius L (1986) Chronic haloperidol and clozapine differentially affect dynorphin peptides and substance P in basal ganglia of the rat. Brain Res 380: 34–41

Onali P, Olinas MC, Gessa GL (1984) Selective blockade of dopamine D 1 receptors by SCH 23390 discloses striatal dopamine 2 receptors mediating the inhibition of adenylate cyclase in rats. Eur J Pharmacol 99: 127–128

Quirion R, Gandrean P, Martel JC, St. Pierre S, Zamir N (1985) Possible interactions between dynorphin and dopaminergic systems in rat basal ganglia and substantia nigra. Brain Res 331: 358–362

Romano GJ, Shivers BD, Harlan RE, Howells RD, Pfaff DW (1987) Haloperidol increases proenkephalin mRNA levels in the caudate-putamen of the rat: a quantitative study at the cellular level using in-situ hybridization. Mol Brain Res 2: 33–41

Sabol S, Yoshikawa K, Hong JS (1983) Regulation of met-enkephalin precursor mRNA in rat striatum by haloperidol and lithium. Biochem Biophys Res Commun 113: 391–399

Shivers BD, Harlan RE, Romano GJ, Howells RD, Pfaff DW (1986) Cellular localization of proenkephalin mRNA in rat brain: gene expression in the caudate-putamen and cerebellar cortex. Proc Natl Acad Sci 83: 6221–6225

Sivam S, Strunk C, Smith DR, Hong JS (1986) Proenkephalin-A gene regulation in the rat striatum: Influence of lithium and haloperidol. Mol Pharmacol 30: 186–191

Tandon R, Day R, Kelsey JE, Watson SJ, Akil H (1986) The effect of haloperidol on prodynorphin end products in the rat striatum and substantia nigra. I.R.N.C. proceedings NIDA Monograph 75: 50

Tang F, Costa E, Schwartz JP (1983) Increase of proenkephalin content of rat striatum after daily injection of haloperidol for 2 to 3 weeks. Proc Natl Acad Sci USA 80: 3841–3844

Taquet H, Javoy-Agid F, Givand P, Legrand JC, Agid Y, Cesselin F (1985) Dynorphin levels in parkinsonian patients: Leu-enkephalin production

from either proenkephalin aA or prodynorphin in human brain. Brain Res 341: 390–392
Yoshikawa K, Williams C, Sabol SL (1984) Rat brain preproenkephalin mRNA. cDNA cloning, primary structure and distribution in the central nervous system. J Biol Chem 259: 14301–14308
Young WS, Bonner TI, Brann MR (1986) Mesencephalic dopamine neurones regulate the expression of neuropeptide mRNAs in the rat forebrain. Proc Natl Acad Sci 33: 9827–9831

Correspondence: Dr. V. Höllt, Physiological Institute of University Munich, Pettenkoferstrasse 12, D-8000 München 2, Federal Republic of Germany.

The diagnostic relevance of Lewy bodies and other inclusions in Parkinson's disease

W. R. G. Gibb

Department of Neuropathology, National Hospitals for Nervous Diseases, Maida Vale, London, U. K.

Summary

Seventy-four of 78 patients with Parkinson's disease (PD) had Lewy bodies in the substantia nigra. Two without Lewy bodies had striatonigral degeneration and two had post-encephalitic pathology. Lewy bodies in controls occurred in 3.6% to 11.1% between the sixth and tenth decades, and similar figures were obtained in other common parkinsonian disorders and Alzheimer's disease. Lewy bodies can also occur in corticobasal degeneration and Hallervorden-Spatz disease. The pathological diagnosis of PD depends on finding Lewy bodies in the substantia nigra, but other parkinsonian disorders must be excluded.

Introduction

Many degenerative neurological diseases are distinguished by their type of neuronal inclusion and distribution of neuronal loss. The severity of cell loss and gliosis, and the presence of other deposits (eg. amyloid) are less important. When stained appropriately many inclusion bodies are easily identified, but they tend not to be completely specific for individual degenerative diseases. Often they are considered to be non-specific or of diagnostic use only when interpreted with knowledge of the clinical history and distribution of neuronal loss elsewhere in the nervous system.

The main neuronal morphological changes in Parkinson's disease (PD) are Lewy bodies, other non-staining inclusions (pale bodies) and chromatolytic-like cells. The Lewy body has been associated with PD since it was first described in 1912, but it has been observed in most degenerative parkinsonian disorders, although not necessarily as part of these disease processes (Table 1). In some diseases Lewy bodies could just be a coincidental finding. To what extent are Lewy bodies helpful in the diagnosis of PD, and what are their limitations in diagnosis? These are two of the questions to be addressed in this paper.

An important diagnostic error with parkinsonian disorders is to assume they can be distinguished with sufficient accuracy on clinical grounds to enable the specific associations of neuronal inclusions to be clarified. Clinical diagnostic inaccuracies may be so considerable that only three quarters of patients with the label of PD may turn out to have pathological features of the disease (Forno, 1966). Of the remaining patients approximately 10% to 15% have striatonigral degeneration (SND), 5% to 10% have Steele-Richardson-Olszewski syndrome (SRO), and 1% to 5% have corticobasal degeneration (CBD) and other parkinsonian disorders. Thorough clinical documentation is vital, but pathological diagnoses must be obtained independently, to prevent biased judgements and to promote objective assessment of potential diagnostic criteria. Which parkinsonian disorders features Lewy bodies as an incidental finding, and which are associated with Lewy bodies as cause or effect of the neuronal degeneration (Table 1)?

Inclusion bodies need to be easily recognizable and clearly defined and described if they are to be considered significant. A Lewy body is

Table 1. Some parkinsonian disorders other than Parkinson's disease

+	striatonigral degeneration (multiple system atrophy)
+	Steele-Richardson-Olszewski syndrome (progessive supranuclear palsy)
*	corticobasal degeneration (corticodentatonigral degeneration with neuronal achromasia
*	Hallervorden-Spatz disease
*	familial olivopontocerebellar atrophy
*	juvenile parkinsonian syndrome (hereditary dystonia-parkinsonism syndrome)
+	Creutzfeld-Jakob disease
£	drug-induced
+	post-encephalitic

Lewy bodies have been described in each of these disorders
+ the association with Lewy bodies is coincidental
* the association with Lewy bodies is more specific
£ see text

Most symptomatic parkinsonian disorders have been omitted from this table, as have other degenerative parkinsonian disorders including Pick's disease, juvenile Huntington's chorea, Guam-parkinsonism dementia complex, and motor neurone disease

Table 2. Inclusions in parkinsonian disorders other than Lewy bodies

* pale body—either granular or smooth
* Lewy body-like neuronal degeneration
* pale body-like neuronal degeneration
globose (spool-like) neurofibrillary tangle
corticobasal inclusion
subcortical Pick body
flame-shaped (rope-like, Alzheimer-type) neurofibrillary tangle

* Inclusions seen in Lewy body diseases

an eosinophilic inclusion with a halo (Fig. 1 A), occurring within nerve cells or their processes, or occasionally lying loose in the neuropil. It is typically soft-textured and smooth-edged. Some have a dense eosinophilic core. This morphology gives scope for confusion, but when located within cytoplasm and stained with haematoxylin and eosin (H and E) the Lewy body is distinctive when compared to other inclusions (Table 2), and other morphological changes such as eosinophilic granules, Marinesco bodies, axonal swellings, and colloid inclusions.

In the following study we have investigated the association of Lewy bodies with parkinsonian disorders with the purpose of clarifying their role in diagnosis. Patients with Alzheimer's disease (AD) were included in the study because of the association repeatedly suggested with PD.

Material and methods

Material for investigation consisted of 78 patients with PD, 31 patients with SND, 26 patients with SRO, 15 patients with post-encephalitic parkinsonian syndrome (PEPS), 121 patients with AD, and 307 healthy controls dying with non-neurological disease. The 78 patients with PD were derived from a group of 99 from which exclusions were made if they failed at accepted diagnostic exclusion criteria for PD (eg. the presence of cerebellar signs, early severe dementia). The other diseases were conclusively diagnosed according to their pathological features. Brains were sampled widely with blocks taken from the cerebral cortex, hippocampus, nucleus basalis of Meynert, anterior and posterior striatum, thalamus, midbrain, pons and medulla. Sections were stained with conventional histological stains including H and E, luxol-fast blue, Nissl cresyl violet, Bielchowsky silver impregnation. For the main study one to four horizontal sections of substantia nigra mostly of 7 μm thickness, but some at 5 μm and 12 μm thickness, were stained with H and E. Cell loss was established as more or less than 60%. The sections were methodically examined so that all the

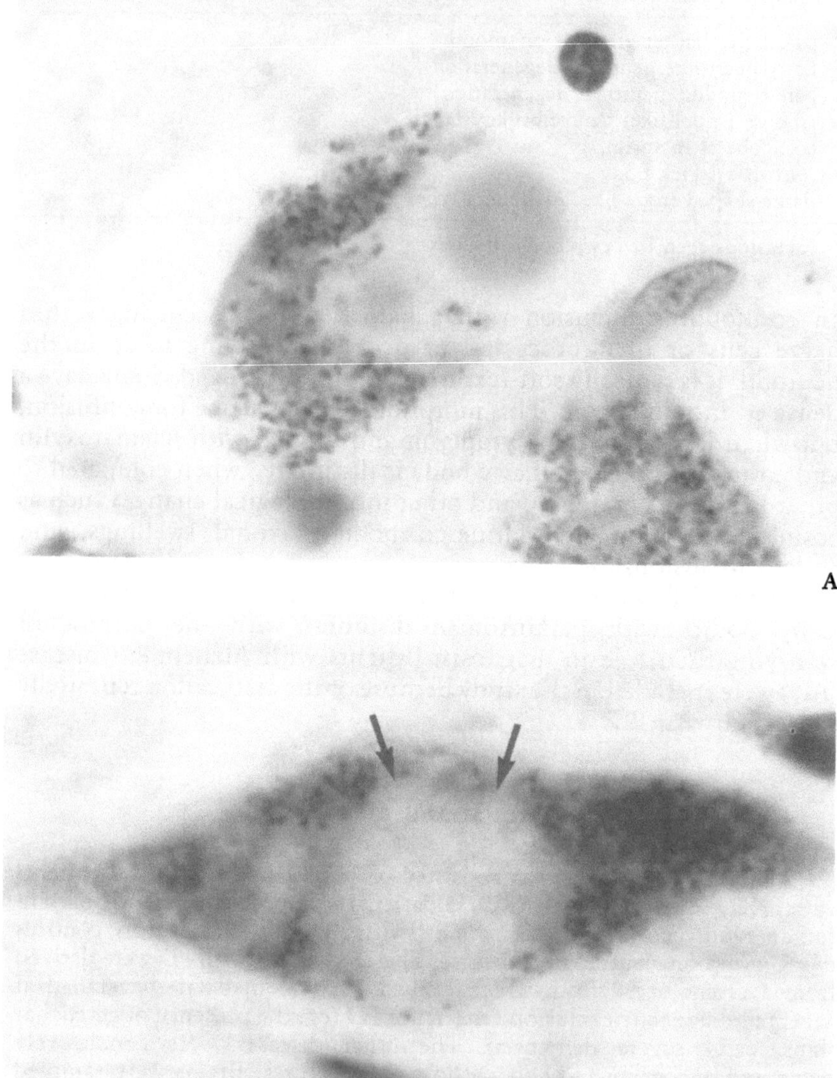

Fig.1. *A* Lewy body in a pigmented nigral cell. H and E, ×4960. *B* Pale body (arrows) with vesicular and granular texture in a nigral cell. H and E, ×6200. *C* Lewy body-like neuronal degeneration in a nigral cell. Soft, cloudy, weakly, eosinophilic matter

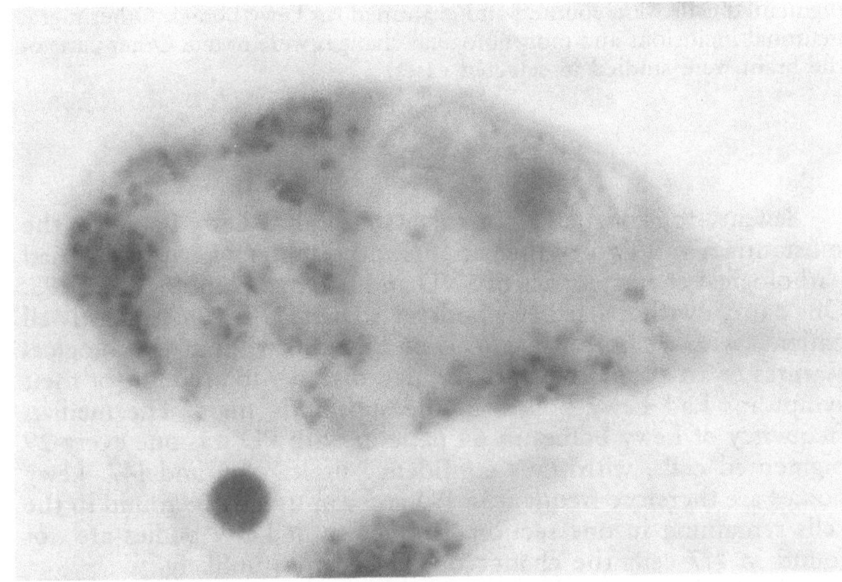

C

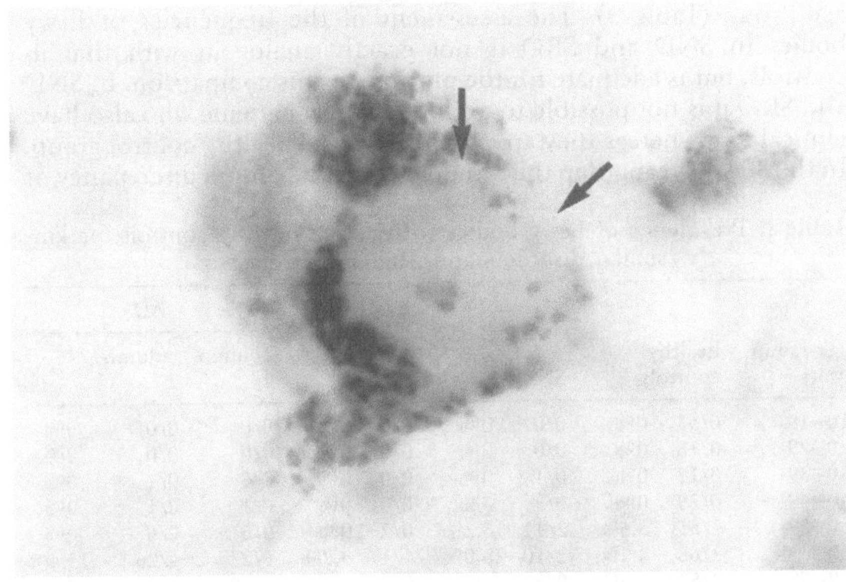

D

separates melanin clumps. H and E, ×6200. *D* Corticobasal inclusion (arrows) which is slightly basophilic with a finely fibrillar component (not seen). A few melanin granules are caught within it. H and E, 4960

pigmented cells were counted and examined for Lewy bodies. Other nigral neuronal inclusions and morphological changes were noted. Other parts of the brain were studied in selected cases.

Results

Seventy-four of the 78 patients with PD had Lewy bodies in the substantia nigra. Two of the patients without nigral Lewy bodies had pathological characteristics of SND and two had pathology of PEPS. One patient with sparse Lewy bodies also had SND. Consequently all patients with a clinical diagnosis of PD, and without pathological features of an alternative parkinsonian disorder to account for their symptoms had Lewy bodies in the substantia nigra. The median frequency of Lewy bodies in 64 patients with PD was one every 29 pigmented cells, with 95% confidence limits of 6 and 147. Lewy bodies are therefore frequent in PD and can usually be found in the cells remaining in one section. Conversely if Lewy bodies are not found in 147 cells the chance of PD becomes unlikely.

The healthy controls showed an increasing prevalence of Lewy bodies from 3.6% in the 50–59 year age group to 11.1% in the 90–99 age group (Table 3). The assessment of the frequencies of Lewy bodies in SND and SRO is not exactly analogous with that in controls, but is adequate for the purpose of this comparison. In SND and SRO it is not possible to exclude the few patients who also have clinical PD, whereas they are absent from the healthy control group. In the 80–89 year group this would create a maximum discrepancy of

Table 3. Prevalence of Lewy bodies with age in healthy controls, parkinsonian disorders and Alzheimer's disease

age group years	healthy controls		SND*		SRO		AD column 1	AD column 2	
10–19	0/5	0%	0/0	0%	0/0	0%	0/0	0/0	0%
20–29	0/18	0%	0/0	0%	0/0	0%	0/0	0/0	0%
30–39	0/17	0%	0/1	0%	0/0	0%	2/3	0/1	0%
40–49	0/29	0%	0/3	0%	0/0	0%	0/1	0/1	0%
50–59	2/55	3.6%	2/11	18.2%	0/2	0%	0/6	0/6	0%
60–69	3/68	4.4%	2/10	20.0%	1/17	5.9%	6/28	4/26	15.4%
70–79	3/54	5.6%	0/5	0%	0/7	0%	9/40	5/36	13.9%
80–89	6/52	11.5%	0/1	0%	0/0	0%	9/36	4/31	12.9%
90–99	1/9	11.1%	0/0	0%	0/0	0%	1/7	1/7	14.3%

* Thirteen of these patients presented with cerebellar signs

2%, equivalent to the community prevalence of PD. One can also argue that the presence of two different pathologies would shorten life expectancy and shift the age-specific prevalence graph leftwards, but this is unlikely to be a significant effect. The frequencies of Lewy bodies in SND and SRO were not significantly different from the healthy control series (Table 3). The series of AD included 11 patients with nigral cell loss in excess of 60%, the threshold required for production of Parkinsonian features (Pakkenberg and Brody, 1965). If these cases are excluded from column 1 prevalence figures closer to healthy controls are found (column 2). None of the patients with PEPS, aged 36–79 years, showed Lewy bodies in the substantia nigra.

The healthy controls and patients without PD carrying Lewy bodies in the substantia nigra, also showed Lewy bodies in other sites such as the locus coeruleus, nucleus basalis, and cerebral cortex. Other morphological changes were also identical to those in PD, except for the mild or moderate cell loss (less than 60%). Pale bodies (Fig. 1 B), Lewy body-like (Fig. 1 C) and pale body-like neuronal degeneration, and chromatolytic-like neurons were also seen, although pale bodies were not as consistently found as Lewy bodies. Occasional chromatolytic-like neurons are quite non-specific among the degenerative diseases. Fragmenting cells and scattered extraneuronal melanin were always present.

Discussion

Lewy bodies are therefore crucial to the diagnosis of PD. Its pathological diagnosis depends on three findings. Firstly, Lewy bodies in the substantia nigra. Secondly, nigral cell loss of 60% or greater. Thirdly, the absence of any other parkinsonian pathology such as SND, CBD or Hallervorden-Spatz disease with which Lewy body disease (usually presymptomatic PD—see below) can coexist. The greatest power of the Lewy body is its role in excluding PD if it cannot be found. It is immensely helpful in the diagnosis of PD, but its diagnostic limitation is its occurrence in other parkinsonian disorders.

Lewy bodies were first clearly described in the brains of healthy persons aged over 50 years (incidental Lewy body disease) by Beheim-Schwarzbach (1952). The frequency of Lewy bodies in normal persons younger than 50 years is likely to be very low. No such case has been found despite documented examination of at least 140 brains of persons aged 40–49, implying that their frequency in this decade is less than 0.7%. Patients with incidental Lewy body disease identified in this study showed modest degrees of nigral cell degeneration and a distribution of Lewy bodies outside the substantia nigra identical to that in PD. Consequently they correspond to a presymptomatic stage of the disease in which nigral cell is not as great as the 60% threshold required for symptoms, and striatal dopamine is less than the 80% threshold (Bernheimer et al., 1973). Similar frequencies of Lewy bodies in SND, SRO, and AD compared with the controls suggest that the occurrence of Lewy bodies in these disorders in coincidental. Another example of this coincidental association with Lewy bodies is Guam-parkinsonism dementia complex, in which 7 of 70 (10%) patients coming to autopsy showed Lewy bodies in the substantia nigra (Hirano, 1973).

These data are particularly important in AD, as they suggest that there is no overlap in pathogenetic mechanisms between AD and PD. There are however exceptions as in two unrelated familial cases of AD included in this study (Table 3). They were both aged 38 years at death and had nigral Lewy bodies. A third sporadic case aged 39 years with dementia and PD had both cortical Lewy body dementia (which includes Lewy bodies in the substantia nigra) and AD (Gibb et al., 1987). A fourth case is that of a mentally retarded man aged 28 years (Popovitch et al., 1987). The coincidental concurrence of these diseases is highly improbable at this young age, because of the rarity of incidental Lewy body disease before the sixth decade, so that some basic pathogenetic interaction is likely. Thus there could be one or more rare diseases, combining AD and PD, possibly inherited in an autosomal dominant pattern. This is unlikely to have relevance to most examples of either AD or PD. Another disorder that has been considered a type of AD—PD overlap syndrome is cortical Lewy body dementia, in which Alzheimer pathology may coexist with PD. While this is not usually the case, some might be examples of the AD—PD disorder cited above (Gibb et al., 1987).

Corticobasal degeneration often presents with focal limb dystonia, an SRO-like syndrome with supranuclear gaze palsies, and parietal lobe features. Two of seven reported patients showed occasional nigral Lewy bodies (Gibb et al., 1988), and we have recently studied yet another case with sparse Lewy bodies giving a prevalence of 3/8 (38%), which is well in excess of an incidental association. The coexistence of Lewy bodies and other inclusions common to PD (Table 2) emphasizes the need to identify the distinctive morphology of CBD. This is a weakly basophilic, very faintly fibrillar body resembling the globose tangle of SRO, probably best called a corticobasal inclusion (Fig. 1 D). In contrast to the globose tangle the skeins are less well-defined and less basophilic. Melanin is characteristically incorporated within the inclusion in bands, streaks, clumps or particles.

Other relevant parkinsonian disorders include Hallervorden-Spatz disease, which is also specifically associated with Lewy bodies. These are reported in 12 of approximately 80 cases (15%). As these patients are usually younger than 30 years this cannot be an incidental association with PD pathology. Lewy bodies are described in two familial cases of olivopontocerebellar atropy and two unrelated patients with Joseph's disease either within the same family or at a relatively young age, suggesting the relationship is not just coincidental (Gibb, 1986). They are also reported in a young patient aged 39 with juvenile parkinsonian syndrome. Again this is a highly significant observation in view of the young age, but more pathological reports are necessary.

The prevalence of Lewy bodies in drug-induced parkinsonian syndrome is unknown. The drug-induced state occurring in older patients, with relatively low doses of neuroleptics, and more typical parkinsonian features (rest tremor) is likely to occur more frequently on a background of mild nigrostriatal dopamine deficiency. Neuroleptic-induced parkinsonian syndrome may thus tend to occur in presymptomatic patients with nigral Lewy bodies and mild cell depletion (Rajput et al., 1982). Therefore a reasonable prediction would be a higher prevalence of Lewy bodies in drug-induced patients than in healthy controls, for example 30%, but the exact frequency would depend on the neuroleptic doses and the prevalence of drug-induced parkinsonian syndrome in the population sampled.

Neuronal inclusions that need to be distinguished from those characteristic of Lewy body disease are the globose neurofibrillary tangle seen in SRO, the corticobasal inclusion seen in CBD, the subcortical Pick body seen in a Pick's disease, and the flameshaped neurofibrillary tangle seen in PEPS and other disorders.

We can conclude that Lewy bodies are plentiful in PD and are a prerequisite for diagnosis. The failure to find them excludes PD. Lewy bodies often occur in other parkinsonian disorders as an incidental or specific feature, so these disorders must be excluded on other pathological criteria before the diagnosis can be confirmed. Nevertheless PD is the only disorder currently known to be invariably associated with Lewy bodies.

Acknowledgement

This work was in part supported by a grant from the Medical Research Council of Great Britain.

References

Beheim-Schwarzbach D (1952) Über Zelleibveränderungen im Nucleus coeruleus bei Parkinsonian-Symptomen. J Nerv Ment Dis 116: 619–632

Bernheimer H, Birkmayer W, Hornykiewicz O, Jellinger K, Seitelberger F (1973) Brain dopamine and the syndromes of Parkinson and Huntington. J Neurol Sci 20: 415–455

Forno LS (1966) Pathology of parkinsonism. A preliminary report of 24 cases. J Neurosurg [Suppl] 24: 266–271

Gibb WRG (1986) Idiopathic Parkinson's disease and the Lewy body disorders. Neuropath Appl Neurobiol 12: 223–234

Gibb WRG, Esiri MM, Lees AJ (1987) Clinical and pathological features of diffuse cortical Lewy body disease (Lewy body dementia). Brain 110: 1131–1153

Gibb WRG, Luthert PJ, Marsden CD (1988) Corticobasal degeneration. Brain (in press)

Greenfield JG, Bosanquet FD (1953) The brain-stem lesions in Parkinsonism. J Neurol Neurosurg Psychiatr 16: 213–226

Hirano A (1973) Progress in the pathology of motor neuron diseases. In: Zimmerman HM (ed) Progress in neuropathology. Grune and Stratton, London, pp 181–215

Pakkenberg H, Brody H (1965) The number of nerve cells in the substantia nigra in paralysis agitans. Acta Neuropathol 5: 320–324

Popovitch ER, Wisniewski HM, Kaufmann MA, Grundke-Iqbal I, Wen GY (1987) Young adult-form of dementia with neurofibrillary changes and Lewy bodies. Acta Neuropathol 74: 97–104

Rajput AH, Rozdilsky B, Hornykiewicz O, Shannak K, Lees T, Seeman P (1982) Reversible drug-induced parkinsonism. Clinicopathological study of two cases. Arch Neurol 39: 644–646

Correspondence: Dr. W. R. G. Gibb, Department of Neuropathology, National Hospitals for Nervous Diseases, Maida Vale, London W9, United Kingdom.

Cytoskeletal pathology of the Lewy bodies

K. Jellinger[1], C. Bancher[2], and H. Lassmann[2]

[1] Ludwig Boltzmann-Institute of Clinical Neurobiology, Lainz Hospital, Vienna

[2] Neurological Institute, University of Vienna, School of Medicine, Vienna, Austria

Summary

Lewy bodies (LB), a major anatomical hallmark of Parkinson's disease (PD), represent abnormal intracytoplasmic neuronal inclusions that show typical antigenic profiles for both the classical and cortical type. Classical or subcortical LB show a strong reaction of the periphery with monoclonal antibodies (mABs) to phosphorylated epitopes of neurofilament proteins (NFP), to micro tubule-associated proteins MAP 2 and MAP 1 (only on cryostate sections), to a mAB to paired helical filaments (PHF) which recognizes ubiquitin, and to polyclonal ABs to ubiquitin, while the core strongly reacts with a mAB to PHF (3.39) that also recognizes determinants of ubiquitin. Cortical LB show diffuse reaction with ABs to NFP and PHF, although no PHF or 15 nm filaments are found in these inclusions. Ultrastructurally, most of the filaments in LB appear to be related to NFP, part of which are uncovered by phosphatase pretreatment indicating that abnormal phosphorylation of NFP plays a role in their pathogenesis. The frequent coexistence of both LB and neurofibrillary tangles (NFT) in the same brain and even in the same neuronal perikaryon, and the sharing of several epitopes have suggested some common pathogenic relations. While both LB and NFT have been shown to contain ubiquitin, a polypeptide required for the ATP-dependent non-lysosomal protein breakdown, LB do not react with tau-protein, the major component of PHF. These data indiciate that ubiquinated proteins in LB and NFT are different: phosphorylated NFP appear to be the major component of LB, while abnormal phosphorylation of tau is suggested to be the first step in the process of NFT formation, and ubiquination may reflect a fruitless attempt of proteolytic degradation. Biochemical changes of cytoskeletal elements may be reflected by the ultrastructural changes in these neuronal inclusions, and reflect the molecular correlate of the different appearance of classical and cortical types of LB, the pathogenesis of which remains to be elucidated.

Introduction

Lewy bodies (LB) are abnormal intracytoplasmic neuronal inclusions occurring in various regions of the central and autonomic system that are considered as the major anatomical hallmark of Parkinson's disease (PD), although they are also observed in various disorders and in normal aging (Forno, 1986). LB are round or oval eosinophilic structures that appear in two characteristic forms: a) the classical or subcortical (brainstem) type consists of a single or multiple round or oval eosinophilic structures with a central core surrounded by a less dense peripheral zone and an outermost pale halo which is sharply demarcated from the neuronal cytoplasm; b) the cortical type is not sharply delineated from the cytoplasm and the outline of the central part is rather indistinct. The LB vary in size from 3 to 20 μm and can be localized within a nerve cell body (classical perikaryal), within nerve cell processes (intraneuronal), and extraneuronally free in the neuropil (Forno, 1986). The LB are composed of proteins, free fatty acids, sphingomyelin and polysaccharides; their core contains alpha amino acids. Electron probe microanalysis revealed a high sulfur content, indicating products of degenerated proteins.

Ultrastructurally, LB are composed of intermediate type filaments, 8 to 10 nm in diameter (with a range from 7 to 20 nm), admixed with vesicular and granular material. In the subcortical type, the central core showing a chaotical arrangement of filaments, is sharply demarcated to the peripheral portions, consisting of radially arranged filaments. The cortical type is rather homogeneous, with random arrangement of the filaments without marked dense core and peripheral radiations. Filaments are admixed with membranous and granular material (Forno, 1986). Similar homogeneous, poorly staining intraneuronal hyaline (colloid) inclusions or "pale bodies" composed of randomly arranged collect ions of 10 nm filaments are found in many neurons in PD. These pale areas, where the cytoplasm appears to be devoid of its normal organelles including the Nissl substance and neuromelanin, are considered to represent the early stages of LB formation, from which they may form as a result of concentration of filaments (Gibbs, 1986).

Immunocytochemical studies have shown that LB react with antisera to neurofilaments (NFP), with monoclonal antibodies (mABs) to phosphorylated and non-phosphorylated epitopes of NFP and to tubulin, various microtubule-associated polypeptides including MAP 1 and MAP 2, with some or most of the mABs to paired helical filaments (PHF) and Alzheimer neurofibrillary tangles

(ANT), and with ubiquitin, a polypeptide required for the ATP-dependent nonlysosomal protein breakdown that recently has been demonstrated to be associated with PHF. LB do not bind antibodies to actin, high molecular weight microtubule-associated proteins, antibodies to 200 kDa components of NFP, mABs against isolated PHF, to anti-PHF that preferentially reacts with phosphorylated tau protein, and the heat-stable microtubule-associated protein tau that strongly reacts with ANT/PHF (see Yen et al., 1986; Kuzuhara et al., 1988; Bancher et al., 1988—Table 1).

The present study presents new data on the antigenic profile of subcortical and cortical LB using antisera and mABs to various cytoskeletal elements in order to further elucidate the cytoskeletal pathology and pathogenesis of LB.

Material and Methods

Immunocytochemistry was performed on paraffin sections of several brains of patients with PD and Diffuse Lewy Body Disease (DLBD) using antibodies against the following cytoskeletal elements: isolated PHF (60 e), isolated PHF (3.39), 150 and 210 kDa NFP, phosphorylated and non-phosphorylated determinants (SM I-31, 34, 33), MAP 1 and MAP 2, tau protein monoclonal (Tau-1) and polyclonal (92 e). Sections were pretreated with alkaline phosphatase. For immune electron microscopy, appropriate areas of LB were selected and, after deparaffination, were osmiated, and gold labeled protein A was used to visualize bound antibodies (see Bancher et al., 1988).

Results

The classical subcortical LB (e. g. in substantia nigra) show intense staining of the peripheral ring with all three mABs to neurofilaments, surrounding a pale central core. This confirms previous data on the presence of epitopes of neurofilaments in the periphery of these inclusions (Forno, 1986; Yen et al., 1986; Dickson et al., 1987). Whereas reactivity with mABs to phosphorylated NFP was found in untreated sections, staining with a mAB to neurofilaments appeared only after phosphatase treatment which led to the appearance of a peripheral ring of reactivity in about one-third of all identifiable subcortical LB (Bancher et al., 1988). This fact indicates that the LB contains epitopes covered with phosphate residues preventing them from being recognized by the antibody.

While antisera against microtubule-associated MAP 2 (recognizing neuronal perikarya and dendrites) were reported to react with

Table 1. Ultrastructure and immunocytochemical features of Lewy bodies and neurofibrillary tangles

Type of lesion	Lewy bodies: Classical		Lewy bodies: Cortical	Neurofibrillary tangles
Ultrastructure	8–10 nm filaments (intermediate type) central condensation peripheral radiation		Random arrangement	10 nm paired helical filaments twisted at 80 nm intervals and few 15 nm straight filaments
Antibodies	Untreated	(PASE)		
PHF 3.39 mAb	++ (Core)	++	+	+++
PHF 60 e pAb	–	–	–	+++
PHF (phosphorylated tau)[a]	–	–	–	+++
P-NFP (150 kDa) SMIK-31 mAb	+ (Rim)	–	+	–
P-NFP (210 kDa) SMI-34 mAb	+ (Rim)	–	+	+
P-NFP (200 kDa) mAb[a]	–/+		–/±	–/±
NFP (150, 210 kDa SMI-33 mAb	–	+ (Rim)	+	–
MAP 1 (52-68 kDa) pAb	–	–	–	–
MAP 1 pAb[c]	+	–	–	+
MAP 2 (200–250 kDa) pAb	–	+/–	–	–
MAP 2 (200–250 kDa) pAb[a]	+ (Rim)		+ (Rim)	++
TAU 1 mAb	–	–	–	+
TAU 92 e pAb	–	–	–	++
TAU (55–62 kDa) pAb[a]	–	–	–	++
DF-2 (5 kDa) (cross-reaction with ubiquitin)[a]	++ (Rim)		++	++
Ubiquitin mAb	++ (Rim)		++	++
Actin[b]	–		–	–
Tubulin[b]	+		+	–
Thioflavin S[b]	+		+	+++

References: [a] Kuzuhara et al. (1988); [b] Yen et al. (1986); [c] Goldman (1987), personal communication (only vibratome slides effective); all others: Bancher et al. (1988). – *mAb* monoclonal antibody, *pAb* polyclonal antibody; *PASE* phosphatase pretreatment

the periphery of LB (Dickson et al., 1987, Kuzuhara et al., 1988), in our material, MAP 2 reactivity occurred only in some subcortical LB after dephosphorylation. Antiserum against the microtubule-associated MAP 1 (recognizing nerve cells and axons) showed no staining of LB on paraffin sections, which, may appear only on glutaraldehyde-fixed vibratome sections (Goldman, personal communication, 1987).

The cores of most subcortical LB were strongly reactive with a mAB to isolated PHF that also recognizes ubiquitin. This reaction remains unchanged after phosphatase treatment. Immunoelectron microscopy confirmed that reaction with this mAB (3.39) was located in the core of typical LB, although no filamentous structures and particularly no PHF or 15 nm filaments were found in this central area (Bancher et al., 1988). This staining pattern differs considerably from the characteristic reaction of the peripheral ring with anti-NFP. Another antiserum to isolated PHF (60 e) did not recognize LB.

The cortical LB often display homogeneous staining pattern, particularly with antibodies to NFP and PHF. No ring forms could be found with these antibodies, while MAP 2 has been reported to stain primarily their periphery (Kuzuhara et al., 1988).

Monoclonal and polyclonal antibodies to microtubule-associated tau protein strongly reacting with ANT showed consistently negative reactions with both cortical and subcortical LB.

Quantitative evaluation revealed that about two-thirds of all subcortical LB were recognized by monoclonal anti-PHF, but only about one-third by anti-NFP antibodies; the number of anti-PHF immunoreactive cortical LB was higher than the number of these lesions recognized in sections stained with hematoxylin-eosin. On the other hand, many LB were immunoreactive with anti-neurofilament antibodies and the cores of reactive LB remained unstained.

Discussion

The results of the present and other studies (Table 1) suggest that most of the filamentous material in LB is related to NFP and contains various antigenic determinants of the neurofilament triplet proteins. In contrast to ANT, all the anti-neurofilament antibodies have been shown to recognize most of the LB, particularly their periphery, while the cores of the reactive LB often remain unstained. A recent study using a sensitive electron microscopic technique has shown weak anti-NFP reaction in the cores, but intense reactivity in the periphery of typical LB (Pappola et al., 1986). This could indicate that the NFP present in the LB undergo biochemical alterations within the inclusions during their development. The uncovering of NF epitopes by treatment with phosphatase indicates that LB contain epitopes covered by phosphate and that phosphorylation of cytoskeletal elements may play a role in the pathogenesis of LB. Phosphorylation of cytoskeletal elements is considered to be a post-translational event rendering neurofilaments more compact and preventing them from proteolytic degradation (Kuzuhara et al., 1988). Under normal circumstances, phosphorylated neurofilaments are chiefly found in axons. The presence of phosphorylated NFP in neuronal perikarya in the form of LB or other inclusions indicated that in affected neurons phosphorylation processes take place at an abnormal lacolization leading to accumulation of abnormal fibrillary proteins within the nerve cell body (Pappola, 1986). This may occur as a cause or consequence of impaired assembling or transport of NF proteins that undergo progressive biochemical and structural changes. These changes may represent the molecular correlate of the different appearance and immunoreactivity of the central core and periphery of the LB and the different reactions of both typical and cortical LB which probably represent different stages of development.

An important finding is the fact that the cores of most LB strongly react with a mAB to PHF, despite the ultrastructurally evident absence of PHF or 15 nm filaments at this location. The immunoreaction of typical LB with this anti-PHF antibody that also recognizes ubiquitin (Perry et al., 1987) differs considerably from the characteristic ring pattern observed with NFP antibodies. On the other hand, another mAB to PHF (DF-2) which recognizes ubiquitin and pABs to ubiquitin have been shown to immunostain virtually all cortical LB in a diffuse manner and the peripheral rim of subcortical LB, whereas the cores of the typical LB and the center of the cortical LB were less intensely stained or remained unstained (Kuzuhara et al., 1988).

Recent studies have demonstrated that ubiquitin, a polypeptide required for the ATP-dependent non-lysosomal protein breakdown is associated with PHF (Perry et al., 1987). The fact that proteins with abnormal structure are more readily conjugated to ubiquitin than are most normal proteins could explain the presence of determinants of ubiquitin in both ANT and LB. The most important function of ubiquitin is in an ATP-dependent proteolytic system responsible for degradation of short-living or abnormal proteins in the cell (see Kuzuhara et al., 1988). The conjugation of ubiquitin with proteins after covalent binding to degradating enzymes is obligatory for the subsequent protein degradation. Recent data provide evidence for the existence of ubiquitin in both PHF and ANT (Perry et al., 1987; Kuzuhara et al., 1988) which have been shown to contain phosphorylated epitopes of the microtubule-associated tau protein as a major antigenic component (see Grundke-Iqbal et al., 1986). In this respect, the filaments of LB are distinct from PHF though both are ubiquinated, since they do not bind antisera to tau and to PHF containinig dominant tau reactivity (Kuzuhara et al., 1988; Bancher et al., 1988). Only the center of LB strongly reacts with a mAB to PHF which recognizes determinants of ubiquitin, while ubiquitin reacts intensely in the periphery of typical and cortical LB, and reaction products are located only in radiating filaments of the periphery, but not in the densely packed filaments of the center of LB (Kuzuhara et al., 1988).

The frequent coexistence of both LB and ANT in the same brain of parkinsonian patients or in Diffuse Lewy Body Disease, the occasional presence of both LB and ANT in the same neuronal perikaryon (Forno, 1986; Dickson et al., 1987; Kuzuhara et al., 1988), and the sharing of several epitopes have led to the speculation that both types of neuronal inclusions might have pathogenic relationship. Although LB and ANT are likely to result from alterations of the neuronal cytoskeleton, some of the basic elements involved in their formation are different. Both LB and NFT have been shown to contain ubiquitin the function of which in both lesions is conjugation to abnormal cytoskeletal proteins, presumably noxious for neurons that should be removed by degradation. However, the ubiquinated proteins in both LB and PHF are apparently different, phosphorylated neurofilaments representing the major component of LB, while phosphorylation of tau proteins is suggested to be the first step in the process of NFT formation (Bancher et al., 1988). Ubiquination occurring in both types of lesions may reflect fruitless attempt of proteolytic degradation of abnormal cytoskeletal proteins. Biochemical changes of cytoskeletal elements may be reflected by the ultrastructural changes in these neuronal inclusion, and reflect the molecular correlate of the different appearance of classical and cortical types of LB, but the molecular basis and principal mechanism for their development are still poorly understood.

The LB have been suggested to represent accumulations of phosphorylated neurofilaments undergoing progressive structural breakdown (Gibb, 1986) or sequestrated neurofilaments after neuronal regeneration or retrograde degeneration. Although ubiquination of abnormal proteins in LB presumably occurs after filament accumulation and ectopic phosphorylation, it is not known why LB are resistent to the ATP-dependent proteolysis system despite ubiquination, since no degradation of LB has ever been observed by light or electron microscopy (Forno, 1986; Gibb, 1986).

LB occurring in catecholaminergic neurons show immunreactivity with antisera to tyrosine hydroxylase (TH), the rate-limiting enzyme of catecholamine synthesis, suggesting that TH enzyme activity is preserved in neurons involved by LB or even may play a role in their formation (Jellinger, 1987). On the other hand, neurons in substantia nigra and locus ceruleus affected by NFT in Parkinson's disease, progressive supranuclear palsy, Parkinson-dementia complex or Alzheimer disease, show positive TH immunoreactivity until the perikarya are entirely replaced by NFT indicating that these neurons still contain TH enzyme protein, except in the final stages of ghost tangle formation (Nakashima and Ikuta, 1984). In Parkinson's disease, most of the remaining melanin-laden dopaminergic neurons in substantia nigra, ventral tegmental area and in the noradrenergic locus ceruleus and dorsal vagal nucleus show substantial reduction or loss of TH immunoreactivity, as do fibers and terminal in the ascending and descending catecholaminergic systems in brain and spinal cord, indicating severe damage to these systems (Jellinger, 1987). The role of neuromelanin in LB formation has been a matter of speculation (Forno, 1986). Ultrastructurally, melanin granules in nigral and locus ceruleus neurons of Parkinson disease brains often cannot be distinguished from those in normal individuals, but there can be a decrease in the very dense component of melanin granules. The substructure of the altered melanin granules including the presence of linear arrays resembles that of lipofuscin, but the relationship between these ultrastructural changes and both the LB formation and degeneration of melanin-containing cells remain unknown. Although the LB frequently affect aminergic neurons and in the cerebral cortex show a distribution pattern partly corresponding to that of dopaminergic axon terminals, they appear not to represent a cytoskeletal abnormality specific for any chemically determined neuronal system. Forno et al. (1986) reported eosinophilic inclusions in brainstem neurons of MPTP treated aged monkeys, suggesting similarity with the pathology of human PD. However, the ultrastructure of these bodies composed of 20 nm filaments appears to differ from human LB (Forno et al., 1988). The use of new scientific strategies will hopefully provide new information about the pathobiological processes involved in the formation of LB and other neuronal cytoskeletal changes in order to enhance our understanding of the basic molecular mechanisms of Parkinson's disease.

References

Bancher C, Lassmann H, Budka H, Jellinger K, Grundke-Iqbal I, Iqbal K, Wiche G, Seitelberger F, Wisniewski HM (1988) Antigenic profiles of Lewy bodies: Immunocytochemical evidence for protein phosphorylation and ubiquination. J Neuropath Exp Neurol 47 (in press)

Dickson DW, Davies P, Mayeux R, Crystal H, Horoupian DS, Thompson A, Goldman JE (1987) Diffuse Lewy body disease: Neuropathological and biochemical studies of six patients. Acta Neuropathol (Berlin) 75: 8–15

Forno LS (1986) The Lewy body in Parkinson's disease. Adv Neurol 45: 35–43

Forno LS, Langston JW, DeLanney LE, Irwin I, Ricaurte GA (1986) Locus ceruleus lesions and eosinophilic inclusions in MPTP-treated monkeys. Ann Neurol 20: 449–455

Forno LS, Langston JW, DeLanney LE, Irwin I (1988) An electron microscopic study of MPTP-induced inclusion bodies in an old monkey. Brain Res 448: 150–157

Gibb WRG (1986) Idiopathic Parkinson's disease and the Lewy body disorders. Neuropathol Appl Neurobiol 12: 223–234

Grundke-Iqbal I, Iqbal K, Tung Y-C, Quinlain M, Wisniewski HM, Binder LI (1986) Abnormal phosphorylation of the microtubule-associated protein (tau) in Alzheimer cytoskeletal pathology. Proc Natl Acad Sci USA 83: 4913–4917

Jellinger K (1987) The pathology of parkinsonism. In Marsden CD, Fahn St (eds) Movement disorders II. Butterworths, London, pp 124–162

Kuzuhara S, Mori H, Izumiyama N, Yoshimura M, Ihara Y (1988) Lewy bodies are ubiquinated. Acta Neuropathol (Berlin) 75: 345–353

Nakashima S, Ikuta F (1984) Tyrosine hydroxylase proteins in Lewy bodies of parkinsonism and senile brain. J Neurol Sci 66: 91–96

Pappola MA (1986) Lewy bodies of Parkinson's disease: Immunoelectron microscopic demonstration of neurofilament antigens in constituent filaments. Arch Pathol Lab Med 110: 1160–1163

Perry G, Friedman R, Shaw G, Chau V (1987) Ubiquination is detected in neurofibrillary tangles and senile plaque neurites of Alzheimer disease brains. Proc Natl Acad Sci USA 84: 3033–3036

Yen SH, Dickson DW, Peterson C, Goldman JE (1986) Cytoskeletal abnormalities in neuropathology. Prog Neuropathol 6: 63–90

Correspondence: Prof. Dr. K. Jellinger, Ludwig Boltzmann-Institute of Clinical Neurobiology, Lainz Hospital, Wolkersbergenstrasse 1, A-1130 Wien, Austria.

Plasma concentrations of endogenous DOPA and 3-O-methyl-DOPA in rats administered benserazide and carbidopa alone or in combination with the reversible COMT inhibitor Ro 41-0960

A. Colzi, G. Zürcher, and M. Da Prada

Pharmaceutical Research Department, F. Hoffmann-La Roche & Co., Ltd., Basle, Switzerland

Summary

In this study in rats the plasma concentration of endogenous 3,4-dihydroxyphenyl-l-alanine (DOPA) and 3-O-methyl-DOPA (3-OMD) have been used as a biochemical index to assess the degree of in vivo inhibition of peripheral L-amino acid decarboxylase (AADC) activity after oral benserazide or carbidopa. Benserazide (170 μmoles/kg p.o.) increased the concentration of DOPA and 3-OMD much more efficiently than equimolar doses of carbidopa, showing that benserazide is the most potent of the peripheral AADC inhibitors presently available. The novel reversible catechol-O-methyltransferase (COMT) inhibitor Ro 41-0960 in combination with benserazide was able, by suppressing the formation of 3-OMD, to markedly increase the plasma concentrations of endogenous DOPA. The non-toxic and potent COMT inhibitor Ro 41-0960 might offer therapeutic advantages in the long-term treatment of Parkinson's disease by substantially improving the DOPA bioavailability of the Madopar-standard or Madopar-HBS formulations.

Introduction

The combination of benserazide or carbidopa with 3,4-dihydroxyphenyl-l-alanine (DOPA) is currently the most effective treatment of Parkinson's disease (Da Prada et al., 1984). Previous studies have shown that, in healthy volunteers and in the rat, the plasma concentrations of endogenous DOPA increased more markedly after peripheral inhibition of the aromatic L-amino acid decarboxylase (AADC) by benserazide than by carbidopa (Da Prada et al., 1987). It was also shown that oral benserazide was more potent than carbidopa in inhibiting the AADC activity in peripheral organs (liver and kidney) of rats and mice. Based on these results we concluded that, at therapeutic doses, carbidopa was less potent than benserazide as peripheral AADC inhibitor (Da Prada et al., 1984, 1987).

In the present study in rats, we examined the effect of very high doses of benserazide and carbidopa either administered alone or in combination with the new catechol-O-methyltransferase (COMT) inhibitor Ro 41-0960 (Zürcher et al., 1988) on the plasma concentrations of endogenous DOPA and 3-O-methyl-DOPA (3-OMD).

The aim of the study was to assess whether the marked increase of the endogenous DOPA level produced by high doses of benserazide or carbidopa (170 μmoles/kg p.o., each) was paralleled by increased plasma concentrations of endogenous 3-OMD. In additional experiments, we studied the effect of the COMT inhibitor Ro 41-0960 given p.o. on the elevation of endogenous DOPA and 3-OMD concentrations induced by benserazide.

Materials and methods

Blood was collected (using as anticoagulant 1 vol. 1% EDTA in saline for 9 vol. blood) from the abdominal aorta of male albino rats (Fü-albino, 150—200 g body weight) under Vetanarcol (0—4 ml/kg i.p., Veterinaria A.G., Zürich) anaesthesia and the plasma separated by centrifugation. Plasma samples were stored at $-80\,^{\circ}$C until analysis.

Endogenous DOPA was measured in plasma containing α-methyldopa as internal standard by high performance liquid chromatography (HPLC) with coulometric detection (almost complete oxidation of the sample) after alumina extraction according to Ito et al. (1984).

3-OMD was measured by an automated column switching HPLC-system with electrochemical detection using a two-column system on-line with the liquid chromatograph. Deproteinized plasma (300 μl $HClO_4$ 1 N/600 μl plasma) was injected into the precolumn dry-packed with strong acid cation-exchange pellicular particules (Dupont). In this purification step the amino acids were retained on the precolumn, whereas the majority of the endogenous plasma components were eluted and discarded. The precolumn carrying the aromatic amino acids was switched in series with the analytical column (Nucleosil 7 μ, C_{18}) through time switches and the retained compounds were rapidly eluted. In this way, the amino acids to be analyzed were quantitatively transferred from the precolumn to the analytical reversed-phase column. While the amino acids were resolved on the analytical column and quantified by electrochemical detection, the precolumn was regenerated for the next sample prepurification.

For the assay of 3-OMD the prepurified plasma extracts were measured using an electrochemical detector (Bioanalytical System) equipped with a glassy carbon electrode (TL-S). The electrode potential was maintained at + 0.75 V (versus an Ag/AgCl reference electrode).

For the assay of DOPA the potential of the two electrodes (glassy carbon electrodes, Analytical cell ESA mod. 5011) was maintained at + 0.45 and – 0.35 V, respectively. The optimized mobile phase consisted of 1 liter water solution containing citric acid. H_2O (13.7 g), Na_2HPO_4. $2H_2O$ (6.2 g), 1-octanesulphonic acid sodium salt. H_2O (9.0 mg), EDTA disodium salt. $2H_2O$ (37.2 mg), methanol (80 ml). The water was glass-distilled and freshly deionised. Before use, the mobile phase was filtered and degassed by ultrasonication. Both for the DOPA and 3-OMD assay the flow rate of the analytical column was 1 ml/min. The retention times for DOPA and α-methyldopa were 4.5 and 7 min, respectively, whereas for 3-OMD the retention time was 10 min. The detection limit of DOPA (about 0.1 ng/ml plasma) was lower than that of 3-OMD (about 5 ng/ml plasma).

Benserazide [1-(DL-seryl)-2-(2, 3, 4-trihydroxybenzyl) hydrazine] and carbidopa [(S)-α-(3, 4-dihydroxybenzyl)-α-hydrazinopropionic acid] and Ro 41-0960 [2'-fluoro-3, 4-dihydroxy-5-nitrobenzophenone] were suspended using all glass Potter-Elvehjem homogenizer in 0.9% NaCl. The volumes for p. o. administration were 10 ml/kg body weight.

Results

In rats we were unable to measure significant differences in the plasma DOPA and/or 3-OMD levels between non-fasting and 24-h fasting animals (not shown). The relatively constant concentrations of DOPA and 3-OMD in plasma during day-time, provide the baseline for a reliable in vivo investigation of the effects of peripheral AADC inhibitors (e. g. benserazide and carbidopa) as well as of the COMT inhibitor Ro 41-0960.

Very high doses of benserazide (170 μmoles [50 mg]/kg p. o.) induced a marked increase (about 10 fold) of the concentration of the endogenous DOPA in rat plasma (Fig. 1 B).

The concentration of DOPA in plasma remained markedly elevated (about 30 ng DOPA/ml plasma) from 2 to 6 h after benserazide dosing (Fig. 2 B). The DOPA level did not completely return to the basal value even 16 h after drug administration. As expected, the marked AADC inhibition induced by benserazide produced a parallel elevation of DOPA and of its O-methylated product 3-OMD (Fig. 1 B). The level of 3-OMD increased linearly for at least 6 h after

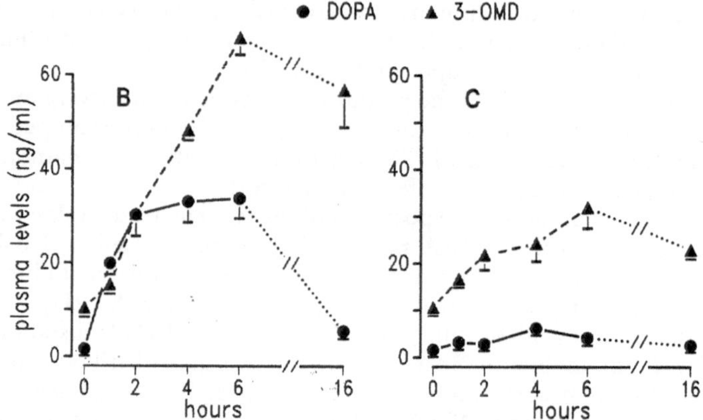

Fig. 1. Time course of the concentrations of endogenous DOPA and 3-OMD in rat plasma after equimolar doses of benserazide *(B)* or carbidopa *(C)* (170 μmoles/kg p. o.). Each point is a mean $\pm$ SEM (N = 6 rats for each point)

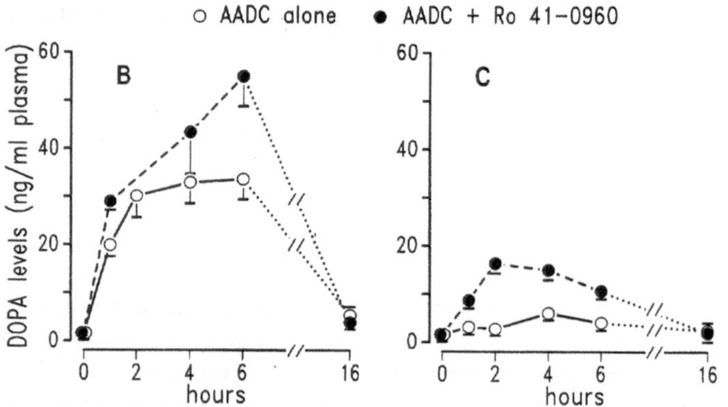

Fig. 2. Time course of the concentration of endogenous DOPA in rat plasma after equimolar doses of benserazide *(B)* and carbidopa *(C)* alone (170 μmoles/kg p. o.) or in combination with the COMT inhibitor Ro 41-0960 (200 μmoles/kg p. o.). Each point is a mean $\pm$ SEM (N = 6 rats for each point)

benserazide (Fig. 1 B), attaining and maintaining very high concentrations (about 60 ng 3-OMD/ml plasma) for more than 16 h. Carbidopa, at an equimolar dose to benserazide (170 μmoles/kg p. o.), produced only a minimal increase of DOPA (Fig. 1 C). The carbidopa-induced increase of 3-OMD in plasma was also much less pronounced than after benserazide (Fig. 1 C).

Ro 41-0960 (200 μmoles [55 mg]/kg p. o.) coadministered with benserazide or carbidopa (170 μmoles/kg p. o., each) almost completely suppressed the formation of 3-OMD down to the detection limit of the assay (about 5 ng/ml). After benserazide plus Ro 41-0960 the concentration of endogenous DOPA attained its highest value (about 50 ng DOPA/ml plasma) 6 h after dosing (Fig. 2 B). When combined with carbidopa the COMT inhibitor Ro 41-0960 produced only a moderate increase of the plasma DOPA concentration, the peak effect (about 18 ng DOPA/ml plasma) being attained 2 h after dosing (Fig. 2 C).

Discussion

The endogenous plasma concentrations of DOPA and 3-OMD in humans and laboratory animals could only recently be precisely measured using highly sensitive and specific assay techniques (Zürcher and Da Prada, 1979; Ito et al., 1984; Da Prada et al., 1987). The present study documents the power of these techniques.

The results in Figs. 1 and 2 clearly demonstrate that the effect of peripheral AADC inhibitors (e. g. benserazide and carbidopa) as well as of the COMT inhibitor Ro 41-0960 can be easily monitored by measuring the changes induced by these compounds on the DOPA and 3-OMD plasma levels. The present study clearly demonstrate that the potent inhibition of the peripheral AADC activity induced by benserazide is reflected by a marked increase of the concentrations of endogenous DOPA and 3-OMD in plasma. These results are consonant with a previous study in which at pharmacological doses (5.1 μmoles/kg p. o.) benserazide was more effective than carbidopa in increasing the plasma level of endogenous DOPA in healthy subjects and in rats (Da Prada et al., 1987).

Other experiments show also that Ro 41-0960 (ED_{50} for COMT inhibition in rat liver = 1.2 mg/kg p. o., 1 h) potently inhibits COMT activity in combination with peripheral AADC inhibitors. Thus, Ro 41-0960 (200 μmoles [55 mg]/kg p. o.) administered together with benserazide or carbidopa completely suppresses the accumulation of 3-OMD and, thereby, further increases the concentration of the endogenous DOPA in plasma.

In the present experiments the 3-OMD level in plasma is maintained at high levels for more than 16 h (Fig. 1 B, C) indicating that, as already shown in humans, 3-OMD has a very long-lasting half-life also in the rat.

In conclusion, this study provides additional support for the view that benserazide is a more potent inhibitor of peripheral AADC than carbidopa. Moreover, we demonstrate that the COMT inhibitor Ro 41-0960 markedly suppresses the increase of endogenous 3-OMD induced by benserazide or carbidopa so that higher concentrations of DOPA are accumulated into the plasma.

Altogether, the present findings strongly indicate that the bioavailability of DOPA will be profoundly improved by its combination with Ro 41-0960 and a peripheral AADC inhibitor (Zürcher et al., 1988). The marked increase of the plasma concentrations of DOPA and 3-OMD after benserazide represents an excellent index for monitoring changes in the synthesis of peripheral catecholamines which could occur in various physiological or pharmacological conditions.

References

Da Prada M, Keller HH, Pieri L, Kettler R, Haefely WE (1984) The pharmacology of Parkinson's disease: basic aspects and recent advances. Experientia 40: 1165–1172

Da Prada M, Kettler R, Zürcher G, Schaffner R, Haefely WE (1987) Inhibition of decarboxylase and levels of DOPA and 3-O-methyldopa: a comparative study of benserazide versus carbidopa in rodents and of Madopar standard versus Madopar HBS in volunteers. Eur Neurol 27 [Suppl] 1: 9–20

Ito S, Kato T, Maruta K, Fujita K, Kurahashi T (1984) Determination of DOPA, dopamine, and 5-S-cysteinyl-DOPA in plasma, urine, and tissue samples by high-performance liquid chromatography with electrochemical detection. J Chromatogr 311: 154–159

Zürcher G, Da Prada M (1979) Radioenzymatic assay of femtomole concentrations of DOPA in tissues and body fluids. J Neurochem 33: 631–639

Zürcher G, Keller HH, Kettler R, Schaffner R, Bonetti EP, Borgulya J, Da Prada M (1988) DOPA-potentiating effects of Ro 41-0960: neuropharmacological profile of a novel reversible, orally active COMT inhibitor. Proceedings, 9th International Symposium on Parkinson's disease, Jerusalem

Correspondence: Dr. A. Colzi, Pharmaceutical Research Department, F. Hoffmann – La Roche & Co., Ltd., CH-4002 Basel, Switzerland.

Catecholamines in urine, blood and cerebrospinal fluid

H. Wisser and D. Ratge

Department of Clinical Chemistry, Robert-Bosch-Krankenhaus, Stuttgart, Federal Republic of Germany

Summary

Quantification of catecholamines in body fluids such as urine, blood and cerebrospinal fluid is one way—beside the determination of receptor densities and/or the stimulation of the cAMP-system—of evaluating the activity of the sympathetic nervous system. In the following discourse the analytical problems and then the clinical importance of the determination of catecholamines in different specimens will be discussed.

Methodology

Methods

Numerous methods for determining catecholamines have been developed. The methods most used during recent years are radioenzymatic assay and high performance liquid chromatography (HPLC) combined with electrochemical or fluorometric detection (Ratge and Wisser, 1983).

The radioenzymatic assay as a single isotope-derivative method is based on the conversion of the catecholamines with ^{3}H-labelled S-adenosyl-L-methionine and catechol-O-methyltransferase into radiolabelled O-methyl derivatives. In our laboratory these compounds are purified by extraction and separated by reversed phase high performance liquid chromatography. The fractions corresponding to labelled normetanephrine, metanephrine and 3-methoxytyramine are collected automatically and assayed radiometrically after oxidation of the metanephrines and acetylation of 3-methyoxytyramine. The quantitative determination is based on internal standards. For determination of the conjugated fraction of catecholamines, hydrolysis of the conjugates before analysis is necessary. Hydrolysis is performed by addition of a sulfatase to the incubation mixture or by deproteinizing with perchloric acid and heating the samples for 20 minutes at 95°C.

Recently applications of high pressure liquid chromatography in combination with amperometric or fluorometric detection have gained widespread use. The first step of this analytical procedure is the extraction of the catecholamines from the different body fluids by chromatography on alumina, cation-exchange resin or boric acid gel. The separation is performed on reversed phase columns with various elution conditions. Using a sample volume of 1 ml plasma this method was not sensitive enough in our hands to determine dopamine concentrations in the normal range.

The radioenzymatic method and the HPLC-method are equally precise. The between run precision for concentrations in the normal range is between 6—10% for both methods, but the radioenzymatic method is 10—20 times more sensitive than the HPLC-method (Ratge and Wisser, 1983). In our experience the long-term stability of the radioenzymatic method is better than that of the HPLC-method. We therefore prefer the single isotope derivative method in spite of the laborious handling and the radioactive waste.

Comparison of the two methods by analyzing 38 different patient samples showed a good correlation of the results for adrenaline ($r = 0.923$) and noradrenaline ($r = 0.921$) (Bauersfeld et al., 1986). These results are not representative if one compares the results of different laboratories. It is not surprising that an inter-laboratory comparison of plasma catecholamines determinations using several different assays for analyzing the same samples revealed sometimes a considerable variability between methods as well as between laboratories using the same method (Hjemdahl, 1984). The measurements of basal noradrenaline concentrations revealed a considerable interlaboratory scatter of the results, demonstrating a poor accuracy, except for the laboratories using HPLC for chromatographic separation. The poor reproducibility of the duplicate determinations suggests that there are frequent problems with intralaboratory reproducibility. If one considers that only laboratories engaged in the development of plasma catecholamine assays and/or in the use of them participated in the survey, the results of the study are disappointing. Further errors often occur during specimen sampling and in addition during the subsequent storage of the samples.

Sample collection

Because of the lability of plasma catecholamines in response to emotional and physical stress, a strongly controlled environment must be ensured during sample collection. Posture, venepuncture, food intake, time of day, physical activity, stress, medications, concomitant illness, age, sex and, especially for CSF, ventriculospinal concentration gradients or obstructed CSF circulation are some of the factors which must be considered (Wood, 1980). Basal levels are preferably drawn via indwelling intravenous cannulas while patients are in the supine position for at least 20–30 minutes before sampling and following overnight fasting. The concentration of catecholamines in serum and cerebrospinal fluid and the excretion in urine show a great dependence on the time of day. Therefore a time factor must be considered (Ratge and Wisser, 1983). As shown by our own experiments, the catecholamines in plasma cerebrospinal fluid or in acidified urine (pH <2) frozen at $-25°C$ are stable for 12 months.

Summarizing the methodological aspects of catecholamine determinations, the entire procedure from environment of the patient during sampling to collection, storage and assay must be rigorously standardized (Bravo, 1986).

Clinical importance

Besides the factors described which influence plasma catecholamine concentrations, there are some diseases which are involved in a disturbed release, synthesis or sensitivity of the target cells.

Internal medicine

Patients suffering from postural hyposensitive syndromes may be distinguished on the basis of plasma noradrenaline concentrations and the changes in the plasma concentration induced by standing. In patients with the hyperadrenergic postural hypotension syndrome the basal levels are normal or increased and plasma noradrenaline concentration increases excessively in response to standing, while in patients with the hypoadrenergic postural hypotension syndrome the basal noradrenaline levels are normal or decreased and the response to standing is either blunted or absent (Bravo, 1986).

The results of extensive literature analysis lead to the following conclusions: Plasma noradrenaline levels are increased in a subgroup of hypertensive patients especially in the young and plasma adrenaline levels are elevated independently of age (Goldstein and

Lake, 1984). The most marked increases in noradrenaline and/or adrenaline plasma concentrations and excretion in urine are found in patients with phaeochromocytoma.

Several signs and symptoms associated with hyperthyroidism suggest that there is a sympathetic nervous system overactivity. However, as found by different authors, noradrenaline and adrenaline levels are decreased. In contrast patients with hypothyroidism have increased plasma catecholamine concentrations (Bravo, 1986).

Neurology

A survey of the literature shows that catecholamine and metabolite determinations have been performed for the following indications:

1. as an aid in diagnosis,
2. as an indicator of the duration of disease,
3. as an indicator of the severity of disease,
4. as a predictor of L-DOPA response or side effects,
5. as a pathophysiological explanation of specific symptoms,
6. as an indicator of a failure of the autonomic nervous system.

In the following section some examples of the results of the determination of catecholamines and/or metabolites in urine, blood and CSF are described.

Markianos et al. (1982) showed that an increased urinary excretion of HVA in Parkinsonian patients is found in the early stages of the disease, but only the subgroup with tremor as the main symptom had a greater HVA excretion than the control group. They also showed that a subgroup of Parkinsonians with akinesia as the main symptom excreted more MHPG in urine than the controls. In comparison to schizophrenic patients a decreased excretion of DA, DOPAC and HVA in the Parkinsonians was shown by Hoehn et al. (1977). This result is a confirmation of several publications in the sixties. In the study cited the excretion of free DA correlated negatively with the severity of rigidity and akinesia. As is well known, the biogenic amines do not cross the blood-brain barrier. Therefore the diminished DA-excretion may be the result of a generalized central and peripheral disorder of DA metabolism as postulated by the authors (Hoehn et al., 1977). Thus the urinary excretion of free DA, although it is primarily of peripheral origin, may reflect a disturbance of central metabolism.

The results of the following study of Turkka et al. (1986) support this hypothesis. The basal concentrations of NA in the controls and

Table 1. Homovanillic acid (HVA) levels in lumbar spinal fluid* (Chase et al., 1976)

	Number of patients	HVA
Controls	25	36 ± 4.0
Parkinson's disease	30	10 ± 1.2[a]
Parkinsonism-dementia	15	9 ± 2.7[a]
Huntington's chorea	14	16 ± 3.3
Dystonia musculorum deformans	15	20 ± 3.6[b]
Down's syndrome	9	56 ± 11.0
Amyotrophic lateral sclerosis	21	13 ± 2.4[a]

* Values are the means ± SEM, expressed in ng per ml for untreated patients
[a] $p < 0.001$ [b] $p < 0.01$

the Parkinsonian patients were 3.19 ± 0.93 nmol/l and 2.64 ± 0.94 nmol/l respectively and were not stastistically significantly different. The mean increments of serum NA levels in response to standing were 38% and 5.9% respectively, indicating a significant failure of the autonomic nervous system of the Parkinsonians. The serum NA responses of the patients treated with levodopa or without levodopa treatment were similar, suggesting that levodopa does not markedly influence the results.

Finally some results of noradrenaline and catecholamine metabolite determinations in CSF are discussed. Instead of dopamine, most authors prefer HVA-determination when studying dopamine metabolism. The HVA in CSF originates from brain parenchyma adjacent to the lateral ventricle. Most of noradrenaline and its major metabolite MHPG in CSF originate from brain. A study of Turkka et al. (1987) showed, that the concentrations of NA and MHPG in Parkinsonian patients did not differ from those of controls, but a significant correlation was found between the CSF MHPG-concentration and the severity of autonomic failure.

HVA levels in lumber spinal fluid of patients with different diseases are given in Table 1 and Table 2. There is a significant depression of the HVA-concentration in CSF, but statistically significant decreases of the HVA concentration are also shown for other diseases (Table 1). The diminished probenecid-induced accumulation of HVA, shown in Table 2 was also not specific for patients with Parkinson's disease.

The diminution in the rate of probenecid-induced HVA accumulation in CSF is significant also for Huntington's chorea and amyotrophic lateral sclerosis. A recent publication of Gibson et al. (1985) in patients with Parkinson's and Alzheimer's disease showed the

Table 2. Effect of probenecid on homovanillic acid levels in lumbar spinal fluid* (Chase et al., 1976)

	Number of patients	Baseline	Treatment	Differences
Controls	8	28 ± 6.8	194 ± 24	166 ± 24
Parkinson's disease	20	12 ± 1.7	73 ± 12	62 ± 11[a]
Huntington's chorea	12	14 ± 5.0	115 ± 14	100 ± 14[b]
Dystonia musculorum deformans	8	14 ± 4.9	168 ± 22	154 ± 19
Amyotrophic lateral sclerosis	7	23 ± 7.7	129 ± 22	104 ± 16[b]

* Values are the means ± SEM, expressed in ng per ml. Probenecid (2 g) was administered orally immediately after obtaining the baseline spinal fluid sample and again 3 and 6 hours later. The second lumbar puncture was performed 9 hours after the initial one

[a] $p < 0.001$ [b] $p < 0.01$

same result—decreased basal HVA concentration in both patient groups, but the difference was not statistically significant. MHPG concentrations in both groups were similar. Probenecid administration significantly increased the HVA and MHPG concentration. The increase was significantly higher in patients with Alzheimer's than in Parkinson's disease. A similar result was found by Pinessi et al. (1987)—decreased DA, HVA and DOPAC concentrations in CSF and in plasma compared with a control group. A correlation was found between decreased DA and HVA levels and duration of the disease.

Summing up, it may be said, that CSF and plasma concentrations of DA and metabolites are lower in Parkinson's disease, but there is a considerable overlap between patients and controls. The diminished increase of HVA and MHPG in CSF after probenecid application further reduce the value of such determinations for the early diagnosis of Parkinson's disease.

References

Bauersfeld W, Ratge D, Knoll E, Wisser H (1986) Determination of catecholamines in plasma by HPLC and amperometric detection. Comparison with a radioenzymatic method. J Clin Chem Clin Biochem 24: 185–188

Bravo LE (1986) Plasma catecholamines: their measurement and clinical utility. Lab Med 17: 512–516

Chase ThN, Eng N, Gordon EK (1976) Biochemical aids in the diagnosis of Parkinson's disease. Ann Clin Lab Sci 6: 4–10

Gibson CJ, Logen M, Growdon JH (1985) CSF monoamine metabolite levels in Alzheimer's and Parkinson's disease. Arch Neurol 42: 489–492

Goldstein DS, Lake RC (1984) Plasma norepinephrine and epinephrine levels in essential hypertension. Fed Proc 43: 57–61

Hjemdahl P (1984) Inter-laboratory comparison of plasma catecholamine determinations using several different assays. Acta Physiol Scand [Suppl] 527: 43–54

Hoehn MM, Crowley ThJ, Rutledge ChO (1977) The Parkinsonian syndrome and its dopamine correlates. Adv Exp Med Biol 90: 243–254

Markianos M, Hadjikonstantinou M, Bistolaki E (1982) Urinary noradrenaline and serotonin metabolites in drug free Parkinson patients and the effect of L-dopa treatment. Acta Neurol Scandinav 66: 267–275

Pinessi L, Rainero I, de Gennaro T, Gentile S, Portaleone P, Bergamasco B (1987) Biogenic amines in cerebrospinal fluid and plasma of patients with dementia of Alzheimer type. Function Neurol 2: 51–58

Ratge D, Wisser H (1983) Praeanalytik und Analytik der Katecholamine. Ärztl Lab 29: 209–214

Turkka JT, Juujärvi KK, Lapinlampi TO, Myllylä VV (1986) Serum noradrenaline response to standing up in patients with Parkinson's disease. Europ Neurol 25: 355–361

Turkka JT, Juujärvi KK, Lapinlampi TO, Myllylä VV (1987) Correlation of autonomic dysfunction to CSF concentrations of noradrenaline and 3-methoxy 4-hydroxyphenylglycol in Parkinson's disease. Europ Neurol 26: 29–34

Wood JH (1980) Neurochemical analysis of cerebrospinal fluid. Neurology 30: 645–651

Correspondence: Prof. Dr. Dr. H. Wisser, Department of Clinical Chemistry, Robert-Bosch-Krankenhaus, Auerbachstrasse 110, D-7000 Stuttgart 50, Federal Republic of Germany.

^{3}H-spiperone binding to lymphocytes is increased in schizophrenic patients and decreased in parkinson patients

B. Bondy[1], F. X. Dengler[1], W. H. Oertel[2], and M. Ackenheil[1]

[1]Psychiatric and [2]Neurological Clinic of University Munich, Federal Republic of Germany

Summary

The binding parameters (B_{max} and K_D) for the dopamine antagonist ^{3}H-spiperone were investigated in psychiatric patients and in patients with Parkinsonian diesease. B_{max} values were found to be increased in acute, unmedicated schizophrenics (n = 46), during neuroleptic treatment (n = 25) and in chronic schizophrenics (n = 6) being under long term neuroleptic treatment. In all other psychiatric patients, including endogenous depressives, manics, neurotics, borderline and alcoholic patients the binding capacity did not differ from that of healthy control persons (n = 50). This finding suggests, that increased ^{3}H-spiperone binding could be valid as marker for schizophrenia.. In contrast to that, ^{3}H-spiperone binding was found to be decreased in drug free (n = 5) and in MAO-B (n = 2) or short term L-DOPA (n = 3) treated Parkinsonian patients (mean decrease 80%) as compared to age and sex matched controls. The significance of the later finding has to be further investigated.

Introduction

Amongst receptor alterations in neuropsychiatric disorders, changes in dopamine (DA) receptors are of significant importance. The "DA hypothesis" of schizophrenia suggests that increased central dopaminergic activity is in part responsible for some of the clinical abnormalities associated with this disease. In idiopathic Parkinson's disease (PD) motor symptoms of the illness are a consequence of the degeneration of mesencephalostriatal neurones. Loss of these neurones results in DA deficiency in the striatum with functional overactivity of the cholinergic system. Using radioreceptor assays in postmortem brains, it has been shown, that the density of DA 2-receptors is elevated in brain of at least part of schizophrenic patients in support of the dopamine hypothesis of schizophrenia (Rev. Seeman, 1987). Concerning DA receptors in the striatum of patients with PD, only recently consensus has been reached, that striatal DA receptor density is either normal or increased in drug free patients and normal or slightly decreased during L-DOPA therapy (Guttman and Seeman, 1986).

Besides postmortem investigations the functional state of receptors as well as the mode of transmembrane signalling are increasingly studied on cultured cell lines, including those of neuronal origin, and in peripheral blood cells. Although the functional significance of neurotransmitter receptors on blood cells is often questioned, and some of these specific binding sites even lack the conformation of being a receptor, investigations with blood cells offer an opportunity to study disease- or drug-related alterations in the periphery during life time. In respect to neuropsychiatric disorders binding of the DA-antagonist ^{3}H-spiperone to lymphocytes is decreased in patients with unmedicated PD (LeFur et al., 1980) but increased in schizophrenics (LeFur et al., 1982, Rotstein et al., 1983) and might even be valid as vulnerability marker for schizophrenia (Bondy and Ackenheil 1987). This report demonstrates results obtained for ^{3}H-spiperone binding to lymphocytes of patients with various psychiatric disorders and preliminary results for patients with PD.

Material and methods

Subjects

The ^{3}H-spiperone binding parameters (B_{max} and K_D) to lymphocytes were investigated in acute, nonmedicated schizophrenics (n = 46), during neuroleptic treatment (n = 25), in chronic schizophrenics, being under longterm neuroleptic treatment (n = 6), in a psychiatric control group comprising endogenous depressed patients (n = 7), manic patients (n = 5), alcoholics (n = 5), borderline patients (n = 4) and neurotics (n = 10). Patients were diagnosed according to the ICD criteria (9th Revision). The clinical status of psychiatric patients was determined using the AMDP and BPRS Rating Scales. Binding assays were also performed in 10 patients with idiopathic PD (7 males, 3 females, mean age 57 + 7 years, mean duration of disease between 1 and 5 years). 5 patients were classified free of dopamimetic drugs: 2 were drug-naive, 1 took occasionally benzodiazepines, 1 had taken

150 mg L-DOPA for 4 weeks, eight months before the assay and 1 patient had taken a MAO-B-inhibitor for two months and stopped this medication 2 weeks before the assay. 2 patients were on MAO-B-inhibitor therapy, with one of them receiving anticholinergics in addition. 3 patients were on L-DOPA, one for several years and two for several weeks. Plasma HVA-concentrations were determined in the Parkinsonian patients and sex and age matched controls together with the radioligand bin ding assays. 50 healthy persons, aged from 18 to 65 years and without any history of psychiatric or neurological diseases were investigated as control group.

Laboratory assay

The lymphocyte preparation and ^{3}H-spiperone binding assay were performed as previously described in detail (Bondy et al., 1985), using ^{3}H-spiperone binding (20-100 Ci/mmol) as ligand and (+)-butaclamol (1 M) as displacing agent. The maximum number of binding sites (B_{max}) and the affinity constants (K_D) were calculated using a computer program for non-linear Scatchard analysis and are expressed as fmoles/10^6 cells and nM respectively.

Plasma HVA concentrations in Parkinson patients were performed with high performance liquid chromatography on a reversed phase column (C 18), followed by quantitation based on electrochemical detection, according to a method of Chang et al. (1983).

Results

In Table 1 the parameters for ^{3}H-spiperone binding to lymphocytes of healthy controls and psychiatric patients with different diagnoses are listed. An increase in maximum binding capacity (B_{max}) could be found only in schizophrenic patients, independently if they were drug naiv, drug free for more than 1 year or under neuroleptic treatment for several weeks or years. All other psychiatric patients (neurotics, borderline patients, alcoholics, depressives, manics) had B_{max} values not significantly different from those of healthy controls. In schizophrenics also statistical significant, although small increase in the K_D values could be observed. The degree of increase was less than that of of the binding capacity.

In preliminary experiments ^{3}H-spiperone binding was investigated in lymphocytes from Parkinsonian patients (Fig. 1) who were either free from anti-parkinsonian medication ($n = 5$) or under treatment with MAO-B-inhibitors ($n = 2$). The binding capacity was found to be decreased as compared to sex and age matched healthy controls with a mean decrease of about 80% ($p < 0.0001$), without

Table 1. ^{3}H-spiperone binding to lymphocytes of psychiatric patients and controls

	B_{max} fmol/10^6 cells	K_D nM
Drug free schizophrenics (n = 46)	8.32 ± 5.34*	0.35 ± 0.16
During NL-therapy (n = 25)	7.98 ± 4.32*	0.31 ± 0.09
Chronic schizophrenics (n = 6)	7.72 ± 5.31*	0.34 ± 0.21
Depressives (n = 7)	2.45 ± 0.36	0.17 ± 0.06
Manics (n = 5)	3.75 ± 1.25	0.22 ± 0.07
Alcoholics (n = 5)	2.78 ± 0.86	0.17 ± 0.07
Neurotics (n = 10)	2.83 ± 0.86	0.13 ± 0.09
Borderline psychosis (n = 4)	2.12 ± 0.85	0.16 ± 0.10
Controls (n = 50)	2.13 ± 0.99	0.17 ± 0.07

Data are present as mean ± S_D; * $p < 0.0001$

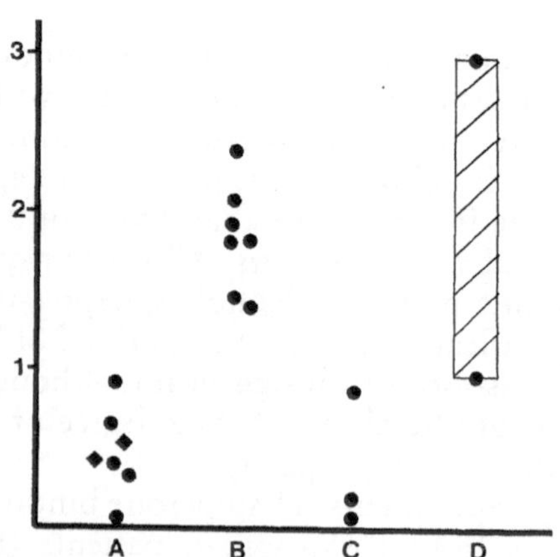

Fig. 1. Binding of ^{3}H-spiperone to lymphocyters of Parkinsonian patients and controls (B_{max}: fmoles/10^6 cells). *A* ● drug free patients, ✦ patients under MAO-B medication, *B* sex and age matched controls, *C* patients under continuous L-DOPA therapy, *D* total control group with maximum and minimum values

significant change in K_D values (controls: K_D 0.17 ± 0.07 nM; patients: K_D 0.13 ± 0.1 nM). In patients being under L-DOPA therapy (n = 3), binding capacity as a mean was again below that of controls. Determination of the plasma HVA concentrations revealed no differences between the patients, who were not treated with dopamimetic drugs and healthy controls (patients: 12.5 ± 4.4 ng/ml; controls: 13.9 ± 6.0 ng/ml). A correlation between the density of ^{3}H-spiperone binding sites and HVA concentration was not observed.

Discussion

Inspite of their important function in neurotransmission limited information is available about the premortem state of postsynaptic DA-receptors either in psychiatric or in Parkinsonian patients. The new in vivo data, obtained with positron emission tomography (PET) in schizophrenic patients are still controversial and prone to confuse more than to clarify. One of the disadvantages of postmortem studies is the fact, that alterations of both, the receptors and neurotransmitters can be monitored only after several years of ongoing disease, thus representing a final and a possibly drug-modified state. An alternative may be premortem investigations in the periphery, as e. g. using blood cells as peripheral model. Although a direct correlation between brain and periphery is sill questioned, platelets and lymphocytes are gaining significance for these studies. Several neurotransmitter receptors have been identified on these cells. In addition, the interaction of the nervous and the immunological system attracts increasing interest (Biziere et al., 1985).

The finding that ^{3}H-spiperone binding capacity is increased in lymphocytes of schizophrenic patients (LeFur et al., 1983; Rotstein et al., 1983; Bondy et al., 1985) strongly suggests that this phenomenon is a marker for schizophrenia. Elevated binding appears to be associated with schizophrenia, since it could not be found in other psychiatric patients. The fact, that elevated binding remained on a constant level during neuroleptic therapy and in drug free remission period is further supportive for a specific trait phenomenon in schizophrenia, worth being investigated as possible vulnerability marker in family studies (Bondy and Ackenheil, 1987).

With our preliminary investigations in patients with PD we can confirm the findings reported by LeFur et al. (1981), that ^{3}H-spiperone binding is decreased in drug-free patients as compared to sex and age matched controls. Binding capacity was also decreased in patients being under treatment with MAO-B inhibitors. In contrast to the results of LeFur et al. (1981) treatment with L-DOPA did in our investigation not clearly elevate binding to normal control values. These results are however very preliminary, since so far we did not follow up possible alterations of the binding parameters before and during L-DOPA treatment. Thus, at present we are unable to confirm, if ^{3}H-spiperone binding sites on lymphocytes of PD patients might be upregulated during L-DOPA therapy, this phenomenon would be the opposite to the changes in brain DA receptors under L-DOPA therapy (Guttman and Seeman, 1986), or if the number of binding sites remains on a constant level as observed with schizophrenic patients (Bondy and Ackenheil, 1987). The observation, that plasma HVA concentrations were similar in Parkinson patients without L-DOPA therapy and healthy controls is in good agreement with findings obtained with cerebrospinal fluid, where only a slight, non significant reduction of HVA levels was observed in Parkinson patients (Cramer et al., 1984).

Our observations raise the problem of the significance of the altered numbers of ^{3}H-spiperone binding sites in lymphocytes of neuropsychiatric patients. The initial finding by LeFur et al. (1980), suggesting these binding sites being peripheral DA receptors could not be replicated by other investigators (Fleminger et al., 1982; Maloteaux et al., 1983) and it was supposed, that binding is a trapping phenomenon into acidic intracellular components. However, trapping phenomena are rather nonspecific cellular events and the stereospecifity of these ^{3}H-spiperone binding sites, although rather low (Bondy et al., 1985; Shaskan et al., 1983) is not supportive for nonspecific trapping phenomena. In a recent report using rat thymocytes, it was again claimed that DA receptors, being insensitive to neuroleptic drugs, are present on these cells (Ovadia et al., 1987). This demonstrates that there is still an ongoing discussion if DA receptors, valid as peripheral model, are present on lymphoid cells.

As long as the significance of these specific binding sites in lymphocytes is not claryfied, we can only describe the results as observations found in different laboraties. For schizophrenia, the elevated binding appears to become extremely important as peripheral marker for genetic and high-risk studies. The significance and diagnostic importance of the decreased binding in lymphocytes of patients with PD has to be further investigated.

Acknowledgement

The authors are very grateful to the excellent laboratory work of G. Hölzl, A. Johnson, B. Kuhl, U. Thüringer. — This work was supported by the Deutsche Forschungsgemeinschaft.

References

Biziere K, Guillaumin JM, Degenne D, Bardos P, Renoux M, Renoux G (1985) Lateralized neocortical modulation of the T-cell lineage. In: Guillemin R (ed) Neural modulation of immunity. Raven Press, New York, pp 81–93

Bondy B, Ackenheil M, Elbers R, Fröhler M (1985) Binding of ^{3}H-spiperone to lymphocytes: a biological marker in schizophrenia? Psychiat Res 15: 41–48

Bondy B, Ackenheil M (1987) ^{3}H-spiperone binding sites in lymphocytes as possible vulnerability marker in schizophrenia. J Psychiat Res 21: 521–529

Chang WH, Scheinin M, Burns RS, Linnoila M (1983) Rapid and simple determination of homovanillic acid in plasma using high performance liquid chromatography with electrochemical detection. Acta Pharmacol Toxicol 53: 275–279

Cramer H, Warter JM, Renaud B (1984) Analysis of neurotransmitter metabolites and adenosine 3', 5'-monophosphate in the CSF of patients with extrapyramisal motor disorders. In: Hassler RG, Christ JF (eds) Advances in neurology, vol 40. Raven Press, New York, pp 431–435

Guttman M, Seeman P (1986) Dopamine D 2 receptor density in parkinsonian brain is constant for duration of disease, age and duration of L-DOPA therapy. Adv Neurol 45: 51–57

Fleminger S, Jenner P, Marsden CD (1982) Are dopamine receptors present on human lymphocytes? J Pharm Pharmacol 34: 658–663

LeFur G, Zarifian E, Phan T, Couche H, Flamier A, Bouchami F, Burgevin MC, Loo H, Gerard A, Uzan A (1980) Identification of stereospecific ^{3}H-spiroperidol binding sites in mammalian lymphocytes. Life Sci 26: 1139–1148

LeFur G, Meininger VA, Baulac M, Uzan A (1981) Recepteurs dopaminergiques lymphocytaires et maladie de parkinson idiopathique. Rev Neurol 137: 89–96

LeFur G, Zarifian E, Phan T, Couche H, Flamier A, Bouchami F, Burgevin MC, Loo H, Gerard A, Uzan A (1983) ^{3}H-spiperone binding in lymphocytes: changes in two different groups of schizophrenics and effect of neuroleptic treatment. Life Sci 32: 249–254

Maloteaux JM, Gossuin W, Waterkyn C, Laduron PM (1983) Trapping of labelled ligand in intact cells: pitfalls in binding studies. Biochem Pharmac 23: 2543–2548

Ovadia H, Lubetzky-Korn I, Abramsky O (1987) Dopamine receptors on isolated membranes of rat thymocytes. Ann NY Acad Sci 496: 211–216

Rotstein E, Ram RK, Singal DP, Barone D (1983) Lymphocyte ^{3}H-spiroperidol binding in schizophrenia: preliminary findings. Progr Neuro-Psychopharmac Biol Psychiat 7: 720–723

Seeman P (1987) Dopamine receptors and the dopamine hypothesis of schizophrenia. Synapse 1: 133–152

Shaskan EG, Ballow M, Oreland L, Wadell G (1983) Is there a functional significance for dopamine antagonist binding on lymphoid cells? Biol Psychiat 12: 123–145

Correspondence: Dr. B. Bondy, Psychiatric Clinic of University Munich, Nussbaumstrasse 7, D-8000 Munich, Federal Republic of Germany.

Platelet MAO-B activity in humans and stumptail monkeys: in vivo effects of the reversible MAO-B inhibitor Ro 19-6327

R. Kettler and **M. Da Prada**

Pharmaceutical Research Department, F. Hoffmann-La Roche & Co., Ltd., Basle, Switzerland

Summary

MAO activity in platelets of six different animal species was compared with that of healthy volunteers. Only stumptail monkey (macaca arctoides) have similar high MAO-B and corresponding extremely low MAO-A activity in platelets as man. The highly selective, reversible inhibitor of MAO-B, Ro 19-6327, produced a marked and short-lasting inhibition of MAO-B in platelets of healthy volunteers as well as of stumptail monkey after oral administration. Therefore, platelets from stumptail monkey can adequately replace human platelets as model for the study of the pharmacological characteristics of MAO-B inhibitors.

Introduction

Several clinical studies have shown that circulating platelets represent an excellent model for an easy assessment of the effect of MAO-B inhibitors in extracerebral tissues (periphery). The degree of platelet MAO-B inhibition is usually taken as an index of central MAO-B inhibition. Since small laboratory animals display very low MAO activity which is, moreover, almost exclusively of the A type (Da Prada et al., 1981), these animals are unsuitable for the study of MAO-B inhibitors.

The aim of this study was to identify an animal species with platelet MAO-B activity similar to that found in humans. The effect of the new, reversible MAO-B inhibitor Ro 19-6327 (N-[2-aminoethyl]-5-chloro-2-pyridine carboxamide hydrochloride) (Da Prada et al., 1988) on platelet MAO activity was then investigated in healthy volunteers and stumptail monkeys. For comparison, the effect of Ro 19-6327 on MAO-B activity was also measured in rat liver.

Materials and methods

Blood collection and isolation of blood platelets

A 10-ml venous blood sample was collected via a butterfly cannula in a 10-ml plastic syringe containing 1 ml of an EDTA solution (1% in saline). After gentle shaking the blood was slowly transferred in a polystyrene tube and centrifuged at room temperature for 15 min at 200 × g. The resulting platelet rich plasma (PRP) was transferred using a plastic pipette into a new set of polyethylene tubes. The PRP samples were then centrifuged for 10 min at 3000 × g at 4°C. After removal of the supernatant the pellets were washed with 2 ml ice-cold modified Tyrode. The washed platelets were sedimented by centrifugation. This washing procedure was repeated for a second time. The supernatant was discarded and the platelet pellets stored at −80°C until analysis. Platelet protein, measured spectrophotometrically, was <1 mg protein/ml PRP. In human platelets a value of 0.85 ± 0.07 mg protein/ml PRP was obtained (mean ± SEM, $N = 6$).

Measurement of MAO activity

A pellet of washed platelets sedimented from 1 ml PRP was suspended in 100 μl Tyrode at 4°C. The platelets in suspension were then lysed with 900 μl sodium borate buffer (0.1 M, pH 8.2) and vigorous vortexing. MAO-A and -B activities were measured by adding the following incubation mixture to 15 ml-glass tubes: 50 μl lysed platelets, 100 μl potassium phosphate buffer (1 M, pH 7.4), 70 μl distilled water and 80 μl [^{14}C]5-hydroxy-tryptamine (^{14}C-5-HT) or [^{14}C]2-phenylethylamine (^{14}C-PEA) (final concentration of the labelled substrates 200 or 20 μM, respectively). All other steps of the assay and the measurement of the MAO activity in liver homogenates were performed as described earlier (Keller et al., 1987).

Results

An extremely high variance on human platelet MAO-B activity is found in the literature even when the same substrate (e.g. PEA) is used. This variance is probably due to differences in platelet preparation and/or MAO assay procedures. With our assay procedure of platelet preparation and MAO-B detection, the MAO activity found in human platelets (Table) attained a relatively high value (22.6 ±

Table 1. Platelet monoamine oxidase activity in different mammalian species

Species	Platelet MAO activity (nmoles/h/mg protein)	
	MAO-A (5-HT)	MAO-B (PEA)
Mouse	8.9	0.34
Guinea pig	5.2 ± 0.1	traces
Rabbit	8.2 ± 1.1	2.5 ± 0.9
Squirrel monkey	0.3 ± 0.1	0.8 ± 0.1
Rhesus monkey	traces	6.5 ± 1.2
Macaca arctoides	2.6 ± 0.5	19.7 ± 0.5
Man (healthy volunteers)	traces	22.6 ± 2.9

MAO activity was measured with 200 μM 5-HT (MAO-A substrate) or 20 μM PEA (MAO-B substrate). For details see methods

4 nmoles/h/mg protein). The present results show that in human, rhesus monkey and macaca arctoides (stumptail monkey) platelets, the bulk of MAO activity is of the B-type (Table). In contrast, the total MAO activity in the squirrel monkey was extremely low with a predominance of MAO-B. In other animal species (mouse, guinea pig and rabbit) the total MAO activity was relatively low (about 10 nmol/h/mg protein) and was prevalently of the A-type.

The dose-dependent effect of the reversible MAO-B inhibitor Ro 19-6327 on rat liver and stumptail monkey platelets is shown in Fig. 1. The rat experiment clearly shows that Ro 19-6327 is a very

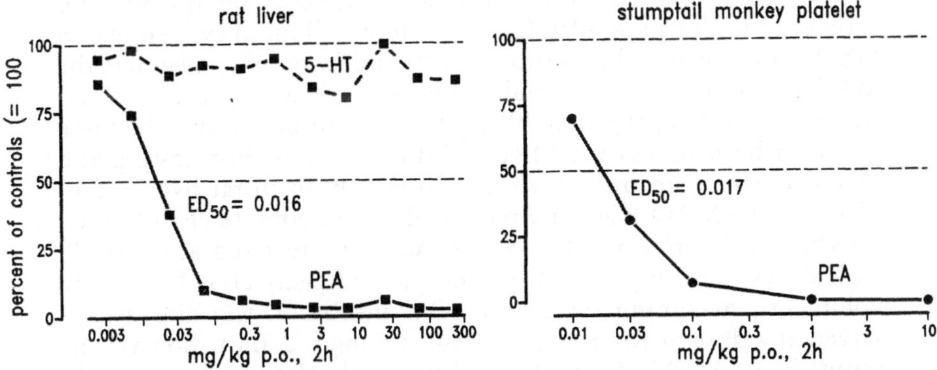

Fig. 1. Effect of different doses of Ro 19-6327 on MAO activity in rat liver (left) and in platelets from stumptail monkeys (right). Data are given in percent of controls. MAO activity was measured with 200 μM 5-HT or 20 μM PEA as substrates. MAO activities in absolute terms in the rat liver were: 19.1 ± 0.6 (5-HT) or 11.9 ± 0.6 (PEA) nmol/h/mg fresh tissue (N = 6). The predrug value in platelets from stumptail monkey (N = 2) was 19.7 ± 0.5 nmol/h/mg protein

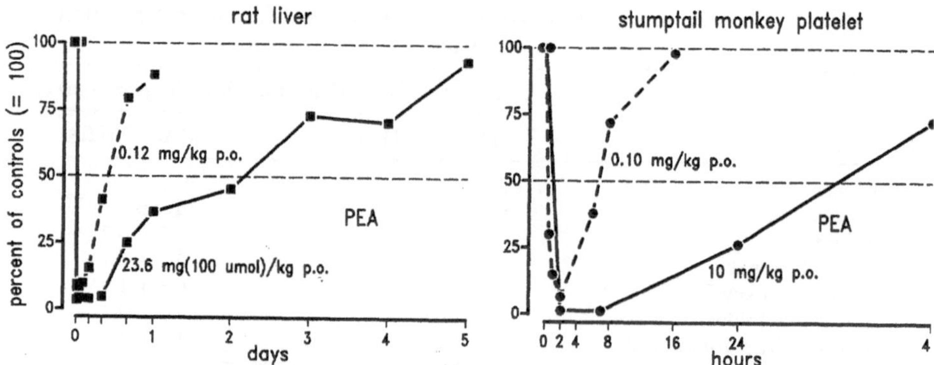

Fig. 2. Time course of MAO-B inhibition after two different doses of Ro 19-6327 in rat liver or in platelets from stumptail monkeys. Data are given in percent of controls (= 100%). MAO-B activity was measured using PEA (20 μM) as substrate. Absolute MAO-B activity for rat liver of untreated controls was to 14.4 $\pm$ 1.2 nmol/h/mg fresh tissue (N = 6). The predrug value of MAO-B activity in monkey platelets was 20.3 $\pm$ 0.8 nmol/h/mg protein (N = 2)

potent and an extremely selective MAO-B inhibitor (ED_{50} for rat liver: 16 μg/kg p.o., 2 hours). On the other hand, virtually no liver MAO-A inhibition was measured even at extremely high doses of Ro 19-6327 (300 mg/kg p.o., 2 hours). It is worthy of note that the ED_{50} value for platelet MAO-B was virtually the same in monkeys as for rat liver (17 μg/kg p.o., 2 hours).

The time-course of the effect of Ro 19-6327 on the MAO-B activity of rat liver and platelets from stumptail monkey is shown in Fig. 2. In rat liver or brain (not shown) and in monkey platelets the MAO-B activity was maximally inhibited already one hour after Ro 19-6327 (120 μg/kg and 100 μg/kg p.o., respectively). At these doses, in both tissues the MAO-B inhibition was short-lasting and MAO-B activity recovered within 8 hours. Both in rat liver and in platelets, the MAO-B activity remained almost completely inhibited for about 8 hours when Ro 19-6327 was administered at relatively high doses (23.6 mg and 10 mg/kg p.o., respectively). In healthy volunteers, maximum platelet MAO-B inhibition was already observed at a dose of Ro 19-6327 as low as 5 mg p.o. (not shown). The results in Figure 2 indicate that in human platelets (and most probably also in brain) the duration and the degree of the MAO-B inhibition increased with the dose of Ro 19-6327 used. Maximum platelet MAO-B inhibition lasted for about 4 hours after 10 mg and 12 hours after 80 mg of Ro 19-6327 p.o.; platelet MAO-B activity rapidly recovered even after high doses of the compound (Fig. 3).

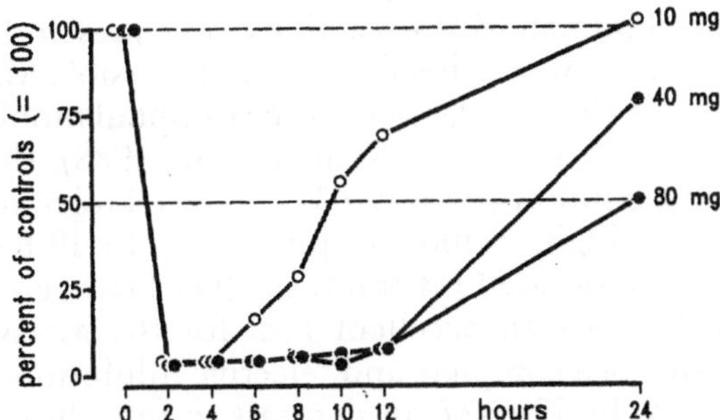

Fig. 3. Time course of MAO-B inhibition in human platelets using three different doses of Ro 19-6327. Each dose was administered to 6 healthy volunteers. Platelet MAO-B activity was measured using PEA (20 μM) as the substrate. Predrug values were 17.9 ± 0.6 nmol/h/mg protein

Discussion

An extremely wide range of human platelet MAO-B activities are reported in the literature even when considering only studies in which PEA was used as the substrate. These marked variances are probably not due to factors such as gender or age but most probably to the method used in the preparation of the platelets and to the MAO assay procedure employed. Several sources of error in the determination of human platelet MAO-B activity have already been discussed by others (Wise et al., 1980). The values obtained in this study for platelet MAO-B activity in healthy subjects are relatively high and are virtually the same as the highest values reported by other laboratories. Using PEA as substrate, we found virtually the same MAO-B activity in platelets from healthy volunteers and stumptail monkeys (Table). However, using benzylamine as the substrate, Murphy et al. (1978) reported a platelet MAO-B activity 20 times higher in human than in stumptail monkeys.

Platelets from stumptail monkeys were used in the present study for the biochemical characterization of the new selective and reversible MAO-B inhibitor Ro 19-6327. The potency of Ro 19-6327 as inhibitor of MAO-B activity in rat liver and in platelets from stumptail monkeys (Fig. 1) as well as the duration of the inhibition were virtually the same for both tissues (Fig. 2). Experiments under way will clarify whether stumptail monkeys are a suitable model also for investigating the level and the distribution of MAO-B in the CNS. Thus, stumptail monkeys could serve as a model for peripheral as well as central human MAO-B activity. The present findings show that Ro 19-6327, given p.o., is a potent and reversible MAO-B inhibitor in rat liver as well as in platelets from stumptail monkeys and man (Fig. 3). As recently observed (Zimmer et al., 1988), the potent and reversible MAO-B inhibitor Ro 19-6327 was well tolerated in humans even at doses as high as 100 mg p.o. Since Ro 19-6327 is a very selective and specific MAO-B inhibitor (Da Prada et al., 1987) the compound represents an excellent tool for the assessment of the therapeutic value of a complete and selective inhibition of MAO-B in the human CNS. Ro 19-6327, in contrast to selegiline (L-deprenyl), does not release catecholamines and, hence, represents a better choice as adjunct to L-dopa plus peripheral decarboxylase inhibitors in the therapy of Parkinson's disease. Ro 19-6327 prevented the MPTP-induced depletion of DA and its metabolites in the striatum of black mice (Da Prada et al., 1988). Ro 19-6327 offers therefore an unique opportunity for clarifying whether a chronic inhibition of the MAO-B activity in the CNS would prevent or block the degeneration of the nigrostriatal dopaminergic neurons leading to Parkinson's disease. Due to its high affinity for MAO-B (Cesura et al., 1988) Ro 19-6327 will be also very useful for in vivo studies of the distribution of the MAO-B in the human CNS by positron emission tomography.

References

Cesura AM, Imhof R, Takacs B, Galva MD, Picotti GB, Da Prada M (1988) [^{3}H]Ro 16-6491, a selective probe for affinity labelling of monoamine oxidase type B in human brain and platelet membranes. J Neurochem 50: 1031–1043

Da Prada M, Kettler R, Keller HH, Burkard W (1988) Ro 19-6327: a reversible, highly selective inhibitor of type B monoamine oxidase completely devoid of tyramine potentiating effect. A comparison with selegiline. In: Proceedings of the 6th international catecholamine symposium. Raven Press (in press)

Da Prada M, Kettler R, Keller HH, Kyburz E, Burkard WP (1987) Ro 19-6327: a novel highly selective and reversible MAO-B inhibitor. Acta Pharm Toxicol 60 [Suppl] 1: 10

Da Prada M, Richards JG, Kettler R (1981) Amine storage organelles in platelets. In: Gordon G (ed) Platelets in biology and pathology. Elsevier/North-Holland Biomedical Press

Keller HH, Kettler R, Keller G, Da Prada M (1987) Short-acting novel MAO inhibitors: in vitro evidence for the reversibility of MAO inhibition by moclobemide and Ro 16-6491. Naunyn-Schmiedeberg's Arch Pharmacol 335: 12–20

Murphy DL, Redmond DE, Baulu J, Donelly CH (1978) Platelet monoamine oxidase activity in 116 normal Rhesus monkeys: relations between enzyme activity and age, sex and genetic factors. Comp Biochem Physiol 60 C: 105–108

Wise CD, Potkin SG, Bridge TP, Phelps BH, Cannon-Spoor HE, Wyatt RJ (1980) Sources of error in the determination of platelet monoamine oxidase: a review of methods. Schizophrenia Bull 6: 245–253

Zimmer R, Kettler R, Bauer K, Thiede HM (1988) First pharmacological investigation in humans of Ro 19-6327, a new reversible and selective MAO-B inhibitor. In: Proceedings of the 9th international symposium on Parkinson's disease, Jerusalem, Israel, June 5–9, 1988

Correspondence: Dr. R. Kettler, Pharmaceutical Research Department, F. Hoffmann-La Roche & Co., Ltd., CH-4002 Basel, Schwitzerland.

Aspartate, glutamate, and glutamine in platelets of patients with Parkinson's disease

L. H. Rolf[1], Th. Klauke[1], E. W. Fünfgeld[2], and G. G. Brune[1]

[1]Klinik und Poliklinik für Neurologie der Westfälischen Wilhelms-Universität, Münster

[2]Schloßklinik Wittgenstein, Bad Laasphe, Federal Republic of Germany

Summary

Aspartate, glutamate and glutamine were determined in platelets of 29 patients with Parkinson's disease (PD) and the results were compared with those of an age matched control group consisting of 24 healthy persons (group A). According to the predominant clinical symptomatology the patients with PD were divided into three groups: group B = patients with tremor (n = 12), group C = patients with akinesia without "on/off" phenomena (n = 11), and group D = patients with akinesia with "on/off" phenomena (n = 6).

Significant decreases of all investigated amino compounds were observed in all groups of patients with PD in comparison with the control group. In addition, aspartate and glutamate were found to be significantly decreased in the group with akinesia and with "on/off" phenomena as compared with the tremor-group. Aspartate was also significantly decreased in patients with akinesia without "on/off" phenomena as compared with the patients with tremor. Glutamine was found to be significantly different between the akinesia groups without and with "on/off" phenomena.

The results point out that platelet aspartate, glutamate and glutamine are significantly decreased in Parkinsonian patients and that there is a correlation between severity of PD and the degree of decreases of the investigated amino compounds.

Introduction

Parkinson's disease (PD) is characterized by a progressive degeneration of the dopaminergic nigrostriatal pathway, however additional pathways are assumed to be involved in the pathophysiology of that disease (Hassler, 1966). The etiology of PD is still unknown. The depletion of dopamine is considered to represent the main biochemical lesion of PD, but alterations of other neurotransmitters have been described, including acetylcholine, noradrenaline, 5-hydroxytryptamine, aminobutyric acid (GABA) and neuropeptides (Lloyd et al., 1975; Riederer and Jellinger, 1983).

Surprisingly, little attention has been paid to possible abnormalities of the excitatory neurotransmitters aspartate and glutamate. The actions of aspartate are little understood. Glutamate is regarded to participate in the functions of the striatum and other parts of the central nervous system (Hassler et al., 1980; Fagg and Forster, 1983; McGeer et al., 1984). Hassler et al. (1980) postulate a complex interaction of various types of neurotransmitters and their respective synaptic terminals in the striatum, and that increased striatal concentrations of dopamine, noradrenaline and glutamate accelerate locomotor and specialized movements and reduce muscle tone, while on the other hand other neurotransmitters such as acetylcholine, 5-hydroxytryptamine and GABA have opposite effects: slowing down of movements and enhancing muscle tone.

This paper reports on amino compounds in platelets of patients with PD. Platelets are considered to represent "multifunctional cells" which are involved in a variety of processes, i.e. thrombosis and haemostasis. In addition, they exhibit a spectrum of cellular reactions which can be readily studied. Thus, they are well suited for the investigation of various biological processes (Gordon and Milner, 1976), and may serve as a model for glutamatergic neurons (Mangano and Schwarzc, 1981).

Patients and methods

The control group consisted of 24 healthy persons without any medication (Group A: mean age [$\pm$ SD] = 68 $\pm$ 8.9 years). The investigation includes 29 patients with PD. According to the predominant clinical symptomatology the patients were divided into 3 groups. In 12 patients tremor was predominant (group B: mean age [$\pm$ SD] = 62.1 $\pm$ 13.0 years; duration of PD [$\pm$ SD] = 8.9 $\pm$ 4.5 years) while in 17 patients the clinical symptomatology was characterized by akinesia. Eleven of these patients showed akinesia without "on/off" phenomena (group C: mean age [$\pm$ SD] = 65.4 $\pm$ 6.6 years; duration of PD [$\pm$ SD] = 6.5 $\pm$ 2.8 years), while 6 patients demonstrated akinesia with "on/off" phenomena (group D: mean age [$\pm$ SD] = 61.5 $\pm$ 7.5 years; duration of PD [$\pm$ SD] = 12.2 $\pm$ 3.8 years). All patients with PD of the various groups obtained comparable therapies. The medication was stopped 24 hours prior to the start of the study.

Kalium-EDTA blood was drawn by venous puncture. Platelets were separated by fractional centrifugation, washed and thrombolized as described by Rolf et al. (1983). The determination of aspartate, glutamate and glutamine was performed by a HPLC-method according to Lenda et al. 1981). The values were calculated to pm amino compound/10^8 platelets. The statistical analyses were performed according to Student's t-test.

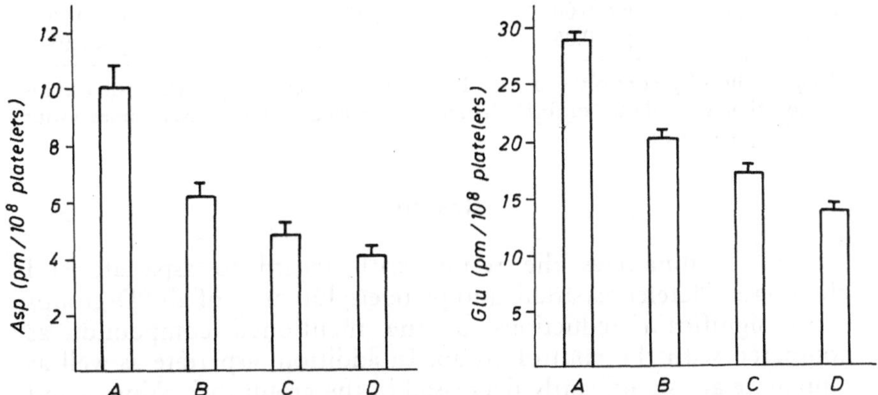

Fig. 1. Aspartate and glutamate in platelets of patients with Parkinson's disease. *Group A:* healthy controls (n = 24); *group B:* Parkinsonism, tremor-type (n = 12); *group C:* Parkinsonism, akinesia without "on/off" phenomena (n = 11); *group D:* Parkinsonism, akinesia with "on/off" phenomena (n = 6). The results are represented as mean values ± S.E.M.

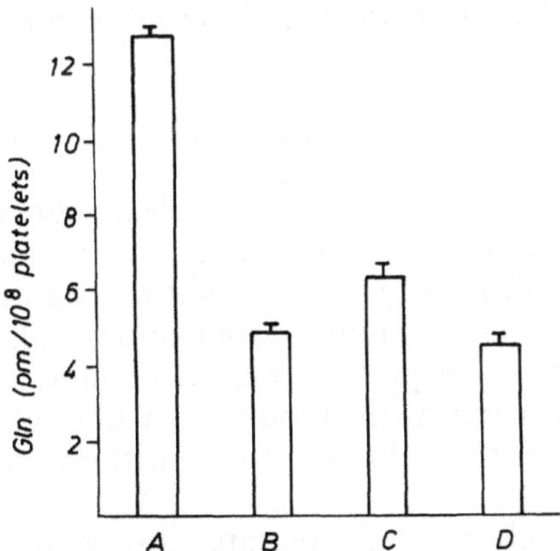

Fig. 2. Glutamine in platelets of patients with Parkinson's disease. *Group A:* healthy controls (n = 24); *group B:* Parkinsonism, tremor-type (n = 12); *group C:* Parkinsonism, akinesia without "on/off" phenomena (n = 11); *group D:* Parkinsonism, akinesia with "on/off" phenomena (n = 6). The results are represented as mean values ± S.E.M.

Table 1. Aspartate, glutamate and glutamine in platelets of healthy controls and patients with Parkinson's disease—Statistical analyses

Group	Asp	Glu	Gln
A/B	$p<0.0005$	$p<0.005$	$p<0.0005$
A/C	$p<0.0005$	$p<0.0005$	$p<0.0005$
A/D	$p<0.0005$	$p<0.0005$	$p<0.0005$
B/C	$p<0.005$	$p>0.05$	$p>0.05$
B/D	$p<0.005$	$p<0.05$	$p>0.05$
C/D	$p>0.05$	$p>0.05$	$p<0.01$

Group A: healthy controls; *group B:* Parkinsonism, tremor-type; *group C:* Parkinsonism, akinesia without "on/off" phenomena; *group D:* Parkinsonism, akinesia with "on/off" phenomena

Results

Fig. 1 summarizes the results with regard to aspartate and glutamate. Platelet-aspartate and platelet-glutamate of all PD-groups show significant reductions of the mentioned compounds as compared with the control group. In addition, aspartate as well as glutamate are significantly decreased in the group with akinesia and with "on/off" phenomena as compared with the tremor-group. Furthermore, aspartate but not glutamate shows significant differences between groups B and C. Fig. 2 demonstrates the results concerning glutamine. The figure demonstrates significant differences between controls and all groups of patients with PD and between groups C and D. The statistical analyses are summarized in Table 1.

Discussion

The results of the study show clear cut decreases of all investigated compounds in all groups of patients with PD as compared with healthy controls. In addition, significant differences were observed between the tremor-group and the akinesia-group with "on/off" phenomena. Aspartate was also found to be reduced in group C as compared with group B, while glutamine showed differences between the akinesia groups without and with "on/off" phenomena.

The described changes of aspartate, glutamate and glutamine in Parkinsonian patients are unexplained, but various mechanisms may play a role, i.e. transport, insufficient production and/or storage. The question may arise as to whether or not medication might be involved in the observed alterations of amino compounds. Recently,

Manyam et al. (1988) determined amino compounds in the CSF of patients with PD including aspartate, glutamate and glutamine without finding any significant differences as compared with controls. Carbidopa/Levodopa was not observed to have significant effects on the levels of the above compounds. Berl and Nicklas (1976) observed significant decreases of glutamine in brain slices of rats after administration of L-dopa while aspartate and glutamate remained unchanged. The supposition that the anti-Parkinson therapy is not responsible for the observed alterations of aspartate, glutamate and glutamine in platelets is supported by the observation that significant differences are not only found between the control group and the Parkinson-groups, but also between different groups of patients with PD. The various Parkinson-groups obtained comparable therapies. On the other hand, the results show a correlation between the severity of PD and the degree of decreases of the determined amino compounds. Nevertheless, the results may suggest but do not prove similar decreases of aspartate, glutamate and glutamine in the central nervous system of patients with PD.

Amyotrophic lateral sclerosis (ALS) represents another progressive degenerative disease of the motor system. Quantitative amino acid analysis of autopsied brains of patients with the sporadic form of ALS revealed decreases of glutamic acid contents in most regions of the brain. Aspartate was only decreased in the thalamus while glutamine was found to be unaltered (Perrry et al., 1987). Preliminary studies on platelets of patients with the sporadic form of ALS showed similar results with regard to aspartate and glutamate. Both amino acids were decreased but also glutamine was found to be reduced (Ludolph and Rolf, 1988). Disruption of the glutaminergic system also has been considered for some of the clinical manifestations of Alzheimer's disease (Rogers and Morrison, 1985). Thus, deficiencies of amino acid neurotransmitters and in particular glutamate may play a role in various disorders of the nervous system.

In addition, abnormally enhanced glutamatergic neurotransmission is considered to cause excitotoxic cell damage and lead to neuronal death (Olney et al., 1971). Such etiological mechanism has been postulated to exist in a special Western Pacific form of ALS combined with Parkinsonism and dementia (Spencer et al., 1987). The hypothesis centres on the consumption of the toxic seed of Cycas circinalis containing the glutamate analogue β-N-methylamino-L-alanine (BMAA). BMAA has been found to be a motor system toxin in primates. Repeated oral, administration of synthetic BMAA to male macaques, over weeks or months, induced signs of corticospinal dysfunction in the extremities, limb weakness, wrist drop and tremor, bradykinesia, masked facies and behavioral changes, with conduction deficits in central and peripheral motor pathways (Spencer et al., 1986 a).

On the whole, the presented study indicates that amino acid neurotransmission and in particular glutamatergic neurotransmission merits further investigations.

References

Berl S, Nicklas WJ (1976) Metabolism of glutamate and related amino acids in brain: effect of drugs that alter catecholamine metabolism. In: Birkmayer W, Hornykiewicz O (eds) Advances in parkinsonism. Editiones Roche, Basel, pp 193–204

Fagg CE, Foster AC (1983) Amino acid neurotransmitters and their pathways in the mammalian central nervous system. Neuroscience 9: 701–719

Gordon JL, Milner AJ (1976) Blood platelets as multifunctional cells. In: Gordon JL (ed) Platelets in biology and pathology. North-Holland, Amsterdam, pp 11–21

Hassler R (1966) Thalamus regulation of muscle tone and the speed of movements. In: Purpura DP, Yahr MD (eds) The thalamus. Columbia University Press, New York, pp 418–438

Hassler R, Nitsch C, Lee HL (1980) The role of eight putative transmitters in the nine types of synapses in rat caudate-putamen. In: Rinne UK, Klingler M, Stamm G (eds) Parkinson's disease—current progress, problems and management. Elsevier/North-Holland Biomedical Press, Amsterdam, pp 61–91

Lenda K, Svenneby G (1981) Rapid high performance liquid chromatographic determination of amino acids in synaptosomal extracts. J Chromatogr 198: 516–519

Lloyd KG, Möhler H, Hertz P, Bartholini G (1975) Distribution of choline acetyltransferase and glutamic acid decarboxylase within the substantia nigra and other brain regions from control and parkinsonian patients. J Neurochem 25: 785–789

Ludolph AC, Rolf LH (1988) Unpublished observations

Mangano RM, Schwarzc (1981) The human platelets as a model for the glutaminergic neuron. Platelet uptake of L-glutamate. J Neurochem 36: 1067–1076

Manyam BV, Ferraro TN, Hare TA (1988) Cerebrospinal fluid amino compounds in Parkinson's disease. Arch Neurol 45: 48–51

McGeer EG, Staines WA, McGeer PL (1984) Neurotransmitters in the basal ganglia. Can J Neurol Sci 11: 89–99

Olney JW, Ho OL, Rhee V (1971) Cytotoxic effects of acidic and sulphur containing amino acids on the infant mouse central nervous system. Exper Brain Res 14: 61–76

Perry TL, Hansen S, Jones K (1987) Brain glutamate deficiency in amyotrophic lateral sclerosis. Neurology 37: 1845–1848

Riederer P, Jellinger K (1983) Morphologie und Pathobiochemie der Parkinson-Krankheit. In: Gänshirt (Hrsg) Pathophysiologie, Klinik und Therapie des Parkinsonismus. Editiones Roche, Basel, S 31–50

Rogers J, Morrison JH (1985) Quantitative morphology and regional laminar distributions of senile plaques in Alzheimer's disease. J Neurosci 5: 2801–2808

Rolf LH, Schlake HP, Brune GG (1983) Plasmafaktoren und Migräne. In: Soyka D (Hrsg) Migräne: Pathogenese - Pharmakologie - Therapie. Enke, Stuttgart, S 79–97

Spencer PS, Nunn PB, Hugon J, Ludolph A, Roy DN (1986) Motor neurone disease on Guam: possible role of a food neurotoxin. Lancet i: 965

Spencer PS, Hugon J, Ludolph A, Nunn PB, Ross SM, Roy DN, Schaumburg HH (1987) Discovery and partial characterization of primate motor-system toxins. In: Ciba Foundation Symposium 126; Selective neuronal death. Wiley, Chichester, pp 221–237

Correspondence: Priv.-Doz. Dr. L. H. Rolf, Klinik und Poliklinik für Neurologie der Westfälischen Wilhelms-Universität, Albert-Schweitzer-Strasse 33, D-4400 Münster, Federal Republic of Germany.

Hypothalamic dysfunction and neuroendocrine research in Parkinson's disease

K.-P. Lesch

Department of Psychiatry, University of Würzburg, Federal Republic of Germany

Summary

The progress of discovery and the expansion of the principal science base of neuroendocrinology to comprise neural science as well as endocrinology include the role of neurotransmitters in the control of hypophysiotropic neurons, the concept of neurosecretion, and the chemical structures of hypothalamic releasing and inhibiting hormones. These advances have corresponded with the emergence of neurobiology of the CNS, with the understanding of the neuropeptides, and some comprehension of their relevance to brain function, their involvement in neuropsychiatric disorders including Parkinson's Disease (PD) and depression, and their introduction into diagnostic strategies and monitoring of treatment.

Introduction

The hypothalamus has historically been considered the pivotal brain region linking experimental neurosciences and clinical neurosciences, namely neurology, psychiatry and endocrinology. In the first place, clinical recognition of specific hypothalamic dysfunction initiated the research that led to the discovery of neuronal control over the pituitary. According to the concept of neurosecretion it was recognized that neurotransmitter-controlled neurosecretory cells terminating in the external layer of the median eminence are characterized by their release of messenger molecules directly into the systemic circulation. The isolation and identification of these hypothalamic hypophysiotropic hormones, primarily representing releasing and inhibiting hormones (e.g. thyrotropin-releasing hormone [TRH], gonadotropin-releasing hormone [GnRH], somatostatin [SRIH], corticotropin-releasing hormone [CRH] growth hormone-releasing hormone [GHRH]), has permitted recognition of functional intercellular regulatory connections at all levels of a neuroendocrine unit. These release and inhibiting hormones are also present in various suprahypothalamic structures as well as in endocrine tissues and the circulation, and have been shown to exert distinctive actions within defined regions of the CNS and spinal cord, thus coordinating complex physiological processes.

Neuroendocrine perturbation interventions with neurotransmitter agonists and antagonists and with synthetic hormones have been used to delineate the pathophysiology of neuropsychiatric disorders such as PD or depression, and to provide laboratory studies as aids in diagnosis and as tools to predict efficacy of pharmacological treatments. The conceptional basis of these studies is that neuroendocrine strategies provide a "window to the brain", since neuropsychiatric disease statuses may be accompanied by various hormonal dysregulations, and limbic-hypothalamic neurotransmitter systems involved in pituitary regulation have been hypothesized to be dysfunctional and to underline the pathogenesis of specific neuropsychiatric disorders (Beckmann and Lesch, 1989). Classical examples include the dopamine (DA) hypothesis of PD and the catecholamine hypothesis of affective disorders.

Hypothalamic-pituitary-lactotropic system

DA released into the hypophyseal portal circulation by the tuberoinfundibular DA (TIDA) neurons tonically inhibits prolactin (PRL) secretion via an action on D_2 receptors. Although inhibitory action of DA is important in determining basal concentrations, supplementary inhibiting (e.g. GnRH-associated peptide [GAP]) and releasing factors (e.g. TRH, vasoactive intestinal peptide [VIP]) have been identified. In addition to its negative "short loop" feedback regulation, PRL secretion may be modulated by central serotonergic, cholinergic, γ-aminobutyric acid (GABA)-ergic and opioidergic mechanisms.

PD is characterized by progressive destruction of DA neurons in the substantia nigra and of their terminals in the caudate and putamen. Correlated with this loss is a depletion of DA, homovanillic acid (HVA), and tyrosine hydroxylase activity in substantia nigra and striatum (Birkmayer and Riederer, 1985). Postmortem studies have shown that postsynaptic DA receptor function is also defective.

Although striatal DA deficiency is crucial to the pathogenesis of PD, neuronal and biochemical lesions are not limited to the nigrostriatal system but involve also the mesolimbic and mesocortical system, and the hypothalamus. Lewy bodies, whose presence in pigmented brainstem nuclei is an invariable feature of PD, are also found in the posterior and lateral hypothalamus. In this respect, the reduction in striatal DA may arise from either selective vulnerability of the nigrostriatal unit, or may be due to a generalized disorder of catecholamine metabolism involving multiple dopaminergic systems. Therefore, evaluation of hypothalamic-pituitary-lactotropic (HHL) system function may be a convenient way to distinguish between localized or generalized dopaminergic dysfunction in PD, and PRL secretion may be used as an indicator of D_2 receptor function.

Eisler et al. (1981) reported that resting levels of PRL, TRH-evoked PRL responses, and suppression of PRL concentration elicited by L-dopa were normal in drug-free patients with PD, while responses to TRH were attenuated after administration of the D_2 receptor agonist bromocriptine (BCT) or L-dopa plus carbidopa. They hypothesized that the enhanced action of DA-mimetic drugs may arise from "up regulation" of D_2 receptors on lactotroph cells as an adjustment to chronically deficient DA release by the TIDA system, thus providing evidence for the existence of extrastriatal dysfunction of dopaminergic systems and supporting reports of decreased DA but unchanged norepinephrine (NE) and serotonin (5-HT) concentrations in the hypothalamus of patients with PD (Conte-Devolx et al., 1985). In a corroborative study by Laihinen and Rinne (1986) a deficient TRH-induced PRL response after BCT treatment in recent-onset parkinsonian patients and increased PRL release following in advanced patients with daily fluctuations in disability as compared with controls was found, while advanced patients without fluctuations exhibited normal PRL responses. In accordance with postmortem studies which demonstrate reduced D_2 receptor concentrations in parkinsonian patients with fluctuations, they suggested that BCT-induced desensitization of D_2 receptors may play a role in the pathogenesis of fluctuations, concluded that HPL function is an indicator of the functional state of central D_2 receptors, and that PRL challenge tests can be used to study these receptors in PD. If the dopaminergic dysfunction in the regulation of the HHL system will be confirmed in future studies, the point at issue in elucidating the etiology of PD will be to determine what makes dopaminergic neurons selectively vulnerable rather than to work out what makes the nigrostriatal pathway selectively susceptible to destruction (Eisler et al., 1981).

Hypothalamic-pituitary-somatotropic system

Growth hormone (GH) release is regulated by a noncompetitive antagonism of GHRH and SRIH, with the effects of other hormones (e.g. glucocorticoids, thyroid hormones, gonadal steroids) on GH secretion apparently mediated via alterations in the secretion of or sensitivity to GHRH and/or SRIH. Feedback inhibition of GH secretion may depend on GH itself or may be mediated by GH-induced somatomedin production. Both GH and somatomedins may feed back at the level of the pituitary directly inhibiting GH release and/or at the level of the hypothalamus stimulating SRIH secretion and possibly suppressing GHRH output.

SRIH influences the release and/or turnover of brain cholinergic and monoaminergic neurotransmitters and, in turn, its release is decreased by NE, 5-HT and GABA, and is stimulated by acetylcholine (ACh) and possibly DA. The physiologic importance of these interactions is supported by the neuronal co-localization of SRIH with a variety of neuroactive substances throughout specific brain structures. In addition to its role as a neuromodulator, SRIH appears capable of inhibiting the secretion of a wide array of hormones.

Recently, peptides with GH-releasing activity isolated from human pancreatic islet cell tumors have been shown to represent the postulated hypothalamic GHRH. Unlike other hypothalamic neuropeptides, GHRH displays a far more restricted central distribution, while its regulatory effects seem to be circumscribed and are specific for GH release. GHRH release is controlled by a complex regulatory mechanism involving several neurotransmitter systems. While NE is stimulatory via α_2- and inhibitory via β_2-adrenoceptors, GHRH secretion is also influenced by serotonergic, dopaminergic, histaminergic and opioidergic innervation.

Observations of decreased CSF SRIH concentrations in severe PD and in depression have been consistently replicated, while in the frontal cortex, hippocampus, entorhinal cortex of PD patients SRIH concentrations are reduced only in those with prominent dementia. However, no variations in SRIH and GHRH contents were found in parkinsonian hypothalami (Conte-Devolx et al., 1985).

Although studies of GHRH in neuropsychiatric disorders and, in particular, of GHRH-stimulated GH secretion in PD are in relatively early stages, aberrant hypothalamic-pituitary-somatotropic (HPS) function as indicated by circadian rhythm disturbances and modified responsiveness to pharmacological challenges have been described. Although normal basal GH concentrations have been reported in untreated PD subjects, decreased episodic secretion as assessed by 24-hour profiles was noted. In PD patients treated with L-dopa a episodic pattern of GH release with secretory pulses, which occur during the daytime hours, was manifest, while only male patients exhibited sleep-associated GH secretion which was not affected by BCT treatment. Challenge tests with dopaminergic agents revealed contradictory findings in parkinsonian patients on chronic DA-mimetic treatment. While Parkes et al. (1976) reported a consistent GH response to L-dopa in 24 out of 26 patients treated with L-dopa for 1 to 5 years, Malarkey et al. (1974) found only 1 of 10 patients treated for several years to be responsive to L-dopa. Further evidence of deranged HHS function was provided by an unexpected suppression of GH following L-dopa in 3 patients, and a paradoxical GH rise after glucose administration in 2 patients. None of these subjects had a defective GH response to insulin-induced hypoglycemia. Martinez-Campos et al. (1981) reported that L-dopa plus benserazide induced an inconsistent rise in GH in patients with PD and controls but a clear-cut and increased GH-releasing effect in PD subjects receiving these drugs chronically. They hypothesized that the more consistent GH response to L-dopa plus benserazide in patients under long-term treatment may be a consequence of "up regulation" of hypothalamic DA receptors.

In conclusion, data appearing in the literature provide no clear evidence for distorted HHS function in PD. L-dopa appears to be a more consistent stimulus for GH release in PD than direct D_2 receptor agonists such as BCT which elicit comparable therapeutic effects (Parkes et al., 1976). However, L-dopa can also be converted to NE thus causing α_2-adrenoceptor-mediated GH releasing effects.

Hypothalamic-pituitary-adrenal system

The function of the hypothalamic-pituitary-adrenal (HPA) axis can be regarded as a model system demonstrating the way that the CNS and the glandular secretory system interact to coordinate and maintain vital homeostatic functions, the response to stress, and regulation of circadian rhythms.

Hypothalamic production and release of corticotropin-releasing hormone (CRH), which has been identified as the primary neuroregulator of the secretion of pro-opiomelanocortin (POMC)-derived peptides, is controlled by various neurotransmitter systems acting through the hypothalamic neuron control network. ACh and 5-HT, possibly via a cholinergic interneuron, are stimulatory, GABA and catecholamines are inhibitory to CRH release, though there may be a stimulating α_1-adrenergic regulation pathway. At the level of the pituitary CRH interacts synergistically with arginine vasopressin (AVP) and NE/E. In addition, numerous neuropeptides (opioid peptides, SRIH, VIP etc.) exert a modulatory influence on ACTH secretion at the level of the hypothalamus and/or pituitary. Opposed to the stimulating effects of CRH, AVP and NE/E is the glucocorticoid-dependent negative feedback regulation which may be mediated by glucocorticoid receptors (GR) either on corticotroph cells or in various brain structures such as hippocampus, septum, and amygdala. Thus, the CNS integrates neural factors and feedback mechanisms controlling circadian and ultradian rhythms (closed loop control) and the open loop control due to stress.

Comparative studies between normal subjects and patients with PD revealed no difference in hypothalamic POMC-derived peptide and CRH concentrations (Conte-Devolx et al., 1985). However, hyperactivity of the HPA system as expressed by 24-hour cortisol secretory patterns, cortisol nonsuppression in response to dexamethasone, and attenuation of ACTH responses to CRH, is a state-dependant biological factor, observable in a substantial subgroup of patients with depression. It has been postulated that the hypercortisolism in depression is due to GR dysfunction which leads to an alteration in the set-point for feedback inhibition of glucocorticoids and/or to neurotransmitter dysregulation-induced hypersecretion of CRH (Lesch and Rupprecht, 1989). Depression has also been reported in 25 to 48% of patients with PD, but the nature of this frequent association remains controversial (Fuchs et al., 1987). Although decreased CSF 5-hydroxyindoleacetic acid (5-HIAA) concentrations which were correlated with psychomotor retardation and loss of self-esteem have been reported, the severity of depression and PD are unrelated (Mayeux et al., 1986). In a large proportion of patients onset of depression may precede the symptoms of PD, an observation that may have therapeutic and prognostic implication. Although the ability of the dexamethasone suppression test (DST) to differentiate central degenerative disease from depression is poor, it has been used to investigate HPA system dysfunction in depression associated with PD. Mayeux et al. (1986) reported a major depressive disorder in 14 and a dysthymic disorder in 7 among 49 consecutive patients with PD. Although 22 patients failed to suppress, the DST was not capable to distinguish between depressed (6/21) and non-depressed patients (11/28) with PD. In addition, no relationship between cortisol nonsuppression to 1 mg dexamethasone and severity of depression was demonstrated in 20 depressed "de novo" parkinsonian patients with 25% nonsuppressors (Fuchs et al., 1987). Since HPA system integrity depends on aminergic-cholinergic interactions, evidence that a reduction in brain NE occurs in PD and that the 24-hour cortisol secretion decreases with aging may account for why suppression fails in nondepressed patients with PD and elderly subjects (Mayeux et al., 1986).

Thus, the DST failed either to be a specific marker for PD-associated depression or to be of diagnostic validity. Although studies of disturbances of CRH-stimulated ACTH and cortisol release and of GR dysfunction in neuropsychiatric disorders are still in their infancy, introduction of these strategies into investigations of HPA system regulation may provide new clues to the pathophysiology of depression associated with PD.

Conclusions

From the above review, evidence emerges that abnormalities in HPL, HPS and HPA system function occur in specific subgroups of PD and may be elicited and studied by a variety of provocative challenge tests. Though multiple interacting variables have to be controlled, neuroendocrine strategies may proof valuable in understanding the pathogenesis of PD and in developing standardized laboratory tests to aid in diagnosis and monitoring of treatment.

Acknowledgements

The authors thank U. Müller for help in compilation of the bibliography, Dr. W. Kuhn for reviewing the manuscript, and S. Groß-Lesch for her secretarial assistance.

References

Beckmann H, Lesch KP (1989) Neurochemische Untersuchungen in der Psychiatrie. In: Kisker KP, et al (Hrsg) Psychiatrie der Gegenwart, Bd IX. Springer, Berlin Heidelberg New York Tokyo (im Druck)

Birkmayer W, Riederer P (1985) Die Parkinson-Krankheit. Biochemie, Klink, Therapie, 2. Aufl. Springer, Wien New York

Conte-Devolx B, Grino M, Nieoullon A, Javoy-Agid F, Castanas E, Guillaume V, Tonon MC, Vaudry H, Oliver C (1985) Corticotropin, somatocrinin and amine contents in normal and parkinsonian human hypothalamus. Neuroscience Lett 56: 217–222

Eisler T, Thorner MO, MacLeod RM, Kaiser DL, Calne DB (1981) Prolactin secretion in Parkinson disease. Neurology 31: 1356–1359

Fuchs G, Maurer K, Kuhn W, Przuntek H (1987) Depression bei Morbus Parkinson. In: Beckmann H, Laux G (Hrsg) Biologische Psychiatrie. Springer, Berlin Heidelberg New York Tokyo, S 185–187

Malarkey WB, Cyrus J, Paulson GW (1974) Dissociation of growth hormone and prolactin secretion in Parkinson's disease following chronic L-dopa therapy. J Clin Endocrinol Metab 39: 229–235

Martinez-Campos A, Giovannini P, Cocchi D, Zanardi P, Parati EA, Caraceni T, Müller EE (1981) Growth hormone secretion in neurological disorder. In: Martin JB, Reichlin S, Bick KL (eds) Neurosecretion and brain peptides. Raven, New York, pp 521–540

Mayeux R, Stern Y, Williams JBW, Cote L, Frantz A, Dyrenfurth I (1986) Clinical and biochemical features of depression in Parkinson's disease. Am J Psychiatr 143: 756–759

Laihinen A, Rinne UK (1986) Function of dopamine receptors in Parkinson's disease: prolactin responses. Neurology 36: 393–395

Lesch KP, Rupprecht R (1989) Psychoneuroendocrine research in depression: hormonal responses to releasing hormones as a probe for hypothalamic-pituitary-endorgan dysfunction. J Neural Transm (in press)

Parkes JD, Debono AG, Marsden CD (1976) Growth hormone response in Parkinson's disease. Lancet i: 483–484

Correspondence: Dr. K.-P. Lesch, Department of Psychiatry, University of Würzburg, Füchsleinstrasse 15, D-8700 Würzburg, Federal Republic of Germany.

The MPTP model: an update

H. Russ[1], W. Mihatsch[1], and H. Przuntek[2]

[1]Department of Neurology, University of Würzburg
[2]Department of Neurology, St. Josef University-Hospital, Bochum, Federal Republic of Germany

Summary

MPTP is a neurotoxic agent which produces a parkinsonlike syndrome in humans and non-human primates by selective destruction of dopaminergic neurons in the nigrostriatal system. The two actual views on the mechanism of MPTP toxicity are the free radical theory and the theory of blocked energy metabolism. The chronic MPTP syndrome in old monkeys is in spite of some disagreements the best model of Parkinson's disease which is available. The main experimental problems of the chronic MPTP model are the nutrition of the monkeys, the homogenity of the groups and the scoring of motor deficiency. The MPTP research results in new aspects for aetiology and treatment of Parkinson's disease.

Introduction

In 1979 Davis et al. reported the case of a young drug abuser who developed a severe and irreversible Parkinson syndrome after self administration of a synthetic meperidine analogue. The analysis of the drug showed that it was contaminated with the toxic 1-methyl-4-phenyl-1,2,3,6-tetrahydropyridine (MPTP) which destroys the dopaminergic neurons of the substantia nigra. Subsequently, similar cases of MPTP intoxicated patients with rigidity, tremor and bradykinesia have been described (Langston, 1983).

The neurotoxic effect of MPTP is based on its metabolism by monoamine oxidase type B (MAO-B) (Chiba, 1984). This enzyme, which is identical with the [^{3}H]MPTP receptor binding site, forms the oxidation product 1-methyl-4-phenyl-2,3-dihydropyridinium ion ($MPDP^+$). In a second step the intermediate product is converted, independently of MAO-B, to the major metabolite 1-methyl-4-phenylpyridinium ion (MPP^+). MPP^+, which is thought to be the principal toxic agent, has a high affinity for the catecholamine uptake carrier and thus accumulates in dopaminergic neurons.

Two major possibilities to block the neurotoxic effect of MPTP result from this mechanism. First: preventing the formation of MPP^+ via MAO-B inhibitors. Second: blocking the transport of MPP^+ into dopaminergic neurons by use of dopamine uptake inhibitors.

Investigations with different animal species showed that the sensitivity to MPTP can only be compared between monkeys and humans. Rats' response to high MPTP doses is unspecific. This could be caused by the melanin-free substantia nigra and the low MAO-B content in rat brain tissue (Przuntek, 1985).

The biochemical induction of neurone degeneration as well as the possible relation to the etiology of idiopathic Parkinson's disease are questions which presently are not definitively clarified. Two actual theories try to explain the biochemical mechanism: formation of cytotoxic superoxid radicals and blocking of energy metabolism.

Free radical theory

It has been suggested that free radicals which have their origin in the autooxidation of dopamine play an important role in the etiology of idiopathic Parkinson's disease.

The metabolism of MPTP via MAO-B also promotes the formation of radicals derived from hydrogen peroxide (e.g. O_2^-, $^{\cdot}OH$) and accelerates dopamine autooxidation, especially in the presence of the transition metal ions and the neuromelanin concentrated in the substantia nigra. Selective damage of these neurons could be explained by the accumulation of MPP^+ in the nigrostriatal dopaminergic system caused by the high density of dopamine uptake sites.

Although the formation of free radicals was proved in vivo by investigation of the redox systems of gluthatione and ascorbic acid (Riederer, 1987), the importance of these effects in the action of MPTP is uncertain, because radical detoxifying agents only partially protect rodents from MPTP toxicity. As far as monkeys (callithrix jacchus) are concerned we found that their sensitivity for MPTP was not altered by long-term application of high doses of antioxidant vitamines (2350 mg alpha-tocopherole plus 100 mg ascorbic acid per kg and day) (Russ, 1987).

Theory of blocked energy metabolism

An other current view on the mechanism of MPTP neurotoxicity is the inhibitory effect on the NADH ubiquinone oxidoreductase (Nicklas, 1985). MPP^+ is actively transported into mitochondria and produces a complete depletion of intracellular ATP by blocking the above mentioned enzyme of aerobic energy metabolism. However, this effect is not specific for nigrostriatal neuron mitochondria. It is assumed that the selectivity of neurodegeneration only results from the ability of dopaminergic neurons to trap MPP^+.

Furthermore, the cytotoxic properties of MPTP could be connected with inhibition of tyrosine hydroxylase (Hirata, 1985) as well as aklylation of specific cellular proteins by electrophilic $MPDP^+$ (Corsini, 1986).

Animal model of Parkinson's disease

At the present time the chronic MPTP syndrome in primates is the best model of Parkinson's disease which is available. The agreement between model and Parkinson's disease is most distinct when considering biochemical features.

Histopathological findings also correspond in a large measure. But the main question is whether pathognomic Lewy bodies appear in MPTP-lesioned brain tissue, and this has not yet been definitively answered. In one human case (Davis, 1979) as well as in subhuman primates intraneuronal inclusions similar to Lewy bodies have been observed. Thus, Forno et al. (1986) found that eosinophilic inclusions especially in locus ceruleus neurons appeared in older squirrel monkeys with MPTP-lesioned substantia nigra if a protracted MPTP regimen was used. In addition, these characteristic lesions arose in substantia nigra, dorsal motor nucleus of the vagus, nucleus basalis of Meynert and in the dorsal raphe nucleus. These are precisely the areas in which Lewy bodies are found in human Parkinson's disease.

Agreement of primate model and Parkinson's disease

The absence of Lewy bodies or similar abnormalities in young monkeys treated by a single dose of MPTP has been one of the two main criticisms of the model. But the reported eosinophilic inclusions suggest that similarities between the neuropathology of the chronic MPTP syndrome and Parkinson's disease are greater than recently thought, if relatively old monkeys are examined.

The other main point of criticism is the difference in motor disturbances between MPTP syndrome and Parkinson's disease. Resting tremor for example has rarely been observed in seven cases of MPTP-intoxicated drug abusers and it seems improbable that a Parkinson-like tremor is induced in subhuman primates. During our investigations with marmosets a resting tremor has never been observed (Mihatsch, 1987). Furthermore, the bradykinesia and rigidity caused by MPTP in monkeys is not progressive as in Parkinson patients, but on the contrary, symptoms improve after MPTP withdrawal.

Experimental problems of the chronic MPTP syndrome

In an experiment with four marmosets we tried to induce a chronic MPTP syndrome with a long series of MPTP doses as represented in Fig. 1. From the 8th MPTP administration on every injection was followed by acute effects such as mydriasis, piloerec-

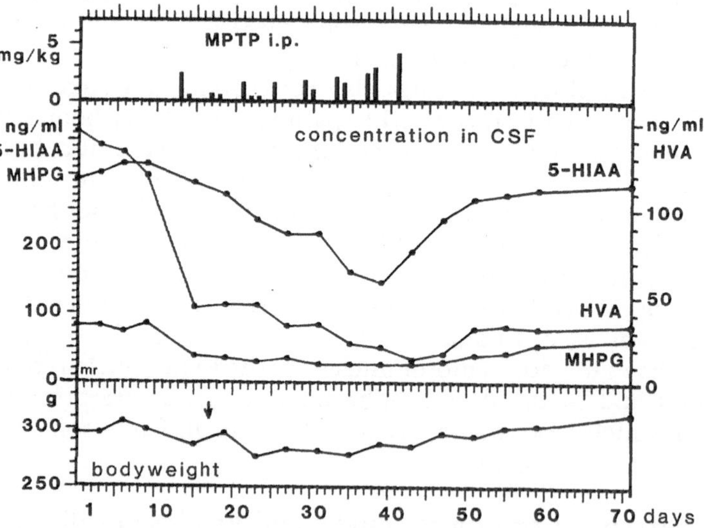

Fig. 1. Cerebrospinal fluid levels of biogenic amine metabolites in relation to MPTP application. From the 12th to the 41th day of the experiment four marmoset were treated intraperitoneally with 15 increasing doses of MPTP as shown in the upper part of the figure (total dose 25 mg/kg). On every fourth day and at the end of the experiment cerebrospinal fluid was obtained and 5-HIAA, HVA and MHPG, the major metabolites of serotonin, dopamine and norepinephrine were determined (each point represents the mean of four subjects). The arrow indicates the start of intragastric feeding

tion, salivation and periods of convulsions. Thereafter a short period of bradykinesia and rigid posture was observed. From the 12th MPTP injection onwards a permanent state of bradykinesia and rigidity developed, but four days after MPTP treatment the marmosets quickly recovered. The only persistant symptom at the end of the experiment seemed to be a reduced coordination of rapid movements, although the cerebrospinal fluid level of HVA remained very low. The post mortem examination of dopamine contents in the brain showed a 75% loss in the caudate nucleus and a more than 90% loss in the putamen compared with an untreated control group.

Another problem of the primate model is that the monkeys stop eating with the beginning of the MPTP application, even when motor deficiencies are still absent. In longterm experiments an intragastric nutrition is necessary to prevent loss of body weight. In Fig. 1 the arrow indicates the start of intragastric feeding with a high caloric diet. The starving of the MPTP treated animals could be caused by an anorectic effect caused by MAO inhibition or by nausea, or possibly by hypothalamic lesions.

A further difficulty is that it is nearly impossible to get a homogeneous group of monkeys (age, weight, breed). Thus, when MPTP is applied in equal doses related to body weight the animals respond with variable degrees of motor impairment. Therefore in therapeutic experiments MPTP should be injected until each monkey has reached a definite level of extrapyramidal motor deficiency. But hitherto no score exists which is based on the observation of motor signs. However, the use of video monitoring (Bakay, 1985) or cages with light sensors allow one to examine changes in locomotor activity. This seems to be a approximate measure of the extrapyramidal affection.

Conclusion

Up to now neither a MPTP-like endogenous nor an exogenous substance has been found which may be operative in idiopathic Parkinson's disease. Nevertheless, a better understanding of the mechanism of MPTP toxicity could show new aspects of the etiology of neurodegenerative processes.

Further research efforts should be directed towards a standardisation of the chronic MPTP syndrome. This is a necessary prerequisite for effective therapeutical investigations and it would allow the comparison of the results from different authors.

References

Bakay RAE, Fiandaca MS, Barrow DL, Schiff A, Collins DC (1985) Preliminary report on the use of fetal tissue transplantation to correct MPTP-induced Parkinson-like syndrome in primates. Appl Neurophysiol 48: 358–361

Chiba K, Trevor A, Castagnoli N (1985) Metabolism of the neurotoxic amine, MPTP by brain monoamine oxidase. Biochem Biophys Res Comm 120: 574–578

Corsini GU, Pintus S, Bocchetta A, Piccardi MP, Del Zompo M (1984) In vitro MAO type B dependent formation of an alkylating metabolite from ^{3}H-MPTP in rat and monkey. In: Markey SP, Castagnoli N, Trevor AJ, Kopin IJ (eds) MPTP—A neurotoxin producing a parkinsonian syndrome. Academic Press, New York, pp 371–376

Davis GC, Williams AC, Markey SP, Ebert MH, Calne ED, Reichert CM, Kopin IJ (1979) Chronic parkinsonism secondary to intravenous injection of meperidine analogues. Psychiatr Res 1: 249–254

Forno LS, Langston JW, DeLanney LE, Irwin I, Ricaurte GA (1986) Locus ceruleus lesions and eosinophilic inclusions in MPTP-treated monkeys. Ann Neurol 20: 449–455

Hirata Y, Nagatsu T (1985) Inhibition of tyrosin hydroxylation in tissue slices of the rat striatum by 1-methyl-4-phenyl-1,2,3,6-tetrahydropyridine. Brain Res 337: 193–196

Langston JW, Ballard P, Tetrud JW, Irwin I (1983) Chronic parkinsonism in human due to a product of meperidine-analog synthesis. Science 219: 979–980

Mihatsch W, Russ H, Przuntek H (1987) The effect of a protracted 1-methyl-4-phenyl-1,2,3,6-tetrahydropyridine (MPTP)-application on marmoset. In: Nappi G, Agnoli A, Fariello R (eds) Neurotoxicology—basic and clinical research. International symposium on neurotoxicology, Turin, 1987. CIC Editioni Internationali

Nicklas WJ, Vyas I, Heikkila RE (1985) Inhibition of NADH-linked oxidation in brain mitochondria by 1-methyl-4-phenyl-1,2,3,6-tetrahydropyridine. Life Sci 36: 2503–2508

Przuntek H, Russ H, Henning K, Pindur U (1985) The protective effect of 1-tert.butyl-4,4-diphenylpiperidine against the nigrostriatal neurodegeneration caused by 1-methyl-4-phenyl-1,2,3,6-tetrahydropyridine. Life Sci 37: 1195–1200

Riederer P, Youdim MBH (1987) MPTP induced dopaminergic neurotoxicity—a useful model in the study of Parkinson's disease. Neurochem Int 11: 379–381

Russ H, Mihatsch W, Kuhn W, Przuntek H (1987) MPTP model of Parkinson's disease in marmoset: radical scarvengers do not prevent neurodegeneration. J Neurochem 48 [Suppl]: 159

Correspondence: Dr. H. Russ, Department of Neurology, University of Würzburg, Josef-Schneider-Strasse 11, D-8700 Würzburg, Federal Republic of Germany.

Histochemistry of MAO subtypes in the brainstem of humans: a relation to the radical hypothesis of Parkinson's disease?

C. Konradi, P. Riederer, and H. Heinsen

Clinical Neurochemistry and Morphological Brain Research,
Department of Psychiatry, University of Würzburg, Federal Republic of Germany

Summary

The distribution of monoamine oxidase subtypes in the human brainstem is described. Results are discussed in relation to the MAO-B inhibiting therapy of Parkinson's disease (PD). Additionally, attention is focused onto the pathogenesis of PD, especially on radical mechanisms which might be mediated via exotoxins and amplified by endogenous radicals.

Introduction

The main pathological mechanism in Parkinson's disease (PD) is characterized by a loss of neurons in the substantia nigra together with a loss of dopamine in this region and its striatal projection areas. Monoamine oxidase (MAO) existing in two subtypes named MAO-A and B, is the main intracellular metabolizing enzyme of dopamine. Blockade of MAO-B by the selective and safe inhibitor l-deprenyl has been shown to be beneficial in the treatment of PD (Birkmayer et al., 1977). It has been concluded from these clinical observations that MAO-B is located in the neurons of the substantia nigra and their axons and terminals projecting to the striatum. Inhibition of this MAO-subtype followed by an increase in neuronal dopamine may account for the positive effect in PD. There is, however, little experimental evidence for this assumption. Therefore, we investigated the distribution of both MAO-subtypes and the cellular site of action of MAO inhibitors in the human brain stem by use of a histochemical technique. Additional investigations were performed to elucidate pathogenic effects of endogenous radicals. Such experiments together with knowledge about the neurotoxin 1-methyl-4-phenyl-1, 2, 3, 6-tetrahydropyridine (MPTP) make it worthwhile to discuss possible preventive therapies of PD.

Materials and methods

Ten human brain stems (6 male/4 female) were used for histochemical investigations, the age ranging from 3 months to 70 years (mean ± SD was 27.6 ± 26.6 years). Patients were without neurological disorders and died by traffic accidents [3], murder [1], sudden infant death [3], suicide by drug-overdose [1], suicide by motor exhaust [1] or by drowning [1]. The post mortem time ranged from 24 to 48 hours.

The histochemical method is based on the generation of hydrogen peroxide via MAO metabolism. The assay consisted of MAO substrate, horseradish peroxidase, 3,3'-diaminobenzidine, sodium azide, MAO inhibitors (if desired) and buffer. The oxidation of MAO substrate is accompanied by a generation of hydrogen peroxide, which is metabolized by added horseradish peroxidase. The latter reaction binds 3,3'-diaminobenzidine, which gives a visible brown precipitate in the respective area (Kishimoto et al., 1983).

The effect of hydrogen peroxide on MAO activity (Table 2; Table 3) was studied in human brain homogenates of parietal cortex and putamen. Tissue was obtained at autopsy from 3 patients (2 female; 1 male). Mean age was 77 ± 5 years (mean ± SD), post mortem time ranged from 4 to 6 hours. Diagnosis was confirmed at autopsy, patients died without neurologic disease. Dissected brain areas were frozen at −70° C for several weeks. Estimation of MAO activity is described elsewhere (Konradi et al., 1986).

Statistics were performed using a Wilcoxon rank sum test, two-tailed approach.

Results and discussion

In contrast to a significant staining and response to respective inhibitors of LC, DNR and GC (Table 1), the most surprising finding was the lack of MAO staining in neurons of the substantia nigra. This could be due either to only a small amount of MAO in those neurons (below the detection limit of the method) or to a total lack of MAO. In contrast, glial cells of this area stained well for MAO, predominantly MAO-B. The question, which arises is about the mode of action of the MAO-B inhibitor l-deprenyl, which is known to be of therapeutic benefit in PD (Birkmayer et al., 1977) and which enhances the probability for an increased life expectancy in Parkinsonian patients (Birkmayer et al., 1985).

Table 1. Human brain areas with stained neurons in presence of the respective substrates and inhibitors

	Serotonin	β-phenylethylamine	Tyramine
Control	LC	DNR, GC	LC, DNR, GC
l-deprenyl	LC	No staining	LC
Clorgyline	No staining	DNR, GC	DNR, GC

DNR dorsal nucleus of raphe; *LC* Locus coeruleus; *GC* glial cells; Inhibitor concentration for the irreversible inhibitors l-deprenyl and clorgyline was 100 μmol, the substrate concentration was 1 mmol

Table 2. Effects of hydrogen peroxide on human brain MAO activity

	PEA oxidation (%)	5-HT oxidation (%)
Homogenates from:	(MAO-B)	(MAO-A)
Parietal cortex	335 ± 67*	82 ± 4*
Putamen	210 ± 17*	79 ± 3*

mean ± SD; data show activity as percent of controls without hydrogen peroxide; $n = 3$; each determination in duplicate; for homogenates see materials and methods; * $p < 0.05$

Several answers for this question are possible:

1) l-Deprenyl inhibits MAO in glial cells, dopamine is accumulated and has hormone-like effects. Nevertheless, an important and limiting factor for this assumption could be the extracellularly located enzyme catechol-O-methyltransferase (COMT).

2) Other monoamines like β-phenylethylamine (PEA), which are not metabolized by COMT, are accumulated and might have stimulatory actions on dopaminergic cells (Riederer et al., 1984).

3) An antidepressant effect of l-deprenyl by inhibition of intraneuronal serotonin (5-HT) decomposition e.g. in the dorsal nucleus of raphe might play an additional role.

4) The increase in life expectancy might be caused by

a) an inhibition of MAO-derived generation of hydrogen peroxide or by

b) a prevention of the degeneration of striatal dopaminergic neurons caused by uptake and accumulation of toxins generated via MAO metabolism, as could be shown for the neurotoxin MPTP which is able to produce Parkinson-like symptoms (Davis et al., 1979).

The last mentioned possibility is closely linked to radical metabolism. Therefore we studied the action of hydrogen peroxide on MAO activity. The results are shown in Table 2.

Table 3. Kinetics of MAO-A and MAO-B in presence of hydrogen peroxide in homogenates of human parietal cortex

	PEA (MAO-B)		5-HT (MAO-A)	
	K_m	V_{max}	K_m	V_{max}
Controls	1.62	0.71	135	2
H_2O_2 10^{-3} M	12.5*	2.6*	1016*	7.6*

n = 2; each determination in duplicate; for homogenates see materials and methods; * $p < 0.01$

Table 4. MAO-B activity in platelets of Parkinsonian patients

Parkinsonian patients	0.275 ± 0.101
Controls	0.210 ± 0.056*

Data in nmol · min^{-1} · mg $protein^{-1}$; mean ± SD; Both groups were age and sex matched; n = 18; 14 female/4 male; age Parkinsonian patients: 73.9 ± 6.2 (mean ± SD); age control group: 73.9 ± 10.0; *$p = 0.05$

Surprisingly, hydrogen peroxide stimulated MAO-B activity and decreased MAO-A activity (Table 2). Lineweaver-Burk-Plots for both MAO subtypes showed an increase in activity (V_{max}) and a decrease in substrate affinity (increase in K_M), which has different consequences for the entire activity given as nmol · min^{-1} · mg $protein^{-1}$ (Table 3). Enhancement in the entire activity of MAO-A in presence of hydrogen peroxide is given only at substrate concentrations > 1 mmol. This means, that in vivo defects in radical scavenging mechanisms could be able to enhance MAO-B activity but not MAO-A.

A radical hypothesis for PD could therefore be forced by a change in MAO activity in those patients. In fact, measurement of MAO-B activity in platelets of PD patients which were not treated with l-deprenyl (but which were treated with l-DOPA), shows enhancement when compared to controls (Table 4; data from Danielczyk et al., 1988).

These data may account for defects in radical scavenging processes in PD and they are forced to our attention by studies with the neurotoxin MPTP. In PD, glutathione has been shown to be decreased (Perry et al., 1982); Riederer et al., 1988). Furthermore, the activity of glutathione peroxidase is extremely dependent on its substrate glutathione.

Corresponding to PD in humans, mice treated with a single dose of MPTP showed significantly decreased glutathione content in the brainstem (Yong et al., 1986). Pretreatment with antioxidants proofed a partial protection from dopamine reduction in presence of MPTP (Perry et al., 1985).

MPTP per se is not able to destroy dopaminergic neurons, but it is metabolized by MAO-B to 1-methyl-4-phenylpyridine (MPP^+), a radical, which acts as neurotoxic agent. MAO-B inhibitors as well as dopamine uptake inhibitors are able to prevent the neurotoxic effects of MPTP (Heikkila et al., 1985; D'Amato et al., 1987). Therefore it is supposed that MPTP is converted via MAO-B in glial cells to MPP^+ which enters the neuronal cells via a high affinity dopamine uptake system (for review see: Kinemuchi et al., 1987). Within neurons, MPP^+ binds to neuromelanin (D'Amato et al., 1987). Thus, by providing a storage of the toxin in dopaminergic, neuromelanin-containing cells, neurons of the substantia nigra are gradually destroyed.

MPP^+ may be able to stimulate MAO-B activity similar to hydrogen peroxide. The oxidation of MPTP by MAO-B is surprisingly rapid, while tertiary amines are thought to be oxidized only very slowly by either MAO subtype (Singer et al., 1986). Thus, in this Parkinson model the MAO-B inhibitor l-deprenyl may have an additional action in preventing MAO-B activation.

It is not clear whether idiopathic PD is caused by an exotoxin of unknown origin. In addition, endogenous radicals may be responsible to trigger PD or may facilitate PD caused by exogenous toxins. There is a lack of evidence how realistic these hypotheses are. In any case a genetic theory seems less realistic. A change in MAO activity which can be observed in the presence of hydrogen peroxide and in platelets of Parkinsonian patients may account for a radical theory.

Therefore, a preventive therapy could be an administration of antioxidants with or without l-deprenyl. A long-term study in healthy volunteers receiving antioxidants, l-deprenyl or both and a comparison of the incidence rate of PD in these groups with the normal population might confirm a hypothesis of an environmental toxin similar to MPTP. In any case, a study of these proportions with regard to time and number of volunteers seems not to be feasible at present.

References

Birkmayer W, Riederer P, Ambrozi L, Youdim MBH (1977) Implications of combined treatment with "Madopar" and l-deprenyl in Parkinson's disease. Lancet i: 439–443

Birkmayer W, Knoll J, Riederer P, Youdim MBH, Hars V, Marton J (1985) Increased life expectancy resulting from addition of l-deprenyl to

Madopar® treatment in Parkinson's disease: A long-term study. J Neural Transm 64: 113–127
D'Amato RJ, Alexander GM, Schwartzmann RJ, Kitt CA, Price DL, Snyder SH (1987) Neuromelanin: A role in MPTP-induced neurotoxicity. Life Sci 40: 705–712
Danielczyk W, Streifler M, Konradi C, Riederer P, Moll G (1988) Platelet MAO-B activity and the psychopathology of Parkinson's disease, senile dementia and multiinfarction dementia. Acta Psychiat Scand (in press)
Davis GC, Williams AC, Markey SP, Ebert MH, Calne ED, Reichert CM, Kopin IJ (1979) Chronic parkinsonism secondary to intravenous injection of meperidine analogous. Psychiat Res 1: 249–254
Heikkila RE, Hess H, Duvoisin RC (1985) Dopaminergic neurotoxicity of 1-methyl-4-phenyl-1,2,3,6-tetrahydropyridine (MPTP) in the mouse: Relationships between monoamine oxidase, MPTP metabolism and neurotoxicity. Life Sci 36: 231–236
Kinemuchi H, Fowler CJ, Tipton KF (1987) The neurotoxicity of 1-methyl-4-phenyl-1,2,3,6-tetrahydropyridine (MPTP) and its relevance to Parkinson's disease. Neurochem Int 11/4: 359–373
Kishimoto S, Kimura H, Maeda T (1983) Histochemical demonstration for monoamine oxidase (MAO) by a new coupled peroxidation method. Cell Mol Biol 29: 61–69
Konradi C, Riederer P, Youdim MBH (1986) Hydrogen peroxide enhances the activity of monoamine oxidase type-B but not of type-A: A pilot study. J Neural Transm [Suppl] 22: 61–73
Perry TL, Godin DV, Hansen S (1982) Parkinson's disease: A disorder due to nigral glutathione deficiency? Neurosci Lett 33: 305–310
Perry TL, Yong VW, Clavier RM, Jones K, Wright JM, Foulks JG, Wall RA (1985) Partial protection from the dopaminergic neurotoxin N-methyl-4-phenyl-1,2,3,6-tetrahydropyridine by four different antioxidants in the mouse. Neurosci Lett 60: 109–114
Riederer P, Jellinger K, Seemann D (1984) Monoamine oxidase and Parkinsonism. In: Tipton KF, Dostert P, Strolin-Benedetti M (eds) Monoamine oxidase and disease. Academic Press, London, pp 403–415
Riederer P, Sofic E, Rausch W-D, Schmidt B, Reynolds GP, Jellinger K, Youdim MBH (1988) Transition metals, ferritin, glutathione and ascorbic acid in Parkinsonian brains. (Submitted)
Singer TP, Salach JI, Castagnoli N, Trevor A (1986) Processing of MPTP to toxic metabolites by monoamine oxidases. In: Markey SP, Castagnoli N, Trevor AJ, Kopin IJ (eds) MPTP: A neurotoxin producing a parkinsonian syndrome. Academic Press, London, pp 235–251
Yong VW, Perry TL, Krisman AA (1986) Depletion of glutathione in brainstem of mice caused by N-methyl-4-phenyl-1,2,3,6-tetrahydropyridine is prevented by antioxidant pretreatment. Neurosci Lett 63: 56–60

Correspondence: Dr. C. Konradi, Clinical Neurochemistry, Department of Psychiatry, University of Würzburg, Füchsleinstrasse 15, D-8700 Würzburg, Federal Republic of Germany.

Importance of dopaminergic and GABAergic neurones of the nucleus accumbens and the caudate nucleus for motoricity

K. Kuschinsky[1] and **U. Havemann**[2]

[1] Institute for Pharmacology and Toxicology, Faculty of Pharmacy, University of Marburg
[2] Max Planck-Institute for Experimental Medicine, Göttingen, Federal Republic of Germany

Summary

In the present brief review, the pharmacological induction of akinesia and muscular rigidity in an animal model (rats) is discussed. Methods measuring these symptoms are described and the possible role played by GABAergic mechanisms in efferent pathways of the basal ganglia is discussed.

Introduction

In Parkinson's disease, poverty of voluntary movement (hypo- or akinesia), slowness of voluntary movement (bradykinesia), muscular rigidity and "resting" tremor appear to be among the most prominent symptoms. The experimental neuropharmacologist's task is to apply animal models which are sufficiently analogous to the pathophysiological processes found in humans in Parkinson's disease. Although primates or cats can and should be used for the detailed studies of some specific problems in this field, most studies must be performed (for ethical and economic reasons) in lower mammals. Most experiments, therefore, have been performed in rats. It seems evident that motoricity of humans and rats differ considerably. One aspect is, for instance, that humans are bipeds and rats quadrupeds. Nevertheless, it seems likely that at least some basic mechanisms are similar in both species, since the structure of the basal ganglia shows a considerable homology, in spite of some differences. The present paper will mainly concentrate on experimental pharmacological models of akinesia and muscular rigidity in rats.

Experimental models for studying akinesia and rigidity in rats

Relatively selective lesions in the central nervous system can be produced by neurotoxins, such as 6-hydroxydopamine (6-OHDA) or MPTP (1-methyl-4-phenyl-1, 2, 3, 6-tetrahydropyridine). Both substances induce irreversible lesions. Since 6-OHDA affects both dopaminergic and noradrenergic neurones, it has to be injected into areas containing either dopaminergic cell bodies or terminals, when selective lesions of dopaminergic neurones are intended. MPTP, when administered systemically, is very potent in destroying dopaminergic neurones in humans and primates, but much less potent in rats. For these reasons, 6-OHDA only seems to be of some use for studying pathophysiological processes in rats resembling those occurring in Parkinson's disease in man. It should be considered, however, that irreversible lesions produce not only direct losses of structure and function of the neurones involved, but also induce compensatory mechanisms which try to circumvent the primary deficits. In this case, the interference of both phenomena, both the direct deficiencies and the compensation mechanisms, is measured. These studies can be helpful, since they try to evaluate chronic processes of irreversible lesions, similarly as they occur in Parkinson's disease. On the other hand, pharmacological studies using shortly and reversibly acting drugs can lead to selective functional alterations with a minimum of compensatory mechanisms and by this may mimick direct deficiencies in Parkinson's disease. Therefore, studies of this kind were prefered by various groups including ourselves.

Reserpine and relatively selective blockers of dopamine receptors produce Parkinson-like symptoms, such as hypokinesia and muscular rigidity in man, which appear after repeated administration of the drug, but at a relatively early stage of the treatment. In rats, hypo- or akinesia can be produced by these drugs as well, even after a single dose. In contrast, rigidity is not prominent under this condition, although it can be induced under special conditions. Administration of dopaminomimetic drugs can antagonize these effects. Opioid-like drugs, such as morphine, can also produce akinesia and a pronounced muscular rigidity, when given in relatively large doses. These effects can be antagonized both by naloxone and by dopaminomimetic drugs, such as L-DOPA and apomorphine.

Methods of measuring akinesia and muscular rigidity

Hypo- or akinesia can be easily measured in various motility meters which work either by interruption of light or infrared beams or by disturbing an electromagnetic field. When akinesia is very pronounced, it becomes manifest as catalepsy: when the animals are brought into an abnormal position, they remain in it for a while, although they are neither atactic nor paralysed and, by appropriate stimuli, can be induced to perform coordinated movements. In the usual tests, a positive score is given, when a defined time is exceeded, in which the rat remains in its position; in many tests 30 sec are used as a limit. Muscular rigidity can be measured and quantified by mechanographic analysis of muscle rigidity (Kolasiewicz et al., 1987) so that the hindlegs are passively straightened and bent and the resistance of flexors and extensors is successively recorded. Alternatively, rigidity can be measured by using an electromyogram recorded either from needle electrodes (Wand et al., 1973) or by stainless steel fine wire electrodes inserted percutaneously (Havemann et al., 1983) into the gastrocnemius-soleus muscle or another muscle to be studied.

Localization of receptors mediating these effects

One possible way to approach this problem is to place selective lesions into a defined brain area and to study a possible decrease or abolishment of the pharmacological effect after systemic administration of the drug. Even if this is the case, however, the results only suggest that the lesioned brain nucleus or pathway is somehow involved in the neuronal circuit mediating the effect. Accordingly, the results are rather ambiguous with regard to the localization of the receptors, from which the effect is triggered.

Another, more promising, but sometimes troublesome way is to inject the drug locally into a definite brain area and to test, if the drug produces the effect expected. By this method, problems of local pharmacokinetics arise leading to a smaller or larger spreading of the drug into adjacent areas. With some hydrophilic drugs, this occurs rather slowly. Another problem is the observation that similar or identical effects can be produced by local injection of the drug into various, distant brain areas. These results demonstrate that drug-specific receptors in the area injected can mediate the effect, at least when the drug is present in a high concentration. However, additional experiments are necessary to verify that a defined brain area is responsible for mediating the effect, when the drug is administered systemically, e.g. by injection of a specific antagonist into this area and to study, whether the drug effect is antagonized.

Local injections of drugs can also be used to study selective effects of neurotransmitter-related drugs in a defined neuronal circuit and to estimate the action of neurotransmitters in the "relais station" of the circuit. In the present studies, both aims were pursued, when drugs were locally injected. The final purpose was to obtain information about the mechanisms responsible for morphine-induced "catatonia" in rats. This syndrome includes a pronounced akinesia, which becomes manifest as a catalepsy (see above) and a muscular rigidity, and it might serve as an animal model of akinesia and rigidity observed in human diseases, with all its limitations discussed above.

Local injections of morphine and other opioids into the striatum produce no akinesia, but a pronounced muscular rigidity. Furthermore, in these studies, evidence was obtained that this brain area is also most probably responsible for rigidity observed after systemic administration of morphine (Havemann and Kuschinsky, 1982). In contrast, akinesia is not produced by injection of opioids into the striatum, but by injection into the nucleus accumbens. Both effects cannot be produced by a blockade of postsynaptic dopamine receptors, since the effects are, in contrast to those of neuroleptics, very easily antagonized by L-DOPA or the agonist at dopamine receptors apomorphine (Wand et al., 1973). Furthermore, an impairment of dopamine release as an explanation of the "catatonia" can also be excluded, since the concentration of 3-methoxytyramine (3-MT), a dopamine metabolite as an indicator of dopamine release, is increased after doses of morphine which produce this syndrome (Möller and Kuschinsky, 1986). Accordingly, the receptors mediating these effects must be located in striatum or nucleus accumbens, respectively, but "downstream" of the dopaminergic synapses. The increase in dopamine release is probably due to an additional action of morphine on dopaminergic neurones in the substantia nigra and the ventral tegmental area (VTA) leading to an activation of these neurones and sometimes antagonizing the rigidity (Havemann and Kuschinsky, 1982) and akinesia. The most relevant efferent pathways from striatum and nucleus accumbens are leading via globus pallidus and substantia nigra pars reticulata (SNR). The striatum is provided by dopaminergic afferents from the substantia nigra pars compacta (SNC), whereas the afferent dopaminergic neurones of the nucleus accumbens arise in the VTA. In further studies, we concentrated ourselves on the SNR, but experiments with local injections of drugs into

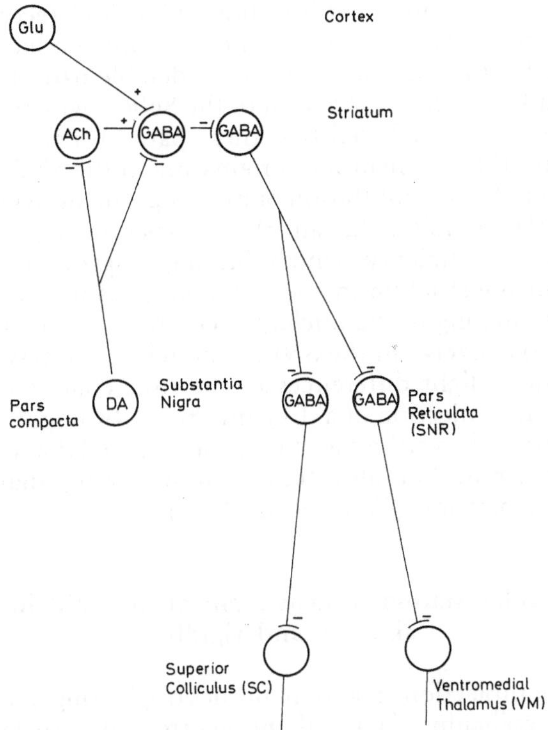

Fig. 1. Simplified scheme of striato-nigral efferent pathways. *ACh* acetylcholine, *DA* dopamine, *GABA* gamma-aminobutyric acid, *Glu* glutamate

nigrofugal target areas will also be briefly described. They were performed by other groups. In Fig. 1, the striato-nigral efferent pathways are shown, but as mentioned above, the SNR relayes efferent pathways of the nucleus accumbens as well.

The role of the SNR in mediating akinesia and rigidity

Injection of a GABAergic agonist (muscimol) into the SNR antagonizes the morphine-induced muscular rigidity, whereas administration of a GABA receptor antagonist (bicuculline) into this brain area enhances this rigidity and produces a rigidity by itself (Havemann et al., 1983). These results suggest that the SNR is an important efferent relay station mediating rigidity by an impairment of GABAergic neurotransmission in this area. This assumption is supported by the finding that injection of a GABA receptor agonist (muscimol) into a ventral area of the striatum produces a similar rigidity (Turski et al., 1984). Since a considerable part of GABAergic pathways leading from the striatum to the SNR originates in the area injected, it can be concluded that these pathways are inhibited by muscimol and that the neurones originating in the SNR, as a result, are dis-inhibited. A dis-inhibition of nigrofugal neurones then results in a muscular rigidity. In all these experiments, rigidity was accompanied by catalepsy, which findings suggest that pathways originating in the striatum (mediating rigidity as mentioned above) and those originating in the nucleus accumbens (mediating akinesia and catalepsy) converge in the SNR, although not necessarily on the same neurones. Slight differences in the time-courses of catalepsy and rigidity after injection of a drug into the SNR might support the latter conclusion. Nevertheless, it must be assumed that the SNR can not only transmit, but also transform incoming signals for its behavioural expression (Cools et al., 1983).

The role of relay stations downstream of the SNR in mediating akinesia and rigidity

Two main projection areas of neurones originating in the SNR are the superior colliculus (SC) and the ventromedial thalamus (VM) (Fig. 1). Both projections are GABAergic (Di Chiara et al., 1979). Injection of the GABA receptor agonist muscimol into the SC produces rigidity, but no catalepsy (Ellenbroek et al., 1984), so that in this efferent pathway a dissociation between both symptoms is apparent. The role of the SC in transmission of striatally-induced rigidity was tested in a further study: the rigidity produced by a haloperidol injection into the rostral part of the striatum was antagonized by administration of the GABA receptor antagonist bicuculline into the SC (intermediate or deep layers) (Ellenbroek et al., 1985).

Injection of muscimol into the VM produces both rigidity and catalepsy (Klockgether et al., 1985). The role of VM in transmission of catalepsy produced by systemic administration of haloperidol or morphine was also studied: whereas haloperidol-induced catalepsy was inhibited by administration of picrotoxin, a GABA antagonist, morphine-induced catalepsy was not affected by this procedure. These results demonstrated that the VM is of relevance for the transmission of haloperidol- but not of morphine-induced catalepsy (Wolfarth et al., 1985), so that even an apparently similar symptom produced by drugs of two different pharmacological groups is transmitted via different pathways. It is not clear from these studies, whether the morphine-induced rigidity may be affected by picrotoxin injections into the VM or not.

In spite of these differences, both nuclei (SC and VM) have in common that they are similar and act in a direction opposite to effects induced in the SNR, when GABAergic drugs are injected into them. This can be easily explained by the fact mentioned above that GABAergic neurones project to both the SC and the VM. An inhibitory action in the SNR must lead to a dis-inhibition in SC and VM and vice versa.

To sum up, these results demonstrate that our knowledge about the relay stations and mechanisms transmitting and transforming at each relay station information necessary for the development of akinesia and muscular rigidity downwards to the motoneurones is still fragmentary. Akinesia/catalepsy on the one hand, muscular rigidity on the other hand can be triggered from different brain areas (nucleus accumbens or striatum, respectively) and show differences in their efferent pathways, although the transmission of both symptoms converges in the SNR. Impairment of dopaminergic neurotransmission, but also processes downstream of dopaminergic nerve terminals can produce these symptoms and GABAergic mechanisms are relevant for passing the information about functional alterations in striatum and nucleus accumbens downwards. The role of the globus pallidus in mediating and transforming these symptoms is as yet rather obscure. Many further studies and a synopsis of clinical and experimental results are necessary and might finally lead to an understanding of the pathophysiological processes resulting in motor disturbances such as akinesia and muscular rigidity.

References

Cools AR, Jaspers R, Kolasiewicz W, Sontag K-H, Wolfarth S (1983) Substantia nigra as a station that not only transmits, but also transforms incoming signals for its behavioural expression: striatal dopamine and GABA-mediated responses of pars reticulata neurons. Behav Brain Res 7: 39–49

Di Chiara G, Porceddu ML, Morelli M, Mulas ML, Gessa GL (1979) Evidence for a gabaergic projection from the substantia nigra to the ventromedial thalamus and the superior colliculs of the rat. Brain Res 176: 273–284

Ellenbroek B, Schwarz M, Sontag K-H, Cools AR (1984) The role of the colliculus superior in the expression of muscular rigidity. Eur J Pharmacol 104: 117–123

Ellenbroek B, Schwarz M, Sontag K-H, Cools AR (1985) The importance of the striato-nigro-collicular pathway in the expression of haloperidol-induced tonic electromyographic activity. Neurosci Letters 54: 189–194

Havemann U, Kuschinsky K (1982) Neurochemical aspects of the opioid-induced "catatonia". Neurochem Internat 4: 199–215

Havemann U, Turski L, Schwarz M, Kuschinsky K (1983) Nigral GABAergic mechanisms and EMG activity in rats: differences between pars reticulata and pars compacta. Eur J Pharmacol 92: 49–56

Klockgether T, Schwarz M, Turski L, Wolfarth S, Sontag K-H (1985) Rigidity and catalepsy after injections of muscimol into the ventromedial thalamic nucleus: an electromyographic study in the rat. Exp Brain Res 58: 559–569

Kolasiewicz W, Baran J, Wolfarth S (1987) Mechanographic analysis of muscle rigidity after morphine and haloperidol: a new methodological approach. Naunyn-Schmiedeberg's Arch Pharmacol 335: 449–453

Möller HG, Kuschinsky K (1986) Interactions of morphine with apomorphine: behavioural and biochemical studies. Naunyn-Schmiedeberg's Arch Pharmacol 334: 452–457

Turski L, Havemann U, Kuschinsky K (1984) GABAergic mechanisms in mediating muscular rigidity, catalepsy and postural asymmetry in rats: differences between dorsal and ventral striatum. Brain Res 322: 49–57

Wand P, Kuschinsky K, Sontag K-H (1973) Morphine-induced muscular rigidity in rats. Eur J Pharmacol 24: 189–193

Wolfarth S, Kolasiewicz W, Ossowska K (1985) Thalamus as a relay station for catalepsy and rigidity. Behav Brain Res 18: 261–268

Correspondence: Prof. Dr. K. Kuschinsky, Institute for Pharmacology and Toxicology, Faculty of Pharmacy, University of Marburg, D-3550 Marburg, Federal Republic of Germany.

Is D-1 receptor stimulation important for the anti-parkinson activity of dopamine agonists?

R. Markstein and J.-M. Vigouret

Preclinical Research, Sandoz Ltd., Basle, Switzerland

Summary

This overview summarises recent experimental findings suggesting that an optimal motor control in Parkinson's disease depends on concurrent activation of D-1 and D-2 receptors. The clinical observation that a combination of bromocriptine (Parlodel; Pravidel) with L-DOPA produces better symptomatic control in Parkinson patients than either agent alone is explained as the result of a synergistic interaction between D-2 receptors activated by Parlodel and D-1 receptors activated by dopamine formed from L-DOPA.

Introduction

The primary disturbance in Parkinson's disease is the progressive degeneration of nigrostriatal dopamine neurons. The disease remains subclinical until the reduction in the striatal dopamine content exceeds 80%. After the onset of parkinsonian symptoms, therapeutic approaches aimed at replacing the lost dopamine ameliorate motor disturbances and improve the quality of life for several additional years. The dopamine precursor L-DOPA is at present the most widely used drug for this purpose. However, after long term treatment its beneficial effects are progressively lost and accompanied by dose limiting side effects. The reasons for these long term problems with L-DOPA are not completely understood but may in part be due to the progression of the disease. Since dopaminergic neurons continue to degenerate, a point is finally reached where the number of remaining neurons is too small to synthesize sufficient dopamine from L-DOPA. Moreover, it has been suggested that

L-DOPA itself contributes to the progression of the degenerative process either by exhausting remaining neurons or by the formation of toxic metabolites. It was therefore hoped that Parkinson patients might benefit from drugs which bypass the degenerating dopaminergic neurons and stimulate postsynaptic striatal dopamine receptors directly. Various dopamine agonists, in particular dopaminergic ergot derivatives, have been shown to be effective in Parkinson's disease (Rinne, 1983). Parlodel (bromocriptine), the first and still the only dopamine agonist approved for the treatment of Parkinson's disease, is effective in patients with mild to moderate symptoms (Calne et al., 1974; Stern and Lees, 1983). However, in patients in an advanced stage it is usually less effective than L-DOPA (Rinne, 1983). Interestingly, a combination of Parlodel with L-DOPA has been reported to produce better symptomatic control of the disease than either agent alone (Bouchard, 1987). Furthermore, early combination of Parlodel and L-DOPA seems not only to improve parkinsonian disability but also to prevent dyskinesias and response fluctuations (Rinne, 1987). This short overview summarises experimental and clinical observations supporting the hypothesis that an optimal therapeutic effect in Parkinson's disease depends on concomitant stimulation of two different classes of dopamine receptors, and that Parlodel stimulates one class and dopamine formed from L-DOPA the other.

Classification and localisation of dopamine receptors

The classification of central dopamine receptors into D-1 and D-2 types, originally based on biochemical criteria, has gained further support by the identification of selective agonists and antagonists for both receptor types, and is now widely accepted. Table 1 summarises the presently accepted classification of central dopamine receptors and Fig. 1 shows the structures of selective drugs.

In the striatum D-1 receptors are exclusively postsynaptically located whereas D-2 receptors occur both on post- and presynaptic elements. Presynaptic D-2 receptors occuring on dopaminergic nerve terminals so called dopamine autoreceptors form part of a negative feedback system by which synthesis and release of dopamine is regulated in an inhibitory manner. In the substantia nigra D-1 receptors seem to occur on terminals of the striatonigral pathway whereas D-2 receptors occur presynaptically on dendrites and the soma of dopaminergic neurons (see Fig. 2).

Table 1. Classification of dopamine receptors

	D-1	D-2
Adenylate cyclase linkage	Stimulatory	Inhibitory or not linked
Selective agonists	SKF 38393 CY 208–243	Bromocriptine LY 171555 RU 24-213
Selective antagonist	SCH 23390	Sulpiride
Prolactin secretion	0	Inhibition
Parathormone release	Stimulation	0
Induction of emesis	No	Yes
Behavioural effect	Non-stereotyped sniffing	Stereotyped sniffing, gnawing

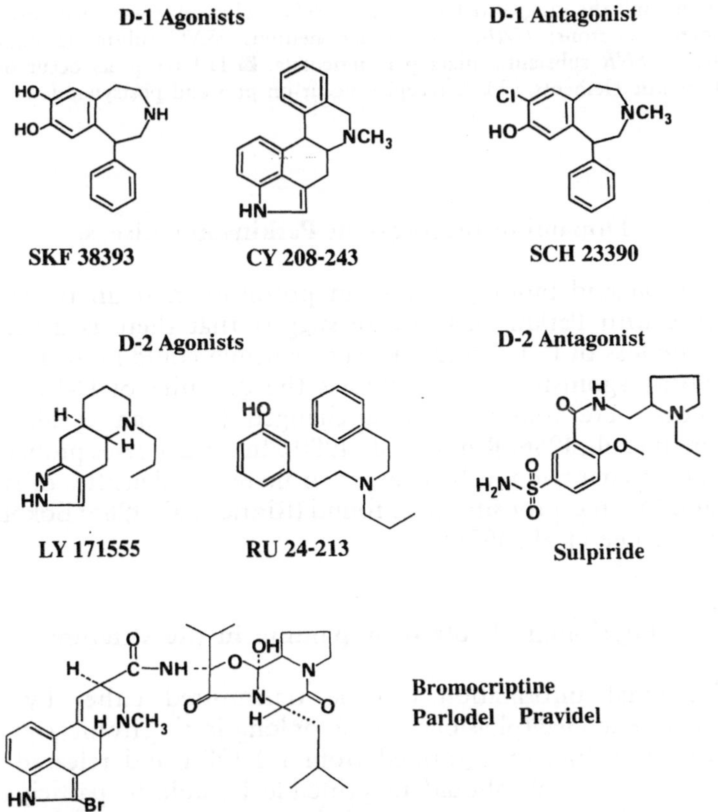

Fig. 1. Selective drugs for D-1 and D-2 dopamine receptors

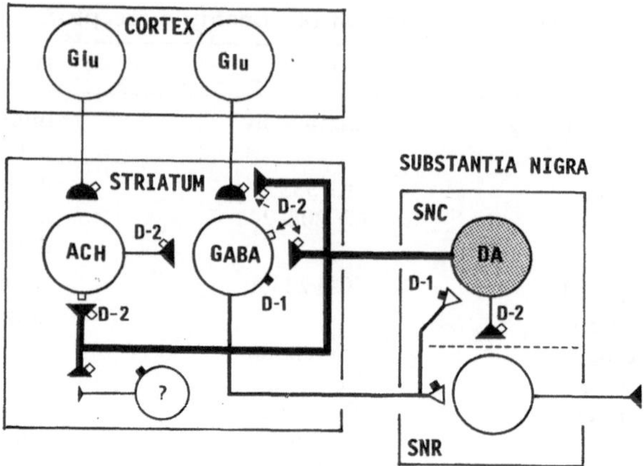

Fig. 2. Schematic representation of known localisation of D-1 and D-2 dopamine receptors in the nigrostriatal system. *GLU* glutaminergic neurons; *ACH* cholinergic neurons; *GABA* GABAergic neurons; *SNC* substantia nigra pars compacta; *SNR* substantia nigra pars reticulata. ■ D-1 receptors occur only on postsynaptic elements. □ D-2 receptor occur on pre- and postsynaptic elements

Dopamine receptors in Parkinson's disease

Radioligand binding studies in postmortem brain tissue from patients with Parkinson's disease suggest that there is usually no dramatic loss of D-1 and D-2 receptors which could limit the use of dopamine agonists. In most studies the densities of D-1 and D-2 receptors were found to be unchanged (Raisman et al., 1985; Guttman et al., 1986; Rinne et al., 1980). In milder cases probably, as a result of adaptive mechanisms even increased densities of striatal D-1 and D-2 receptor sites were found (Rinne et al., 1985; Bokobza et al., 1984; Lee et al., 1978).

Physiological role of dopamine in the striatum

Neuronal information can be transmitted either by rapid frequency encoded (phasic) impulses or long-lasting (tonic) signals. It is clear that dopamine formed from L-DOPA and released from dopamine terminals should in principle be able to participate in phasic and tonic neurotransmission, whereas directly acting dopamine agonists at the postsynaptic level would be expected to produce only tonic effects. However, several experimental observations are compatible with the view that the role of dopamine in motor control is rather that of a tonically acting neurotransmitter. For instance, dopamine neurons innervating the striatum have been shown to exhibit a relatively regular and slow basal discharge rate (3–5 impulses/sec) with very little phasic activity (Chiodo et al., 1984). Furthermore, biochemical and electrophysiological experiments indicate that in particular the D-1 receptor mediated effects of dopamine are long lasting.

Specific roles of dopamine receptor subtypes in motor control

Until recently, the majority of the central effects of dopamine have been linked to D-2 receptors whereas an important physiological function for D-1 receptors has been questioned. However, recent studies with newly available D-1 selective agonists and antagonists strongly suggest that D-1 receptors also participate in motor control. In contrast to D-2 receptor agonists (e.g. bromocriptine, LY 171555) which produce marked behavioural stimulation in experimental animals selective D-1 agonists (e.g. SKF 38393, CY 208-243) induce only subtle behavioural effects consisting mainly of non-stereotyped sniffing and grooming (Molloy and Waddington, 1984). However, in rats with a unilateral 6-hydroxydopamine (6-OHDA) induced dopaminergic denervation of the striatum D-1 and D-2 agonists cannot be differentiated since both induce turning towards the non-lesioned side (Arnt and Hyttel, 1986; Setler et al., 1978). Recently, N-methyl-4-phenyl-1, 2, 3, 6-tetrahydropyridine (MPTP) has been reported to produce motor deficits and neurochemical changes in primates resembling those in Parkinson's disease (Burns et al., 1986). Although in MPTP-treated monkeys a typical marker for idiopathic Parkinson's disease i.e. the Lewy body is absent, it is considered to be the best available animal model for predicting antiparkinson activity of drugs. L-DOPA and several clinically effective dopamine agonists, including bromocriptine, have been shown to counteract parkinsonian symptoms in MPTP-treated monkeys. Surprisingly the selective D-1 receptor agonist SKF 38393 was found to be inactive in this animal model (Nomoto et al., 1985) and also failed to produce clinical improvement in Parkinson patients (Braun et al., 1987). These observations implied that D-1 receptor agonists would be unlikely to be of value in the treatment of Parkinson's disease. However, the novel D-1 agonist CY 208-243 has been shown to be effective in MPTP-treated monkeys (Markstein et al., 1988). Thus, it cannot longer be excluded that also D-1 agonists are effective in Parkinson's disease.

Interaction between D-1 and D-2 receptors

Evidence is now accumulating that D-1 and D-2 receptors do not regulate motor behaviour independently but are functionally linked. In experimental animals doses of D-1 agonists and D-2 agonists which when given alone fail to change motor activity, produce marked behavioural activation when given together (see Table 2). For instance, in unilateral 6-OHDA lesioned rats doses of SKF 38393 and bromocriptine which themselves were inactive, induced a high circling intensity when given together (Robertson and Robertson, 1986). Furthermore, the locomotor stimulating effect of bromocriptine in mice was abolished when central dopamine stores were depleted but was fully restored when the D-1 agonist SKF 38393 was added (Jackson and Hashizume, 1986). Furthermore, the selective D-1 antagonist SCH 23390 has been reported not only to block the effects of D-1 agonists but also to abolish D-2 agonist induced behavioural effects. The most likely explanation for these observations is that D-1 and D-2 receptors interact in a synergistic manner and that D-2 receptor mediated behavioural responses are only expressed when D-1 recepors are activated.

This hypothesis implies that the behavioural effects of selective D-2 agonists depend in some way on the presence of endogenous dopamine which could provide the necessary D-1 background. At the cellular level D-1 and D-2 receptors have been shown to produce opposing biochemical effects. For instance, in striatal preparations D-1 receptors activation enhances cAMP formation and neuronal release of GABA whereas D-2 receptor stimulation produces the opposite effect (Stoof and Kebabian, 1981; Girault et al., 1986). It seems therefore unlikely that the synergistic interplay between D-1 and D-2 receptors occurs at the cellular level. The autoradiographic determination of 2-deoxyglucose (2-DOG) uptake in rat brain, originally described by Sokoloff et al. (1977), is a novel method for differentiating between drug effects on D-1 and D-2 receptors in vivo. In the brains of normal rats, dopamine agonists and antagonists acting on D-2 receptors have been shown to alter regional 2-DOG uptake in a characteristic manner whereas D-1 selective drugs were inactive. By contrast in 6-OHDA lesioned rats D-1 agonists such as SKF 38393 or CY 208-243 induce a marked increase of 2-DOG

Table 2. Evidence for synergistic interactions between D-1 and D-2 receptors. Behavioural studies

Drug treatment		Dose	Animal model	Symptom	Intensity	Reference
D-1	D-2	mg/kg				
SKF 38393	–	6 i.p.	Mouse	Locomotion	+	Jackson and
	Bromocriptine	10 i.p.			++	Hashizume (1986)
SKF 38393	+ Bromocriptine				++++	
SKF 38393	–	10 s.c.	Mouse	Climbing	0	Moore and
	LY 171555	1.2 s.c.			0	Axton (1988)
SKF 38393	+ LY 171555				+++	
SKF 38393	–	15 s.c.	Rat	Stereotypy	0	Mashurano and
	RU 24213	2.5 s.c.			0	Waddington (1986)
SKF 38393	+ RU 24213				+++	
SKF 38393	–	6 i.p.	Reserpin	Locomotion	0	Jackson and
	Bromocriptine	10 i.p.	treated		+	Hashizume (1986)
SKF 38393	+ Bromocriptine		mouse		++++	
SKF 38393	–	0.5 i.p.	6-OHDA	Circling	0	Robertson and
	Bromocriptine	5.0 i.p.	lesioned		0	Robertson (1986)
SKF 38393	+ Bromocriptine		rat		+++	
SKF 38393	–	8 i.p.	AMPT*	Stereotypy	0	Braun and
	LY 171555	3 s.c.	treated		0	Chase (1986)
SKF 38393	+ LY 171555		rat		++++	

* AMPT (alpha-methyl-para-tyrosine): inhibits synthesis of dopamine

uptake in the substantia nigra pars reticulata and the entopeduncular nucleus (Trugman and Wooten, 1987) (Fig. 3). Based on these observation Robertson and Robertson (1987) proposed that the substantia nigra is an important site for the action for D-1 agonists and that the synergistic effect of D-1 and D-2 receptor agonists is mediated by actions in anatomical separate regions. The notion that different brain sites are involved in dopaminergic motor control gains also support from micro-injection experiments. Motor activity was increased when dopamine agonists were injected directly into the striatum, the nucleus accumbens, the substantia nigra and the habenula nucleus. The functional linkage between D-1 and D-2 receptors however can be lost after extensive and prolonged depletion of dopamine. Under such circumstances D-2 agonists have been reported to produce marked behavioural stimulatory effects in the absence of a tonic D-1 background. However, it seems that the loss of dopamine in Parkinson's disease is not sufficient to uncouple the interplay between D-1 and D-2 receptors. In summary, there is now considerable evidence that many behavioural and motor effects considered to be regulated by central dopaminergic systems depend on the concurrent activation of D-1 and D-2 receptors. It appears that D-2 agonist are capable of producing motor stimulating effects because endogenous dopamine provides the necessary tonic D-1 background.

Mechanism of action of L-DOPA

L-DOPA, the immediate precursor of dopamine in the biosynthetic pathway, is usually taken up by dopaminergic nerve terminals. Although there is marked loss of dopamine neurons in Parkinson's disease there is obviously still enough dopa decarboxylase present to catalyse the formation of dopamine from exogeneous L-DOPA. Since this step is not rate limiting dopamine neurons can synthesise almost unlimited amounts of dopamine in the presence of L-DOPA. Although it is likely that dopamine formed from L-DOPA activates both dopamine receptor types, the question arises as to

Fig. 3. Modification of 2-[^{14}C]deoxyglucose (2-DOG) uptake in various brain areas of rats with unilateral 6-OHDA-induced lesion of the substantia nigra following treatment with L-DOPA or selective D-1 agonists (SKF 38393; CY 208-243) or selective D-2 agonists (LY 171555; bromocriptine). Data from Trugman and Wooten (1986), Markstein et al. (1988) and J. M. Palacios and K. H. Wiederhold (unpublished). * $p < 0.01$ (Student t-test)

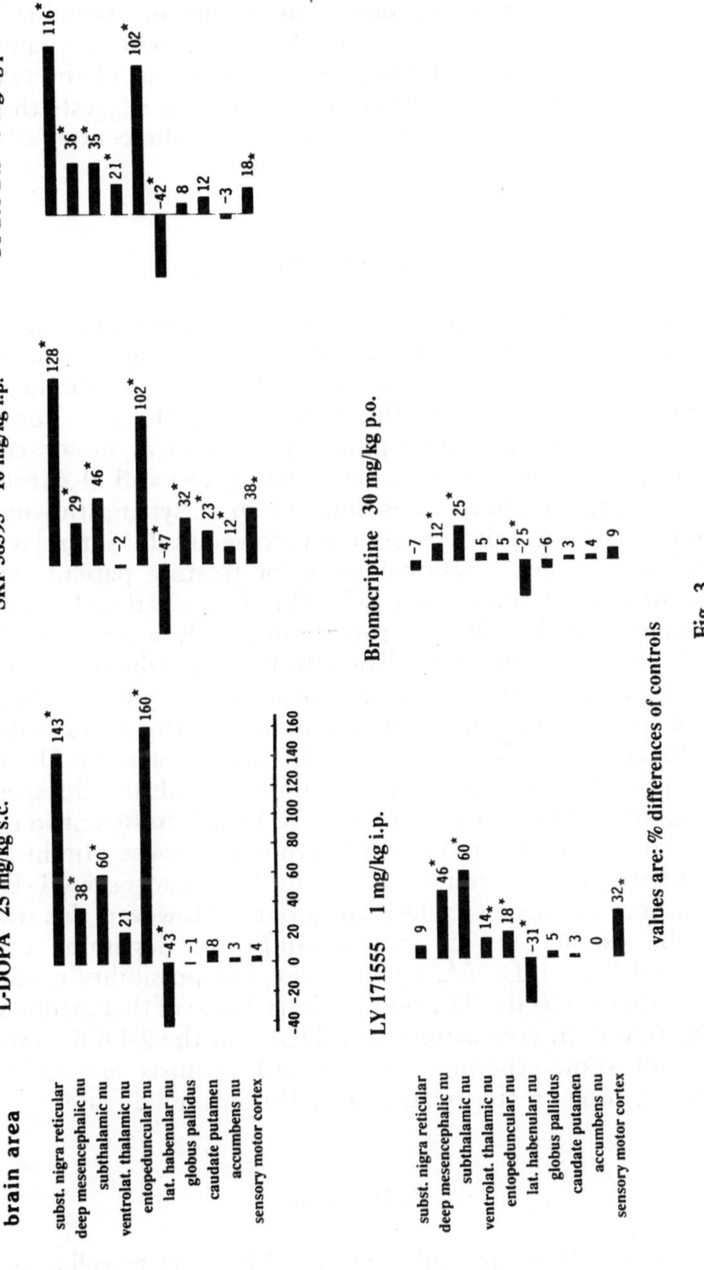
brain area
subst. nigra reticular
deep mesencephalic nu
subthalamic nu
ventrolat. thalamic nu
entopeduncular nu
lat. habenular nu
globus pallidus
caudate putamen
accumbens nu
sensory motor cortex
L-DOPA 25 mg/kg s.c.
143* 38* 60* 21 160* -43* -1 8 3 4
SKF 38393 10 mg/kg i.p.
128* 29* 46* -2 102* -47* 32* 23* 12* 38*
CY 208-243 3 mg/kg p.o.
116* 36* 35* 21* 102* -42* 8 12 -3 18*
LY 171555 1 mg/kg i.p.
9 46* 60* 14* 18* -31* 5 3 0 32*
Bromocriptine 30 mg/kg p.o.
-7 12* 25* 5 5 -25* -6 3 4 9
-40 -20 0 20 40 60 80 100 120 140 160
values are: % differences of controls

Fig. 3

whether one subtype plays a dominant role in the central effect of L-DOPA. In 6-OHDA lesioned rats, L-DOPA produces a pattern of 2-DOG uptake almost indistinquishable from that of the two D-1 agonists SKF 38393 or CY 208-243 (Fig. 3). This suggests that D-1 receptor activation plays an important role in the central effect of L-DOPA.

Conclusion

A prerequisite for the effectiveness of a dopamine agonist in Parkinson's disease is, that postsynaptic receptors are available and that the agonist mimics the physiological role of dopamine. Both conditions appear to be fulfilled. Moreover, there is now considerable evidence that an optimal physiological motor control depends on concomitant tonic activation of D-1 and D-2 receptors and that both receptor types interact in a synergistic manner. Different anatomical sites seem to be involved in this interplay. This hypothesis provides a rational basis for treating patients with a combination of Parlodel with L-DOPA. Thus, Parlodel stimulates D-2 receptors and L-DOPA provides indirectly a tonic D-1 background. Because of the synergistic effects, dose reductions for both agents should be possible. The therapeutic efficacy of selective D-2 receptor agonists as a primary treatment in the early stages of Parkinson's disease probably depends on the degree of tonic D-1 background provided by endogenous dopamine. In advanced stages, the endogenous dopamine content seems to be not sufficient to enable expression of a D-2 receptor mediated motor responses. In this situation, enhanced formation of dopamine from exogenous L-DOPA could restore the necessary D-1 background. However, this strategy would be useful only as long as sufficient dopamine can be synthesised from L-DOPA. Furthermore, the possibility cannot be excluded that a selective D-1 agonist alone is also of therapeutic value since it mimics in vivo actions of L-DOPA in the 2-DOG test. The future will show whether selective D-1 agonists are additional therapeutic tools in the treatment of Parkinson's disease.

References

Arnt J, Hyttel J (1986) Inhibition of SKF 38393- and pergolide-induced circling in rats with unilateral 6-OHDA lesion is correlated to dopamine D-1 and D-2 receptor affinities in vitro. J Neural Transm 67: 225–240

Bokobza B, Ruberg M. Scatton B, Javoy-Agid F, Agid Y (1984) [^{3}H]spiperone binding and HVA concentrations in Parkinson's disease and supranuclear palsy. European J Pharmacol 99: 167–175

Bouchard S (1987) Optimum symptomatic control of Parkinson's disease with dopaminergic therapy. Canad J Neurol Sci 14 [Suppl]: 460–465

Braun AR, Chase TN (1986) Obligatory D-1/D-2 receptor interaction in the generation of dopamine agonist related behaviors. European J Pharmacol 131: 301–306

Braun AR, Fabbrini G, Mouradian MM, Serrati C, Barone P, Chase TN (1987) Selective D-1 dopamine agonist treatment of Parkinson's disease. J Neural Transm 68: 41–50

Burns RS, Phillips JM, Chiueh CC, Parisi JE (1986) The MPTP-treated monkey model of Parkinson's disease. In: Markey SP, Castagnoli N, Trevor AJ, Kopin IJ (eds) MPTP: A neurotoxin producing a Parkinsonian syndrome. Academic Press, Orlando, pp 23–42

Calne DB, Teychenne PF, Leigh PN, Bamji AN, Greenacre JK (1974) Treatment of Parkinsonism with bromocriptine. Lancet ii: 1355–1356

Chiodo LA, Bannon MJ, Grace AA, Roth RH, Bunney BS (1984) Evidence for the absence of impulse regulating somatodendritic and synthesis modulating nerve terminal autoreceptors on subpopulations of mesocortical neurons. Neuroscience 12: 1–16

Girault JA, Spampinato U, Glowinski J, Besson MJ (1986) In vivo release of [^{3}H]gamma-aminobutyric acid in the rat neostriatum. II. Opposing effects of D-1 and D-2 dopamine receptor stimulation in the dorsal caudate putamen. Neuroscience 19: 1109–1117

Guttman M, Seeman P, Reynolds GP, Riederer P, Jellinger K, Tourtellote WW (1986) Dopamine D-2 receptor density remains constant in treated Parkinson's disease. Ann Neurol 19: 497–492

Jackson DM, Hashizume M (1986) Bromocriptine induces marked locomotor stimulation in dopamine-depleted mice, when D-1 dopamine receptors are stimulated with SKF 38393. Psychopharmacology 90: 147–149

Lee T, Seeman P, Rajput A, Farley IJ, Hornykiewicz O (1978): Receptor basis for dopaminergic supersensitivity in Parkinson's disease. Nature 273: 59–61

Markstein R, Seiler MP, Vigouret JM, Urwyler S, Enz A, Palacios JM, Wiederhold KH, Dixon K, Hofmann A, Wuetherich H (1988) CY 208-243, a novel D-1 receptor partial agonist. European J Pharmacol (in press)

Mashurano M, Waddington JL (1986): Stereotyped behaviour in response to the selective D-2 receptor agonist RU 24213 is enhanced by pretreatment with the selective D-1 agonist SKF 38393. Neuropharmacology 25: 947–949

Molloy AG, Waddington JL (1984) Dopaminergic behaviour stereospecifically promoted by the D-1 agonist R-SK&F 38393 and selectively blocked by the D-1 antagonist SCH 23390. Psychopharmacology 82: 409–410

Moore NA, Axton MS (1988) Production of climbing behaviour in mice requires both D-1 and D-2 receptor activation. Psychopharmacology 94: 263–266

Nomoto M, Jenner P, Marsden CD (1985) The dopamine D-2 agonist LY 141865, but not the D-1 agonist SKF 38393, reverses parkinsonism induced by 1-methyl-4-phenyl-1,2,3,6-tetrahydropyridine (MPTP) in the common marmoset, Neurosci Lett 57: 37–41

Raisman R, Cash R, Ruberg M, Javoy-Agid F, Agid Y (1985) Binding of [^{3}H]SCH 23390 to D-1 receptors in the putamen of control and parkinsonian subjects. European J Pharmacol 123: 467–468

Rinne UK, Koskinen V, Loenberg P (1980) Neurotransmitter receptors in the Parkinsonian brain. In: Rinne UK, Klinger M, Stamm G (eds) Parkinson's disease—Current progress, problems and management. Elsevier/North-Holland Biomedical Press, pp 93–107

Rinne UK (1983) Dopamine agonists in the treatment of Parkinson's disease. Adv Neurol 37: 141–150

Rinne JO, Rinne JK, Laasko K, Loenberg P, Rinne UK (1985) Dopamine D-1 receptors in Parkinsonian brain. Brain Res 359: 306–310

Rinne UK (1987) Early combination of bromocriptine and levodopa in the treatment of Parkinson's disease: A 5-year follow up. Neurology 37: 826–828

Robertson GS, Robertson HA (1986) Synergistic effects of D-1 and D-2 dopamine agonists on turning behaviour in rats. Brain Res 384: 387–390

Robertson GS, Robertson HA (1987) D-1 and D-2 dopamine agonist synergism: separate sites of action? Trends Pharmacol Sci 8: 295–299

Setler PE, Sarau HM, Zirkle CL, Saunders HL (1978) The central effects of a novel dopamine agonist European J Pharmacol 50: 419–430

Sokoloff L, Reivich M, Kennedy C, DesRosiers MH, Patlak CS, Pettigrew KD, Sakurada O, Shinohara M (1977) The [14C] deoxyglucose method for the measurement of local cerebral glucose utilisation: theory, procedure, and normal values in the conscious and anesthetised albino rat. J Neurochem. 28: 897–916

Stern GM, Lees AJ (1983) Sustained bromocriptine therapy in 50 previously untreated patients with Parkinson's disease. Adv Neurol 37: 17–21

Stoof JC, Kebabian JW (1981) Opossing roles for D-1 and D-2 receptors in efflux of cyclic AMP from rat neostriatum. Nature 294: 366–368

Trugman JM, Wooten GF (1986) The effect of L-DOPA on regional cerebral glucose utilisation in rats with unilateral lesions of the substantia nigra. Brain Res 379: 264–274

Trugman JM, Wooten GF (1987) Selective D-1 and D-2 dopamine agonists differentially alter basal ganglia glucose utilisation in rats with unilateral 6-hydroxydopamine substantia nigra lesions. J Neurosci 7: 2927–2935

Correspondence: Dr. R. Markstein, Preclinical Research, Sandoz Ltd., CH-4002 Basel, Switzerland.

Pharmacological and clinical-pharmacological aspects of D_1 and D_2 receptors

R. Horowski, I. Runge, P. Löschmann, H. Wachtel, and G. Stock

Clinical and Experimental Research of Schering AG, Berlin/Bergkamen

Summary

In contrast to the situation in some peripheral systems, the role of D_1 receptors for clinical effects or side effects of dopaminergic therapies in Parkinson's disease is still unclear.

This is due to the lack of clinical studies with D_1 agonists as well as to quite opposing results of animal neurobiochemistry and neuropharmacology depending on species, receptor state and other factors. For therapy, dopaminergic actions can be classified by the dopamine systems involved, the presence or absence of a blood-brain barrier (i.e. inhibition by domperidone) and by their evolution on chronic oral treatment: some dopaminergic effects remain unchanged, others increase or decrease with time in the same patient = "functional tolerance / supersensitivity development". New therapeutic strategies should involve more selective stimulation of the affected dopaminergic system (e.g. by dopamine partial agonists) or by new drug delivery systems avoiding high first-pass variations and "yo-yoing" of receptor stimulation as it occurs with all available dopaminergic therapies. Positive results of such a strategy are shown by continuous dopaminergic infusion.

Additional strategies involve selective inhibition of some dopaminergic side effects (e. g. by decarboxylase inhibitors, domperidone and clozapine-like drugs), but only research on events preceeding dopamine loss will help to find possibilities of arresting the progression of the disease. For early diagnosis, except PET studies only a combined strategy involving a drug challenge and a refined measurement of motor ability can be imagined.

I. Can we predict the clinical significance of D_1 receptors in Parkinson's disease?

The discovery and development of dopaminergic therapy of Parkinson's disease has been a breakthrough in neurology which came as a result of close and successful cooperation of clinicians, neuropathologists, neurochemists and neuropharmacologists (e.g. W. Birkmayer, O. Hornykiewicz, N. A. Hillarp, and A. Carlsson) whose ideas, however, first met considerable scepticism (for review see Carlsson, 1987). As a consequence of this work, dopamine no longer could be considered to be a mere precursor of noradrenaline, and the establishment of Parkinsonism as a consequence of reduced dopaminergic activity within the basal ganglia led to the development of a substitution therapy by dopamine resp. its precursor levodopa.

The acceptance of this concept was favoured by the clinical observation that reserpine which depletes dopamine in rat brain can cause reversible Parkinsonism in man, and also the further refinement of dopaminergic therapies was mostly based upon clinical observation and sometimes serendipity.

At the same time, neurobiochemists developed new models and methods; however, the extrapolation from studies with acute parenteral treatment in young intact rats ot the chronic oral treatment of elderly patients has been less successful, and there is a "mismatch" between biochemical theories and clinical improvements (see Table 1).

This also applies to the suggestion that dopamine acts by increasing cAMP levels as a second messenger in a similar way as described for adrenaline by Sutherland and colleagues. At a time, there was a plethora of different dopamine receptor subtypes and many different classifications so that the proposal by Kebabian and Calne (1979) of a dopamine receptor type acting by increased cAMP (D_1) and of another type independent of or even decreasing cAMP (D_2) met general acceptance. There were and there are, however, few pharmacological correlates for the D_1 receptor in brain so that this receptor subtype could be called "a receptor in search of a function" by John Kebabian himself, and it was stated indeed for patients "that the motor and the endocrine effects of L-DOPA, bromocriptine, lisuride and also of different dopamine antagonists in Parkinsonism can be attributed to the effects of these drugs on D_2 (non-adenylate-linked-) receptors ..." (Schachter et al., 1981).

In the meantime, animal research with D_1-agonists such as SKF 38393 or fenoldopam and the D_1-antagonist SCH 23390 has caused further complications because it has been shown that depending on the system tested and its state, D_1 and D_2 receptors can have antagonistic or synergistic effects. It may well be that the cAMP system acts as an integrating system for many simultaneous transmitter and modulator effects and that the interaction of D_1 receptors with the D_2-mediated effects is only a minor part of their function.

Table 1. Dopaminergic treatment of Parkinson's disease: biochemical theories and clinical situation

Theoretical claim from rat neurobiochemistry	Clinical observation
DA ist an independent neurotransmitter; its depletion in the basal ganglia causes akinesia which is reversed by L-DOPA as a DA precursor	L-DOPA is a most effective antiparkinson drug
Presynaptic DA receptors and autoreceptors mediate negative feed-back	Clinical efficacy improves with time, "presynaptic" effects such as sedation develop rapid tolerance
Denervation causes enhanced DA receptor binding	No evidence for greatly enhanced DA receptor binding in patients
DA receptor subtypes (D_1, D_2, "D_3", "D_4")	So far only D_2 receptor of clinical relevance, clinical function of D_1 still unclear
Biochemical basis established only subsequently (DCI not passing BBB)	Use of peripheral decarboxylase inhibitors to reduce side effects and increase activity
Partial DA agonists or selective DA agonists: same efficacy with less side effects?	Clinical studies ongoing, no clear answer so far
Partial agonism of ergot-derived DA agonists?	Positive effects of DA agonists on akinesia and dyskinesia
No biochemical explanation available so far	Slowly increasing oral dosage for inducing tolerance to side effects (dissociated functional tolerance/supersensitivity)
No biochemical explanation available so far	Improved tolerance by intake with food
No biochemical explanation available so far	Motor fluctuations after long-term L-DOPA treatment
No biochemical explanation available so far	Development of psychosis after long-term dopaminergic therapy
No biochemical explanation available so far	Enhanced antiparkinson effects with less peripheral side effects with continuous dopaminergic stimulation

It also could be conceived that the D_1 receptors are a part of a general receptor complex (similar to the GABA-Benzodiazepine receptor complex with its various binding sites on only two receptor proteins whose allosteric changes, again in an integrated way, regulate an ion channel and thus the state of a neuron). Anyway, in a recent review, Clark and White (1987) propose a model which permits both opposing and enabling forms of interaction between D_1 and D_2 receptors controlling animal behaviours. For animal behaviour, they enumerate more than 20 different effects where syn- or antagonism of both receptor subtypes has been proposed. Also these authors state that the "speculation regarding the utility of D_1-agonists in Parkinsonism ... now, necessitates a careful, cautious and throughout evaluation of the effectiveness of D_1 receptor agonists in Parkinsonian patients".

The search of a clinical function for D_1 receptors was further complicated by the lack of specific agonists and antagonists which could be tested in patients, and only very recently, the first clinical studies have been undertaken (Braun et al., 1987). It now seems clear that a peripheral D_1 receptor linked in a positive way to cAMP and stimulated by, e.g., fenoldopam, has a significant role in renal and cardiovascular function as well as in the regulation of parathormone secretion; central effects, however, observed in rats such as increased grooming or orofacial movements so far have not found a clinical correlate. In view of the variable interaction between D_1 and D_2 receptors (synergistic or antagonistic) which may also depend on the structures investigated and the state of the system, clinical results for D_1 agonists are eagerly awaited.

It had been proposed (e.g. by Ruggieri et al., 1987) that the difference observed between the acute effects of levodopa and lisuride infusion in a group of parkinsonian patients (where lisuride during the first hours induces a less "brilliant" state) is due to a lack of D_1 receptor stimulation by lisuride (in contrast to levodopa-derived dopamine). Thus, the relevant clinical test would be to add a D_1-agonist to the initial lisuride infusion on a double-blind basis in such patients. It must be stressed, however, that especially in the case of ergot-derived D_1-agonists, metabolites need to be investigated as well because many ergot derivatives have D_2-agonistic properties, and we have observed significant dopaminergic properties with a lisuride derivative (proterguride) at oral dosages as low as 3—10 μg in humans. Only if no prolactin-lowering effects can be found, such a drug is unlikely to influence D_2 receptors directly or via a metabolite.

II. Effects of dopaminergic therapy in patients: empirical classification and functional tolerance/supersensitivity

For nowadays clinical purposes, it seems much more important to classify effects elicited by all dopaminergic treatments so far in a different way. For clinicians (and for patients), it is of great importance what changes in these effects can be expected with effects and side effects when dopaminergic drugs are used on a chronic basis. It may also be relevant whether they occur outside the blood-brain barrier (even including some CNS systems, e.g. in the medulla and the hypothalamus). This type of side effects can be treated or prevented by peripheral dopamine antagonists such as domperidone without disturbance of the central antiparkinson effects.

In Table 2 we present such a classification of effects which are caused by dopaminergic drugs (mostly, if not exclusively, related to D_2 receptor stimulation). It has to be stressed that in the same patients, some effects remain unchanged whilst others undergo tolerance development and still others occur more frequently with chronic oral treatment (so-called functional tolerance/supersensitivity development, Horowski, 1987). Knowing these changes with time may be helpful for the use of dopaminergic therapies which is difficult anyway.

There is not only a patient factor (severity and type of disease, concomitant diseases) but also the state of the patient's dopamine system (e.g. pretreatment, concomitant treatment, availability of endogenous dopamine) has to be taken into account, as well as, of course, the drug factor itself (highly variable bioavailability of all dopaminergic drugs, interaction with absorption and L-DOPA transport). The variable functional response to chronic dopaminergic therapies cannot be explained simply by receptor changes (even if the rapid tolerance developed by "phasic" systems such as the CTZ or the autoreceptors reminds neurophysiological models of fast adaption). It therefore has to be postulated that events occuring behind the receptor (e.g. in the postreceptorial signal transduction) or simultaneous involvement of other, antagonistic or synergistic transmitter systems must play an important role (in long-term studies with dopaminergic drugs, P. Jenner and his colleagues did not observe relevant changes in dopamine biochemistry, but found changes in several peptides, e.g. neurotensins (Jenner et al., 1988).

It is very likely that the mechanisms for the "plasticity" of dopaminergic effects differ from one system to another depending on its anatomy and organisation: it is, e.g., conceivable that a dopaminergic increase of growth-hormone-releasing factor und subsequent growth-hormone increase causes a rapid compensatory increase in somatostatin and thus a normalization; on the other hand, simple threshold changes known from neurophysiology may be the reason why the chemoreceptor trigger zone develops tolerance so fast (as a "phasic" system).

Table 2. Proposed classification of relevant dopaminergic effects in man

Dopaminergic effect	Pathway and substrate	Inhibition by domperidone	Tolerance on chronic treatment	Supersensitivity on chronic treatment
Prolactin-lowering effect GH-lowering effect in 50% of acromegalics	tubero-infundibular system (DA receptors on PRL cells in ant. pit.)	++	–	–
GH-increase; inhibition of TRH-induced TSH-increase	hypothalamus	++	++[a]	–
Nausea, emesis	CTZ in medulla oblongata	++	+[b]	–
Reduction of sympathetic tone	diencephalo-spinal pathway?	++	+[b]	–
Orthostatic hypotension	medulla oblongata?	++	++[a]	–
Sedation	?	–	++[a]	–
Headache	?	–	++[a]	–
Antiparkinson effects	nigro-neostriatal system	–	–	(+)
Dyskinesias*	"basal ganglia"	–	–	+[c]
Psychosis*	mesolimbic and mesocortical system	–	–	+[c]

* Especially with high-dose therapy in P.D.
[a] Within days.
[b] Within weeks.
[c] After years.

III. New therapeutic strategies (continuous dopaminergic stimulation, partial agonists, early diagnosis and combination, selective antagonists)

It has to be reminded, however, that also the way of application of a dopaminergic drugs plays an important role. Side effects such as nausea and emesis or orthostatic hypotension do not occur on a given drug plasma level but depend clearly on the rat of change of these levels and thus receptor activation; this is the reason why continuous dopaminergic stimulation as caused by the lisuride pump achieves quite higher drug plasma levels without a parallel increase in side effects (for review see Obeso, Horowski, and Marsden, 1988). On the other hand, central effects on motor, limbic and cortical systems seem to increase with time under these conditions so that special care has to be taken to reduce and adjust the dosage quite fast if necessary. These findings have recently been corrobated by animal studies where similar differences were found between continuous and oscillatory application of dopaminergic drugs (Wachtel et al., 1988).

Another important topic in dopaminergic therapy is the concept of "partial agonism", especially in the case of 8-α-amino-ergolines such as terguride, but also lisuride. In situations of dopaminergic hyperactivity, these drugs may even block or "stabilize" dopamine receptors whilst they clearly act as agonists in "empty" systems. Therefore, they have been labelled "high-affinity-low-activity" dopamine agonists by G. L. Gessa (1988) and this, indeed, may explain some anti-dyskinetic properties of the antiparkinson drugs lisuride or terguride.

Similar drugs may be useful, because they will not have many effects or side effects, for early treatment of Parkinson's disease in combination with L-DOPA. Indeed, a controlled prospective randomized study has shown that the early combination of lisuride with L-DOPA is able to postpone the development of fluctuations (U. Rinne, this symposium). Altered pharmacokinetics or an autoreceptor stabilisation might be an explanation for these results but further studies are necessary before we can make a clear statement.

To come to another important question related to this symposium, i.e. the early diagnosis of the disease, maybe even in a presymptomatic stage, and the prophylaxis of its development, it is clear now from the work of D. B. Calne and others that PET will become a powerful instrument to give us early information about the number of dopamine neurons in the basal ganglia in living presymptomatic subjects. However, its usefulness as a mass screening method will be very limited for a foreseeable future due to the very difficult and expensive methodology (especially because no clear and strong risk factors for developing Parkinson's disease are known [except non-smoking] and because it is a quite rare disease). A more simple technique could use motor tests as elaborated by Przuntek and others, maybe including some neurophysiological parameters, and these tests could be performed using functional dopamine antagonists in order to demonstrate an impaired "dopamine reserve". Even in this case, time- and energy-consuming longitudinal studies will be necessary before the predictive value of these tests can be evaluated.

Tetrabenazine might be a candidate for such a test because it reduces dopamine levels and has a short duration of effect; another possibility would be to inhibit tyrosine hydroxylase, the key enzyme of dopamine synthesis which may be impaired very early in the development of the disease (Birkmayer and Riederer, 1985; P. Riederer et al., 1988). This is also a reason why research on the prevention of Parkinson's disease should focus on this enzyme and its performance whilst in advanced disease, a combination of dopaminergic drugs may be supplemented by some postreceptorial manipulation. Ideally, also special antagonists of central side effects (clozapine? dopamine partial antagonists?) have a potential of improving the quality of treatment in a similar way as the introduction of peripheral decarboxylase inhibitors (and now, also domperidone) did by reducing peripheral side effects.

In conclusion, and in order to provoke some discussion, we want to formulate a few theses:

1. A clinical function of D_1 receptors in Parkinson's disease needs to be established by the demonstration of antiparkinson effects by a pure D_1 agonist by its own or synergistic to a D_2 agonist.
2. Most dopaminergic effects can still be related to D_2 agonism; they can be classified by an inhibitory effect of domperidone and by their change with time (functional tolerance/supersensitivity).
3. More interesting new developments are continuous dopaminergic stimulation, the concept of partial agonism and selective interference with special dopaminergic effects or side effects.
4. An early diagnosis of presymptomatic Parkinson's disease will become possible by PET. A more practical approach could be a pharmacological "stress test" on motor function in elderly people.
5. In addition to further refinement of dopaminergic therapy by enzyme inhibitors, dopamine agonists and partial agonists, further research should focus on dopamine synthesis and on events happening before the dopaminergic depletion occurs as well as on other transmitter systems interacting with dop-

aminergic function. For this as for other strategies, close cooperation with clinicians, neuropathologists and clinical pharmacologists is necessary.

References

Birkmayer W, Riederer P (1985) Die Parkinson-Krankheit, 2. Aufl. Springer, Wien New York

Braun A, Fabbrini G, Mouradian MM, Senzti C, Barone P, Chase TN (1987) Selective D_1 dopamine receptor agonist treatment of Parkinson's disease. J Neural Transm 68: 41–50

Carlsson A (1987) Monoamines of the central nervous system. A historical perspective. In: Meltzer HY (ed) Psychopharmacology, the third generation of progress. Raven Press, New York, 39–48

Clark D, White FJ (1987) D_1 receptor–the search for a function. Synapse 1: 347–388

Gessa GL (1988) Agonist and antagonist actions of lisuride on dopamine neurons: electrophysiological evidence. J Neural Transm [Suppl] 27: 201–210

Horowski R (1987) Clinical neuropharmacology of DA agonists in Parkinson's disease. Neuroscience [Suppl] 22: 107

Jenner P, Boyce S, Marsden CD (1988) Receptor changes during chronic dopaminergic stimulation. J Neural Transm [Suppl] 27: 161–175

Kebabian JW, Calne DB (1979) Multiple receptors for dopamine. Nature 227: 93–96

Mogi M, Harada M, Kinchi K, Kojima K, Kondo T, Narabayashi H, Rausch D, Riederer P, Jellinger K, Nagatsu T (1988) Homospecific activity (activity per enzyme protein) of tyrosine hydroxylase increases in Parkinsonian brain. J Neural Transm 72: 77–81

Ruggieri S, Stocchi F, Antonini A, Bellantuono P, Carta A, Agnoli A (1987) Lisuride infusion in chronically treated fluctuating Parkinson's disease: effects of continuous dopaminergic stimulation. Trends Clin Neuropharmacol 1: 55–60

Schachter M, Bedard P, Debono AG, Jenner P, Marsden CD, Price R, Parkes JD, Keenan J, Smith B, Rosenthaler J, Horowski R, Dorow R (1980) The role of D_1 and D_2 receptors. Nature 286: 157–159

Wachtel H, Rettig KJ, Löschmann P-A (1988) Effect of chronic subcutaneous minipump infusion of lisuride upon locomotor activity of rats. J Neural Transm [Suppl] 27: 177–183

Correspondence: Dr. R. Horowski, Clinical Research of Schering AG, Postfach 650311, D-1000 Berlin 65.

Chemical modulation of membrane-bound receptors

W. Wesemann[1], N. Weiner[1], H.-O. Lausch[1], and H.-D. Mennel[2]

[1]Department of Neurochemistry, Institute of Physiological Chemistry
[2]Department of Neuropathology, Institute of Pathology, Philipps-University, Marburg/L., Federal Republic of Germany

Summary

Neurotransmitter receptor binding depends on membrane anisotropy. Endogenous processes such as ageing and exogenous factors such as neurotoxins or drugs can affect membrane anisotropy and receptor binding as shown for dopamine (DA) and serotonin (5-HT) receptors. In M. Parkinson membrane effects have to be considered for the pathogenesis as well as for the treatment of this disease.

Introduction

Two important steps in chemical mediated neurotransmission are the recognition of the signal molecule, i.e. the binding of the neurotransmitter by its receptor, and the transduction of the signal across the membrane. Both processes, the recognition and the transduction of the signal, take place within the membrane. If the membrane is regarded as a medium in which the proteins of the receptors, ionic channels, and second messenger systems are embedded, this medium may play a role with regard to the accessibility and the mobility of the corresponding proteins.

This presentation will focus on the possible influence of the physico-chemical properties of the membrane on the receptor kinetics. Moreover, the question will be addressed whether or not disturbances of neurotransmission may be explained, at least partly, by changes of the membrane characteristics.

Membrane anisotropy

Since according to the Fluid Mosaic Model the biological membrane is a twodimensional fluid, the terms "fluidity" and "microviscosity" had been introduced to illustrate the order and the diffusion aspects of the lipid constituents. These terms can be calculated from the fluorescence polarization P. Today the operational term membrane anisotropy, r = 2 P/3-P, is more widely used to characterize the membrane architecture. Increasing values of the membrane anisotropy are found when the fluidity of the membrane is reduced.

In vitro modulation of binding sites

If membranes isolated from rat brain are "fluidized" by incubation with increasing n-hexanol concentrations (Fig. 1), membrane anisotropy and specific [^{3}H]5-HT binding (5-HT_1 binding sites) are reduced (Wesemann et al., 1986). It could be demonstrated that both kinetic parameters, receptor affinity and the maximal number of binding sites (B_{max}), may be altered by the membrane anisotropy (Shinitzky, 1984). This effect, however, is specific for a given receptor population. Hence, in contrast to 5-HT_1 binding sites the binding of [^{3}H]spiroperidol as a measure of DA receptors is enhanced with decreasing anisotropy of striatal membranes (Henry and Roth, 1986).

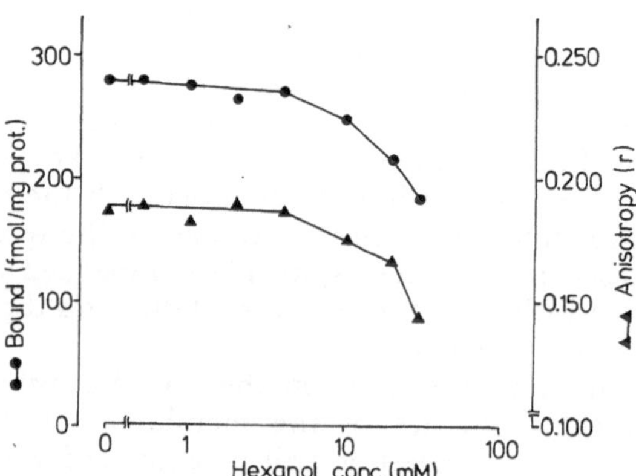

Fig. 1. Effect of n-hexanol on membrane anisotropy and 5-HT_1 binding sites in crude membranes of rat brain. Membrane anisotropy was calculated from the fluorescence polarization P using 2 μM 1,6-diphenyl-1, 3, 5-hexatriene (DPH) as fluorescence probe. Specific 5-HT binding was measured at 20 nM [^{3}H]5-HT by the filtration procedure. Membrane fluidity is significantly ($p < 0.05$) increased by up to 27% at 30 mM n-hexanol. Specific 5-HT binding is significantly ($p < 0.001$) decreased by 36% at 20 mM n-hexanol

In vivo modulation of binding sites

Ageing

From a theoretical point of view also in vivo membrane architecture and consequently membrane anisotropy may be affected by changing metabolic processes. Ageing is one example for this. In the striatum of the rat a small decrease of membrane anisotropy and a significant reduction of the accessibility of DA receptors is observed upon ageing. In addition, the total DA receptor concentration of brain membranes isolated from senescent rats is reduced as compared with that of mature animals (Henry and Roth, 1986).

Fig. 2 comparing two age groups shows that in the hippocampus, putamen, and in four cortical areas of post mortem human brain the membrane anisotropy is decreased with age while the specific [^{3}H]ketanserin binding (5-HT$_2$ binding sites) is enhanced.

With regard to M. Parkinson it has to be considered according to the results obtained with rat and post mortem brain tissues that besides the well-known DA deficit and the reduced number of DA receptors the accessibility of the receptors is modulated by age and/or pathophysiologically induced changes of the membrane anisotropy.

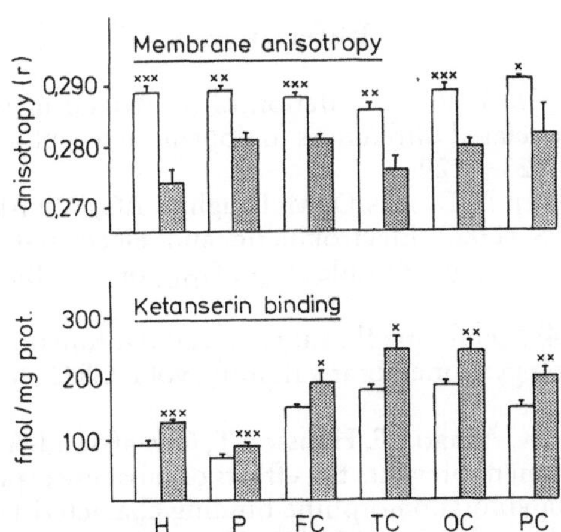

Fig. 2. Age dependence of membrane anisotropy and 5-HT$_2$ binding sites in crude membranes of post mortem human brain. Membrane anisotropy was calculated as described in the legend of Fig. 1. In these experiments 0.2 μM DPH was used. 5-HT$_2$ binding sites were measured at 2 nM [^{3}H]ketanserin. The mean values ± S.E.M. of 7–8 post mortem brains are compared with each other: * $p<0.05$, ** $p<0.01$, *** $p<0.001$. *H* hippocampus, *P* putamen, *FC* cortex frontalis, *TC* cortex temporalis, *OC* cortex occipitalis, *P* cortex parietalis. Age group: □ 52–65 y; ▩ 69–86 y

Environmental neurotoxic substances and drug treatment

The two above mentioned examples of DA and 5-HT binding sites clearly indicate that membrane anisotropy as well as receptor binding can be modulated by metabolic, that means by endogenous processes. The question arises whether or not binding sites may be changed also by exogenous substances like neurotoxins or drugs. If rats were exposed to toluene, the K_D and B_{max} values of N-[^{3}H]propylnorapomorphine ([^{3}H]NPA) binding (D_1 and D_2 receptors) in striatal membranes are increased. It was suggested that this effect is caused by a reduction of membrane anisotropy (Euler et al., 1987). To test this hypothesis, rats were treated with the ganglioside GM1 prior to toluene exposure and [^{3}H]NPA binding measurement. The ganglioside GM1 is known to increase membrane anisotropy (McDaniel et al., 1986). The toluene effect on [^{3}H]NPA binding was found to be blocked by ganglioside pretreatment which by itself had no significant effect. This is in accordance with the assumption that toluene exposure changes DA receptors by the effect on membrane anisotropy. Moreover, the results indicate that pathophysiological alterations of membrane anisotropy and receptor binding may be prevented by drug treatment.

References

Henry JM, Roth GS (1986) Modulation of rat striatal membrane fluidity: effects on age related differences in dopamine receptor concentrations. Life Sci 39: 1223–1229

McDaniel RV, Sharp K, Brooks D, McLaughlin AC, Winiski AP, Cafiso D, McLaughlin S (1986) Electrokinetic and electrostatic properties of bilayers containing gangliosides G_{M1}, G_{D1a}, or G_{T1}. Biophys J 49: 741–752

Shinitzky M (1984) Membrane fluidity and cellular functions. In: Shinitzky M (ed) Physiology of membrane fluidity, vol I. CRC, Boca Raton, pp 1–51

von Euler G, Fuxe K, Agnati LF, Hansson T, Gustafsson J-A (1987) Ganglioside GMl treatment prevents the effects of subacute exposure to toluene on N-[^{3}H]propylnorapomorphine binding characteristics in rat striatal membranes. Neurosci Lett 82: 181–184

Wesemann W, Weiner N, Hoffmann-Bleihauer P (1986) Modulation of serotonin binding in rat brain by membrane fluidity. Neurochem Int 9: 447–454

Correspondence: Prof. Dr. W. Wesemann, Department of Neurochemistry, Institute of Physiological Chemistry, Hans-Meerwein-Strasse, D-3550 Marburg/L., Federal Republic of Germany.

Transplantation of dopamine-synthesizing cells—new therapy for Parkinson's disease?

Comparison of strategies

W. H. Oertel[1] and **C. D. Marsden**[2]

[1] Department of Neurology, Klinikum Grosshadern, University of Munich, Federal Republic of Germany
[2] Institute of Neurology, University Department of Clinical Neurology, The National Hospital for Nervous Diseases, London, U.K.

Summary

In extensive research on rodents and in preliminary studies on non-human primates, allogeneic embryonic ventral mesencephalic grafts are superior to allogeneic adult adrenal medullary grafts in respect to intra-striatal graft survival and functional effectiveness in animal models of Parkinson syndrome. Intrastriatal grafting of Parkinson patients with their own adrenal medulla has so far given discrepant results. The adrenal medulla itself appears to be affected by Parkinson's disease. Transplantation of embryonic ventral mesencephalic tissue raises ethical concern. Issues such as method of application, target areas and long term immunological graft-host reaction await further research in non-human primates. In the future, catecholamine-synthesizing cell lines may provide a valid alternative to presently considered donor tissue.

Introduction

Parkinson's disease is associated with loss of dopaminergic mesencephalo-striatal/mesencephalo-cortical neurons. Drug therapy of Parkinson's disease includes L-DOPA (the precursor of dopamine), dopamine agonists, MAO-B-inhibitors, amantadine and anti-cholinergics, with L-DOPA being still the most effective agent. On average, in the first five years the neurological deficits are continously and consistently reduced by L-DOPA treatment. Thereafter problems such as L-DOPA-induced associated dyskinesias, wearing off and on-off-fluctuations, and L-DOPA-induced psychosis may emerge. The transplantation of dopamine-synthesizing cells may be a potential alternative to drug therapy. This article will discuss different strategies of transplantation research related to Parkinson's disease.

Criteria for transplanted cells (see Table 1)

Several criteria have to be fullfilled by the donor tissue in order to be suitable for transplantation:

- the grafted cells must survive for an extended period of time in a differentiated state;
- the grafted cells should not introduce an inflammatory response;
- the grafted cells should not be immunologically rejected (without or with immunosuppressive therapy);
- the grafted cells must have the capacity to synthesize and store dopamine;
- the grafted cells must release dopamine;
- the grafted cells must be able to reverse the behavioural deficit.

If transplantation is considered as an alternative to drug therapy, at least two of the following criteria should be met:

- the improvement of behavioural deficit should at least match the improvement obtained by drug therapy;
- the improvement of behavioural deficit should last longer than the improvement obtained with drug therapy;
- transplantation should have less side effects than drug therapy;
- transplantation should allow subsequent and/or concomitant drug therapy.

Table 1. Criteria for transplanted dopamine-synthesizing cells

– Survival for an extended period of time in a differentiated state
– No inflammatory response
– No immunological rejection (without or with immunosuppressive therapy)
– Synthesis and storage of dopamine
– Release of dopamine
– Reduction of behavioural deficit

Table 2

Comparison of	Drug therapy	Transplantation
Availability	yes	?
Selectivity	precursor (L-DOPA) D-1/2-agonist MAO-B-inhibitor amantadine anticholinergics	grafted tissue is a mixture of different cells
Tissue properties	–	embryonic ventral mesencephalon: healthy cells? adrenal medulla: affected by Parkinson's disease?
Distribution	systemic	focal
Frequency of treatment	daily	once multiple?
Duration of continous benefit	ca. 5 years (example: L-DOPA)	?
Problems	compliance – wearing off – on/off – dystonia – psychosis	immunology: delayed rejection? side effects of immunosuppression postoperative unknown
Peripheral side effects	yes	unlikely
Progression of disease retarded	no effect (MAO-B-inhibitors?)	?
Controllable	yes	?

These criteria have to be demonstrated in non-human primates (Hefti et al., 1985; Kordower et al., 1987).

Some advantages and disadvantages of the presently available drug therapy for Parkinson's disease and transplantation procedures are compared in Table 2.

In the following paragraphs, experiments in rodents, non-human primates and experimental therapy in human beings are discussed. Embryonic ventral mesencephalic cells and adrenal medullary tissue are compared as sources for dopamine synthesizing grafts.

Embryonic ventral mesencephalic tissue

Rat tissue (donor)—rat (host): allogeneic grafts

Most transplantation studies have so far been carried out in inbred laboratory rat strains. Prior to transplantation, the host rats received a unilateral stereotactic injection of 6-hydroxy-dopamine into the substantia nigra or medial forebrain bundle. This procedure leads to a nearly complete loss of dopaminergic innervation of the dorsal striatum (Ungerstedt and Arbuthnott, 1970). Such animals exhibit asymmetric motor behaviour (both spontaneously and in response to dopamine agonists and dopamine-releasing compounds), sensory inattention (towards stimuli applied to the side of the body contralateral to the dopamine-deficient striatum), and asymmetric posture (for review see: Björklund et al., 1987). Especially the rotational behaviour lends itself to quantification (Ungerstedt and Arbuthnott, 1970). Allogeneic rat embryonic ventral mesencephalic "anlage" has been employed as donor tissue.

Advantages

These studies have convincingly shown:

- embryonic dopaminergic neurons from the ventral mesencephalon survive transplantation in a dopamine-deficient striatum of adult rats for at least 12 months (the average life span of a laboratory rat is 30 months);
- allogeneic (donor and host are from the same species, but from different individual animals) donor tissue survives in the host brain for months without the need for immunosuppression, as the brain is an immunologically relatively privileged structure;
- the transplanted cells in principle form morphologically normal synaptic connections with their natural target cells in the striatum;
- transplanted cells produce dopamine and release dopamine (Strecker et al., 1987);
- the transplanted cells ameliorate or even abolish
  - a motor deficit or
  - a sensory deficit

  depending on the number of surviving dopaminergic neurons and the site of transplantation in the striatum (for review see: Björklund and Stenevi, 1985; Björklund et al., 1987);
- the stereotactic intraparenchymal injection of a suspension of embryonic cells is the method of choice for the application (Schmidt et al., 1983).

Disadvantages in respect to application in human beings

- allogeneic human embryonic tissue is obtained from aborted embryos. Although approved for experimental studies in rodents and non-human primates, ethical objections may not allow the use of human embryonic tissue for transplantation in human beings in many countries;
- a gestation period between about 7 and 12 weeks appears to be optimal for the survival of grafted embryonic dopamine-synthesizing cells (Brundin et al., 1988 b, c). Thus, only embryonic tissue of a particular age is suitable;
- dopaminergic neurons represent only a minority of all cells in the ventral mesencephalon. Thus, a suspension of a brain region is grafted containing non-neuronal cells and neurons with neurotransmitters different from dopamine such as serotonin, GABA and/or neuropeptides (Fine et al., 1988);
- long-term immunological rejection of grafted allogeneic tissue can presently not be ruled out.

Possible mechanisms of dopaminergic graft function

Based on these studies in rats, several mechanisms may influence, how grafted cells interact with the host:

- via a non-specific release of transmitter or/and hormones by diffusion;
- via reinnervation of elements in the host brain through the graft;
- via establishment of reciprocal or even more extensive graft–host connections;
- via a trophic action on the host brain
(for review see: Björklund et al., 1987).

Available data suggest that the first two mechanisms are to a large extent responsible for effects of dopamine-synthesizing grafts. The results with embryonic neural transplants in rodents have led to the suggestion that neural transplants might play a general role in the therapy of neurodegenerative disorders, and especially in the best characterized neurotransmitter deficit syndrome, i.e. Parkinson's disease.

Human tissue (donor)–rat (host): xenogeneic graft

As an initial experimental step human embryonic ventral mesencephalic tissue has been transplanted either as cell suspension (Brundin et al., 1986) or as solid graft (Strömberg et al., 1986). The xenogeneic (donor and host are of different species) graft tissue only reliably survived grafting into the rat dopamine-deficient striatum when the host was treated with immunosuppressives (cyclosporin A: 10 mg/kg).

A study employing mice donor tissue as transplant in rat host striatum supports the necessity of immunosuppression in xenogeneic grafting (Brundin et al., 1985 a). Without concommitant immunosuppressive therapy, xenogeneic donor tissue is rejected (Mason et al., 1986; Brundin et al., 1988 b; Widner et al., 1988).

Reduction of lesion-induced turning asymmetry was seen, when human embryonic ventral mesencephalic tissue of 7—9 (up to 12) weeks of gestation was grafted, but not of 13—19 weeks of gestation. When compared to grafted dopamine neurons of rat origin the grafted dopaminergic neurons of human origin were larger and possessed more prominent dendrites. In addition, reduction of rotational behaviour of rats with human grafts was observed after a longer time interval (12—20 weeks) than with mouse grafts (3 weeks). Thus the human donor tissue appeared to develop with a time course characteristic for human dopaminergic mesencephalic neurons (Brundin et al., 1986; Strömberg et al., 1986).

Despite extensive research, several questions cannot be answered in the rat and have to be solved in a host which is phylogenetically closer to human beings such as non-human primates. These questions are concerned with the target of transplantation, the behavioral benefit to the host, and long term side effects of transplantation.

Non-human primate tissue (donor)—non-human primate (host): allogeneic graft

The discovery of the neurotoxin (MPTP)-induced bilateral primate model of Parkinson's syndrome (Burns et al., 1983; Langston et al., 1983; Ballard et al., 1985) has stimulated transplantation research in non-human parkinsonian primates.

Target

In contrast to the rodent, non-human primates possess a basal ganglia structure very similar to the human. Correlations between motor function and certain nuclei in the basal ganglia in normal and MPTP-treated non-human primates have been found (Alexander et al., 1986; Miller and DeLong, 1987). For example the putamen has been shown to be related to specific aspects of limb movements, including direction, amplitude (and velocity) and load. In addition cells in the primate putamen may participate in the preparation of movements (Alexander et al., 1986).

With appropriate neurosurgical (stereotactic) techniques, dopamine-synthesizing cells can be focally grafted into the parenchyma of the putamen, into the body or into/onto the head of the caudate nucleus. This possibility appears to be of advantage, as dopamine loss in the striatum of patients with Parkinson's disease is uneven (Kish et al., 1988). Improvement of parkinsonian symptoms in non-human primates after transplantation is, therefore, of much greater clinical relevance than those observed in rodents.

Reduction of behavioural deficit

Grafting of allogeneic embryonic ventral mesencephalic cells has been reported in small numbers of rhesus monkeys (Bakay et al., 1988; Bankiewicz et al., 1988) and african green monkeys (Redmond et al., 1986; Sladek et al., 1988) with an MPTP-induced bilateral parkinsonian syndrome. Transplantation was done unilaterally into a cavity carved into the head of the caudate nucleus. The grafted cells therefore had access to the cerebrospinal fluid (CSF) as well as to the parenchyma of the caudate.

Considerable bilateral improvement of rigidity, bradykinesia and tremor has been reported after unilateral transplantation at time intervals varying between a few days and 120 days in rhesus monkey (Bakay et al., 1988; Bankiewicz et al., 1988) and african green monkey (Redmond et al., 1986; Sladek et al., 1988; Sladek and Gash, 1988).

In contrast, animals with control transplants, i.e. with grafted tissue incapable of synthesizing dopamine or other catecholamines failed to reduce the MPTP-induced neurological deficit.

Post-mortem biochemical measurements of dopamine and its metabolite homovanillic acid showed a nearly normal ratio of the metabolite to dopamine in the vicinity of the grafted dopaminergic neurons. In contrast, areas distant from the graft contained a high ratio of the metabolite to dopamine, indicating increased neurotransmitter turnover (Elsworth et al., 1988). Thus a unilateral graft appeared to improve a bilateral neurological deficit and to normalize dopamine turnover in a striatal area which was nearly devoid of its endogenous dopaminergic innervation due to MPTP-exposure.

Only preliminary data are available in common marmosets with MPTP-induced parkinsonian syndrome. They suggest an improvement of the animal's spontaneous locomotor activity up to 6 months after transplantation (Fine et al., 1988).

Technique of application

In rats the intraparenchymal stereotactic injection of suspensions of embryonic cells is widely used (Björklund and Stenevi, 1985). In one publication, however, the reduction of rotational behaviour in rats was reported to be more marked when grafted cells were not only present in the parenchyma, but grew inside the ventricle (Nishino et al., 1986). In non-human primates it is not clear whether a unilateral transplant into/onto the caudate nucleus with contact with the CSF, or a unilateral or bilateral graft into the parenchyma of the putamen or of the nucleus caudatus, or both areas, is the most effective way to ameliorate a bilateral neurological deficit. The observation that a unilateral graft with contact with the CSF reduces a bilateral neurological deficit in parkinsonian non-human primates suggests that, at least in part, a "pharmacological" action via a diffuse release of biologically active molecules such as dopamine into the cerebrospinal fluid and/or the host parenchyma may be responsible for the behavioural effect of the graft. In addition, neural mechanisms such as crossed cortico-striatal projections may play a role.

Immunology

Limited data are presently available on the immune response of non-human primates serving as a host to intracerebral transplantation of allogeneic embryonic tissue. Fiandaca and coworkers (1988) failed to detect antibodies to the grafted tissue in the animals' serum. Rejection was not seen for an observation period of more than twelve months (Sladek et al., 1988) in agreement with data in the rat (Brundin et al., 1985 a; Brundin et al., 1988 b; Widner et al., 1988). It is not known whether a second transplant from an immunologically different allogeneic donor performed several months or years later, will lead to the rejection of the first transplant without immunosuppression. This question may be of importance if such an approach were used in the human, as tissue of one human embryo may only be sufficient for a unilateral transplant into the putamen (Brundin et al., 1988 b).

Side effects

No side effects have been reported in the available literature on transplantation of embryonic ventral mesencephalic tissue in non-human primates.

In summary none of the studies published until April 1988 have been carried out under stringent conditions comparable with drug testing. This is in part due to the high costs and long time period required for evaluation. The latency between induction of a parkinson syndrome with MPTP and the day of transplantation varies from report to report. Only few studies have controlled for the possible degree of partial spontaneous recovery of the MPTP-induced parkinsonian syndrome. Several preliminary reports on a small number of non-human primates await confirmation with larger numbers of experimental animals. Long term follow-up studies are needed in order to compare drug treatment with transplantation therapy in respect to behavioral benefit and side effects, and to study immunological aspects.

Human tissue (donor)—human beings (host): allogeneic graft

In January 1988 the transplantation of human embryonic ventral mesencephalic tissue of 13 weeks of gestation was reported into/onto the head of the right caudate nucleus of a 50 year old Parkinson patient. The neurosurgical method used allowed the grafted tissue to contact both the CSF and the parenchyma of the target area.

The patient was treated with cyclosporin (2 mg/kg/day) and steroids (15 mg/day). Slight improvement of neurological symptoms was reported 8 weeks after the transplantation (Madrazo et al., 1988). Two patients have undergone transplantation of embryonic ventral mesencephalic tissue into the striatum in Sweden and at least three in Great Britain. Patients in Great Britain were claimed to be improved within weeks of the transplantation in March/April 1988 (Vines, 1988). The Swedish patients were transplanted in the autumn of 1987. Full details of the methods and effects for six months thereafter are to be published soon.

Tissue culture of embryonic ventral mesencephalic cell—cryobiology

Availability of embryos at the appropriate gestation date may be limited.

Methods to keep (Berger et al., 1982) and to enrich embryonic mesencephalic cells in culture are under development. New techniques such as immunological identification of cell membrane proteins in living cells and flow cytometry for cell sorting are being used to try to enrich embryonic dopaminergic neurons in tissue culture (Di Porzio et al., 1987). Rat embryonic mesencephalic tissue has been cultured up to 5 days, in one instance even transported for two days, and then grafted. Despite the delay the transplanted tissue was still capable to reduce rotational behaviour in the rat model. As cultivation was initiated on the 14—16th day of gestation, the transplanted tissue was at embryonic day 16—18 on the day of transplantation

(Brundin et al., 1985 b; Brundin et al., 1988 a). This report raises the question whether embryonic ventral mesencephalic tissue may be kept in tissue culture for few days, modified and/or enriched and may still remain in a condition suitable for successful grafting.

Cocultivation of embryonic ventral mesencephalic cells with embryonic striatal target cells has so far failed to increase the number of dopaminergic neurons, although dopamine uptake and synthesis were enhanced. These phenomena are compatible with enhanced maturation of embryonic mesencephalic neurons in culture in the presence of their natural target cells (Prochiantz et al., 1979 a, b; Di Porzio et al., 1980). Finally the survival of embryonic mesencephalic neurons has been studied after prolonged freezing by means of cryobiological techniques (Collier et al., 1988).

Adrenal medullary tissue

In the search for a catecholamine-synthesizing tissue which is available without public ethical concern and does not bear the danger of immunological rejection, the Parkinson patient's own adrenal medulla has been used.

Disadvantages

Not only the central nervous system, but also the adrenal medulla appears to be affected in Parkinson's disease, as

- the occurrence of "adrenal bodies" and Lewy bodies has been described in adrenal medulla (Den Hartog Jager, 1970; Jellinger, this volume),
- the level of activity of the enzyme tyrosine hydroxylase, the rate limiting enzyme for the synthesis of dopamine, has been reported to be decreased (Riederer et al., 1978),
- the concentration of catecholamines in the adrenal medulla is reduced (Carmichael et al., 1988);

Furthermore

- the grafted tissue contains not only chromaffin cells, but vessels, muscle cells, connective tissue etc.;
- the adrenal medulla is obtained in a major abdominal operation under general anaesthesia lasting more than one hours, which may impair the patients neurological condition;
- the adrenal medulla lacks a blood-tissue-barrier:

Thus if host vessels contacted capillaries of the graft, substances which can pass fenestrated vessels should have access to the cerebrospinal fluid and/or parenchyma of the transplanted host area. Provided this were the case in human beings, administration of a DOPA-decarboxylaseinhibitor in combination with L-DOPA to Parkinson patients with an adrenal medullary transplant might reduce the production of dopamine in the adrenal medulla implant, as the "peripheral" inhibitor would block the enzyme dopa-decarboxylase in the graft.

Advantages

- no public ethical concern, as the consent to the procedure is given by the patient itself. Whether this approach presently fullfills the standards of medical ethics is debatable;
- immunological rejection is not possible, as the patients own tissue is used;
- catecholamine production: chromaffin cells of the adrenal medulla produce catecholamines with a preponderance of noradrenaline and adrenaline (Müller and Unsicker, 1986). When isolated and placed into the ventricle or under tissue culture conditions adrenal medullary tissue has been reported to secret dopamine and noradrenaline in about equal proportions (Freed et al., 1983; Patel-Vaidya et al., 1985);
- adrenal medullary tissue can be kept in tissue culture for several days prior to transplantation (Patel-Vaidya et al., 1985).

Rat adult tissue (donor)—rat (host): allogeneic graft

Since 1981 allogeneic adrenal medullary tissue of young or adult rats has been implanted into the dopamine-deficient striatum of adult rats. These grafts were placed either intraventricularly or into the striatal parenchyma. Grafted adrenal chromaffin cells maximally survived 6 months (Freed et al., 1981, 1983, 1986 a). The additional intraventricular infusion of nerve growth factor for 4 weeks via minipumps prolonged the survival time and enhanced differentiation of the grafted cells. In addition it resulted in some neurite outgrowth into the host neuropil. This later phenomen was reported to be dependent on the presence of nerve growth factor. In none of the experiments, however, was a complete recovery of the behavioural deficit achieved (Strömberg et al., 1985), in contrast to reports with embryonic ventral mesencephalic cells.

Non-human primate tissue (donor)—non-human primates (host): allogeneic graft

In only one study adult allogeneic adrenal medullary tissue was transplanted into the dopamine-deficient striatum of non-human primates (Morihisa et al., 1984). Survival of a few hundred chromaffin cells was reported, but no consistent improvement of the unilateral behavioural deficit was observed.

Embryonic human tissue (donor)—rat (host): xenogeneic graft

Embryonic human adrenal medullary tissue of 11 weeks of gestation has been transplanted into a dopamine-deficient striatum of rats. The tissue was cultured for 3 weeks in vitro before transplantation. Cells exhibiting catecholamine histofluorescence survived for 4 weeks without immunosuppression. No histological signs of inflammation or rejection were observed. A longer survival time was not studied nor was the effect of the transplant on the animals behaviour reported (Kamo et al., 1985). Immunological studies were not carried out.

Embryonic human tissue (donor)—non-human primate (host): xenogeneic graft

In a preliminary report xenogeneic embryonic human adrenal medullary tissue between 16 and 18 weeks of gestation was purified and cultured in the presence of nerve growth factor prior to transplantation (Silani et al., 1988). Such pretreated cells were then stereotactically grafted into one putamen of one rhesus monkey rendered parkinsonian with MPTP. Neurological symptoms started to improve 7 days after transplantation. Later on ipsilateral rotation was observed after slight physical stimulation or treatment with apomorphine, compatible with some restoration of dopaminergic function in the grafted striatum (Pezzolli et al., 1988). Immunological studies in this model would be of interest.

Adult human tissue (donor)—human beings (host): syngeneic graft

Parkinson patients have been grafted with their own adrenal medullary tissue as an experimental transplantation therapy since 1981. The results are controversial.

Four patients were operated until the end of 1987 in Sweden by means of a stereotactic procedure. Tissue was applied with a coil. Two patients received adrenal medullary tissue into the parenchyma of the nucleus caudatus (Backlund et al., 1985; Olsen et al., 1986) and two patients into the parenchyma of the putamen (Lindvall et al., 1987). None of the patients showed any longterm improvement. In addition, the progression of the disease did not appear to be influenced by the transplantation. On the other hand, side affects were not observed either.

In contrast dramatic improvement of neurological symptoms were reported in two young individuals with Parkinson syndrome following syngeneic adrenal medullary grafting by a group in Mexico (Madrazo et al., 1987). Their neurosurgical method differed from the Swedish approach, as pieces of adrenal medullary tissue were implanted into a cavity carved into the head of one caudate nucleus during open surgery. This method allows the grafted tissue to be in contact with the parenchyma of the nucleus caudatus as well as with the cerebrospinal fluid.

Since this publication more than 100 patients worldwide have undergone this procedure using methods similar to, but in general less traumatic than the technique employed by the Mexican group. Countries in which this has been done include Mexico, Cuba, the United States of America, the People's Republic of China, Italy, Spain and Portugal. The Mexican group, and also a group from Cuba report successful transplantation therapy in young patients, but not in elderly Parkinson patients. The later group includes at least two Western European patients who underwent adrenal medullary grafting in Mexico without any functional benefit (personal communication).

None of the centers in the United States of America (Goetz et al., 1988; Olanow et al., 1988; Tanner et al., 1988) have so far reported a response comparable to the results described in patients operated in Mexico.

Biochemical data suggest, that observed improvement of neurological symptoms does not relate to a dopaminergic mechanism. At least one post mortem study has been performed in the USA on a patient with an adrenal medullary graft who was operated in Mexico. The graft was devoid of viable cells at the time of death (Peterson et al., 1988).

It is presently unclear why transplantation of syngeneic adrenal medullary tissue in Mexico is reported to result in dramatic improvement of motor and cognitive deficits whereas centers in the USA and Europe can so far not confirm these results. Speculations relate for example to the extent of trauma caused by the open surgery, the age of the host, and thus the capacity of the host brain to vascularize the graft or to provide trophic factors for graft survival. In young patients transplanted tissue may survive grafting better and/or contain more trophic factors than old donor tissue. In addition the grafted tissue may absorb an immunological factor directed against substantia nigra compacta cells (Carvey et al., 1988).

From the presently available scientific information several lines of evidence suggest that the adrenal medulla of Parkinson patients is affected by the disease process. One post mortem study failed to demonstrate survival of grafted catecholamine-synthesizing adrenal medullary cells. As discussed above, the administration of DOPA-decarboxylase inhibitor, however, may prevent dopamine production in the adrenal medullary transplant. A beneficial effect of this procedure might therefore be obscured by concommitant L-DOPA/DOPA decarboxylase inhibitor therapy.

Embryonic human tissue (donor)—human beings (host): allogeneic graft

The first transplant of allogeneic embryonic adrenal medullary tissue into/onto the head of the right caudate nucleus has been performed in a 35 year old Parkinson patient in autumn 1987. The tissue was obtained at 13 weeks of gestation. An improvement of the host's parkinsonian syndrome was described two months post transplantation in a short communication (Madrazo et al., 1988).

Comparison of allogeneic adult adrenal medulla and allogeneic embryonic ventral mesencephalic tissue

Only one comparison between the effect of allogeneic embryonic ventral mesencephalic tissue and adult allogeneic adrenal medullary tissue has been carried out in the unilateral rotation model in rhesus monkeys induced by injection of MPTP into one carotid artery (Bankiewicz et al., 1988). The embryonic ventral mesencephalic cells grafted into/onto the dopamine deficient nucleus caudatus reversed the drug-induced turning for at least six months. Adrenal medullary tissue was effective for 3 months, but at 6 months the transplants showed no effect and the animals showed rotational behaviour not different from control animals.

Catecholamine-synthesizing cell lines

In light of the public ethical concern related to the use of embryonic human cells for transplantation therapy of Parkinson patients, the so-far disappointing results in animal studies with the grafting of allogeneic adrenal medullary tissue and the controversial results of transplantation of syngeneic adrenal medullary tissue in

Parkinson patients, studies on the use of other types of dopamine (catecholamine-)synthesizing cells would be of value. One alternative could be the use of a permanent (tumor) cell line containing the enzymes for the synthesis of catecholamines. These cells would be constantly available, could be well-characterized and manipulated in vitro.

In the case of these cell lines an additional prerequisite for transplantation has to be fulfilled: the cells should stop proliferating once implanted (Hefti et al., 1985).

Particular chemical treatment converts these tumour cells into an amitotic state. Potential candidates include the PC 12 cell line (Hefti et al., 1985; Jaeger et al., 1985, 1987; Freed et al., 1986 b) the neuroblastoma cell line LA-N-1 (Seeger et al., 1977) or the neuroblastoma cell line IMR-32 (Gash et al., 1986).

Studies related to transplantation of these cell lines in animal models of Parkinson's disease are at an early stage. They lack long term follow-up of the behavioural effect of transplanted cell lines combined with biochemical and immunohistochemical assessment. However, this research may be of extreme importance for future clinical application.

Table 3. Comparison of presently used donor tissue

	Adrenal medulla	Embryonic ventral mesencephalic tissue
Public concern	no	yes
Immunology	no	(yes)
Tissue	mixture of cells: likely to be affected by Parkinson's disease	mixture of cells: healthy? contamination?
Effectiveness	animals:	
	rat: 6 months; nerve growth factor prolongs survival	rat: at least 12 months
	monkey: 3 months	monkey: at least 1 year
	human: discrepancy of results	human: unknown
Number of patients operated	>150	>5

Outlook

Molecular genetic techniques and tissue culture methods may help to modify existing catecholamine-synthesizing cell lines or even to design defined cell lines capable of synthesizing, storing and releasing dopamine (Gage et al., 1987). The genetic code for some of the necessary proteins for example for tyrosine-hydroxylase (Craig et al., 1986) is known. The use of "growth or survival factors" (Anderson and Axel, 1986; Rydel and Greene, 1987) for the modification and survival of embryonic and adult catecholamine-synthesizing cells will have to be explored further. The search for mechanisms which may promote dendritic outgrowth (Chamak et al., 1987; Barbin et al., 1988) or axonal outgrowth (Bohn et al., 1987) of dopaminergic mesencephalic neurons will be intensified.

Conclusion

Table 3 summarizes selected problems to be faced in grafting of adrenal medulla or embryonic ventral mesencephalic tissue. Transplantation studies in rodents have provided convincing evidence that embryonic ventral mesencephalic tissue substantially improves a unilateral dopamine deficit syndrome.

Studies in non-human primates are preliminary. Issues such as the type of donor tissue, the technique of application of the graft, the target area and the immunological graft-host reaction await controlled and long term follow-up investigations with a sufficient number of non-human primates. Whether embryonic tissue, either of the adrenal medulla or the ventral mesencephalon of non-human primates, may be used as graft tissue for human beings is an untouched problem.

The major obstacle to the use of human embryonic ventral mesencephalic tissue is the public and medical ethical concern. The same ethical problems remain with the use of human embryonic medullary adrenal tissue. Whether a population of purified adrenal medullary cells converted to catecholaminergic cells by exposure to nerve growth factor is equal or even superior to embryonic ventral mesencephalic neurons remains to be clarified.

Syngeneic or allogeneic human adult adrenal medullary tissue would bypass ethical objections. The available experimental data, however, and the limited information on clinical experimental transplantation therapy show this tissue to be inferior to embryonic ventral mesencephalic tissue in respect to graft survival, functional effectiveness and also its dependence on nerve growth factor.

Most of the studies mentioned dealt with aspects of naturally available donor tissue and the behavioural and immunological response of the host to the graft.

One problem not studied so far at all is the question whether the underlying cause of Parkinson's disease will eventually affect the transplanted tissue. This problem may be answered only in clinical trials (Brundin et al., 1988 c). On the other hand, this question may be of little clinical relevance, if a sufficient number of healthy dopamine-synthesizing cells were grafted, as Parkinson's disease appears to become clinically manifest only after a long subclinical period and it may take years for endogenous or exogenous factors to destroy the transplanted dopaminergic neurons.

In summary numerous problems remain to be solved by basic research, before transplantation of catecholamine-synthesizing cells may be considered as treatment of Parkinson syndrome.

Acknowledgement

Supported by Deutsche Forschungsgemeinschaft grant Oe 95/3-1 and by Medical Research Council grant.

References

Alexander GE, DeLong MR, Strick PL (1986) Parallel organization of functionally segregated circuits linking basal ganglia and cortex. Ann Rev Neurosci 9: 357–381

Anderson DJ, Axel R (1986) A bipotential neuroendocrine precursor whose choice of cell fate is determined by NGF and glucocorticoids. Cell 47: 1079–1090

Backlund E-O, Granberg PO, Hamberger B, Knutsson E, Martensson A, Sedvall G, Seiger A, Olson L (1985) Transplantation of adrenal medullary tissue to striatum in parkinsonism. J Neurosurg 62: 169–173

Bakay RAE, Sweeney KM, Colbassani HJ, Collins DC (1988) Delayed stereotactic transplantation technique in primates. In: Gash DM, Sladek JR Jr (eds) Transplantation into the mammalian CNS: preclinical and clinical studies. Prog Brain Res 74 (in press)

Ballard PA, Tertrud JW, Langston JW (1985) Permanent human parkinsonism due to 1-methyl-4-phenyl-1, 2, 3, 6-tetrahydropyridine (MPTP). Neurology 35: 949–956

Bankiewicz KS, Plunkett RJ, Oldfield EH, Jacobowitz DM, Porrino LJ, Vaidya U, Di Porzio U, Schuette WH, Markowitz A, London WT, Kopin IJ (1988) Transient and long term functional improvement by adrenal and fetal mesencephalic implants into caudate nuclei of MPTP parkinsonian monkeys. In: Gash DM, Sladek JR Jr (eds) Transplantation into the mammalian CNS: preclinical and clinical studies. Prog Brain Res 74 (in press)

Barbin G, Katz DM, Chamak B, Glowinski J, Prochiantz A (1988) Brain astrocytes express region-specific surface glycoproteins in culture. Glia 1: 96–103

Berger B, Di Porzio MC, Daguet M-C, Gay M, Vigny A, Glowinski J, Prochiantz A (1982) Long-term development of mesencephalic dopaminergic neurons of mouse embryos in dissociated primary cultures: morphological and histochemical characteristics. Neuroscience 7: 193–206

Björklund A, Stenevi U (eds) (1985) Neural grafting in the mammalian CNS. Elsevier, Amsterdam

Björklund A, Lindvall O, Isacson O, Brundin P, Wictorin K, Strecker RE, Clarke DJ, Dunnett SB (1987) Mechanisms of action of intracerebral neural implants: studies on nigral and striatal grafts to the lesioned striatum. Trends Neurosci 10: 509–516

Bohn MC, Cupit L, Marciano F, Gash DM (1987) Adrenal medulla grafts enhance recovery of striatal dopaminergic fibers. Science 237: 913–916

Brundin P, Nilsson OG, Gage FH, Björklund A (1985 a) Cyclosporin A increases survival of cross-species intrastriatal grafts of embryonic dopamine-containing neurons. Exp Brain Res 60: 204–208

Brundin P, Barbin G, Isacson O, Mallat M, Chamak B, Prochiantz A, Gage FH, Björklund A (1985 b) Survival of intracerebrally grafted rat dopamine neurons previously cultured in vitro. Neurosci Lett 61: 79–84

Brundin P, Nilsson OG, Strecker RE, Lindvall O, Astedt B, Björklund A (1986) Behavioural effects of human fetal dopamine neurons grafted in a rat model of Parkinson's disease. Exp Brain Res 65: 235–240

Brundin P, Barbin G, Strecker RE, Isacson O, Prochiantz A, Björklund A (1988 a) Survival and function of dissociated rat dopamine neurones grafted at different developmental stages or after being cultured in vitro. Develop Brain Res 39: 233–243

Brundin P, Strecker RE, Widner H, Clarke DJ, Nilsson OG, Astedt B, Lindvall O, Björklund A (1988 b) Human fetal dopamine neurons grafted in a rat model of Parkinson's disease: immunological aspects, spontaneous and drug-induced behaviour, and dopamine release. Exp Brain Res 70: 192–208

Brundin P, Strecker RE, Lindvall O, Isacson O, Nilsson OG, Barbin G, Prochiantz A, Forni C, Nieoullon A, Widner H, Gage FH, Björklund A (1988 c) Intracerebral grafting of dopamine neurons: experimental basis for clinical trials in patients with Parkinson's disease. In: Gash DM, Sladek JR Jr (eds) Transplantation into the mammalian CNS: preclinical and clinical studies. Prog Brain Res 74 (in press)

Burns RS, Chiueh CC, Markey SP, Ebert MH, Jacobowitz DM, Kopin IJ (1983) A primate model of parkinsonism: selective destruction of dopaminergic neurons in the pars compacta of the substantia nigra by N-methyl-4-phenyl-1, 2, 3, 6-tetrahydropyridine. Proc Nat Acad Sci (USA) 80: 4546–4550

Carmicheal SW, Wilson RJ, Brimijoin WS, Melton LJ III, Okazaki H, Yaksh TL, Ahlskog JE, Stoddard SL, Tyce GM (1988) Decreased catecholamines in the adrenal medulla of patients with Parkinsonism. New Engl J Med 318: 254

Carvey PM, Kroin JS, Zhang TJ, O'Dorisio TM, Yaksh TL, McRea A, Dahlstrom A, Kao LC, Penn RD, Goetz CG, Tanner CM, Shannon KM,

Klawans HL (1988) Biochemical and immunochemical characterization of ventricular CSF from Parkinson's disease (PD) patients with adrenal medulla transplants in the patients. Neurology 38 [Suppl] 1: 144

Chamak B, Fellous A, Autillo-Touati, Barbin G, Prochiantz A (1987) Are neuronotrophic neuron-astrocyte interactions regionally specified? Ann NY Acad 495: 528–535

Collier TJ, Gallagher MJ, Sladek CD, Blanchard BC, Daley BF, Foster PN, Redmond DE Jr, Roth RH, Sladek JR Jr (1988) Cryopreservation of fetal rat and non-human primate mesencephalic neurons: viability in culture and neural transplantation. In: Gash DM, Sladek JR Jr (eds) Transplantation into the mammalian CNS: preclinical and clinical studies. Prog Brain Res 74

Craig SP, Buckle VJ, Lamourous A, Mallet J, Craig I (1986) Localization of the human tyrosine hydroxylase gene to 11 p 15: gene duplication and evolution of metabolic pathways. Cytogenet Cell Genet 42: 29–32

Den Hartog Jager WA (1970) Histochemistry of adrenal bodies in Parkinson's disease. Arch Neurol 23: 528–533

Di Porzio U, Daguet MC, Glowinski J, Prochiantz A (1980) Effect of striatal cells on in vivo maturation of mesencephalic dopaminergic neurons grown in serum-free conditions. Nature 288: 370–373

Di Porzio U, Rougon G, Novotny EA, Barker JL (1987) Dopaminergic neurons from embryonic mouse mesencephalon are enriched in culture through immunoreaction with monoclonal antibody to neural specific protein 4 and flow cytometry. Proc Nat Acad Sci (USA) 84: 7334–7338

Elsworth JD, Redmond DE Jr, Deutch AY, Collier TJ, Sladek JR Jr, Roth RH (1988) Dopamine production by fetal neuron transplants in the caudate nucleus of the MPTP-treated primate. In: Gash DM, Sladek JR Jr (eds) Transplantation into the mammalian CNS: preclinical and clinical studies. Prog Brain Res 74 (in press)

Fiandaca MS, Bakay RAE, Sweeney KM, Chan WC (1988) Immunologic response to intracerebral fetal neural allografts in the rhesus monkey. In: Gash DM, Sladek JR Jr (eds) Transplantation into the mammalian CNS: preclinical and clinical studies. Prog Brain Res 74 (in press)

Fine A, Oertel WH, Hunt S, Chong PN, Nomoto M, Waters CM, Temlett J, Dunett S, Annett L, Jenner P, Marsden CD (1988) Transplantation of embryonic marmoset dopaminergic neurons to the corpus striatum of marmosets rendered parkinsonian by MPTP. In: Gash DM, Sladek JR Jr (eds) Transplantation into the mammalian CNS: preclinical and clinical studies. Prog Brain Res 74 (in press)

Freed WJ, Morihisa JM, Spoor E, Hoffer BJ, Olson L, Seiger A, Wyatt RJ (1981) Transplanted adrenal cromaffin cells in rat brain reduce lesion-induced rotational behavior. Nature 292: 351–352

Freed WJ, Karoum F, Spoor HE, Morihisa JM, Olson L, Wyatt RJ (1983) Catecholamine content of intracerebral adrenal medulla. Brain Res 269: 184–189

Freed WJ, Cannon-Spoor H, Krauthammer E (1986 a) Intrastriatal adrenal medulla grafts in rats. Long-term survival and behavioral effects. J Neurosurg 65: 664–670

Freed WJ, Patel-Vaidya U, Geller HM (1986 b) Properties of PC 12 pheochromocytoma cells transplanted to the adult rat brain. Exp Brain Res 63: 557–566

Gage FH, Wolff JA, Rosenberg MB, Xu L, Yee J-K, Shults C, Friedmann T (1987) Grafting genetically modified cells to the brain: possibilities for the future. Neuroscience 23: 795–807

Gash DM, Notter MFD, Okawara SH, Kraus AL, Joynt RJ (1986) Amitotic neuroblastoma cells used for neural implants in monkeys. Science 233: 1420–1422

Goetz CG, Tanner CM, Penn RD, Shannon KM, Klawans HL (1988) Efficacy of intrastriatal adrenal medulla transplant in Parkinson's disease. Neurol 38 [Suppl] 1: 142

Hefti F, Hartikka J, Schlumpf M (1985) Implantation of PC 12 cells into the corpus striatum of rats with lesions of the dopaminergic nigrostriatal neurons. Brain Res 348: 283–288

Jaeger CB (1985) Immunocytochemical study of PC 12 cells grafted to the brain of immature rats. Exp Brain Res 59: 615–624

Jaeger CB (1987) Morphological and immunocytochemical characteristics of PC 12 cell grafts in rat brain. Ann NY Acad Sci 495: 334–349

Kamo H, Kim SU, McGeer PL, Shin DH (1985) Transplantation of cultured fetal human adrenal chromaffin cells to rat brain. Neurosci Lett 57: 43–48

Kish SJ, Shannak K, Hornykiewicz O (1988) Uneven pattern of dopamine loss in the striatum of patients with idiopathic Parkinson's disease. New Engl J Med 318: 876–880

Kordower JH, Notter MFD, Yeh HH, Gash DM (1987) An in vivo and in vitro assessment of differentiated neuroblastoma cells as a source of donor tissue for transplantation. Ann NY Acad Sci 495: 606–621

Langston JW, Ballard P, Tertrud JW, Irwin I (1983) Chronic parkinsonism in humans due to a product of meperidine-analog synthesis. Science 219: 979–980

Lindvall O, Backlund E-O, Farde L, Sedvall G, Freedman R, Hoffer B, Nobin A, Seiger A, Olson L (1987) Transplantation in Parkinson's disease: two cases of adrenal medullary grafts to the putamen. Ann Neurol 22: 457–468

Madrazo I, Drucker-Colin R, Diaz V, Martinez-Mata J, Torres C, Becerril JJ (1987) Open microsurgical autograft of adrenal medulla to the right caudate nucleus in two patients with intractable Parkinson's disease. New Engl J Med 316: 831–834

Madrazo I, Leon V, Torres C, del Aguilera C, Varela G, Alvarez F, Fraga A, Drucker-Colin R, Ostrosky F, Skurovich M, Franco R (1988) Transplantation of fetal substantia nigra and adrenal medulla to the caudate nucleus in two patients with Parkinson's disease. New Engl J Med 318: 51

Mason DW, Charlton HM, Jones AJ, Lavy CBD, Puklavec M, Simmonds SJ (1986) The fate of allogeneic and xenogeneic neuronal tissue transplanted into the third ventricle of rodents. Neuroscience 19: 685–694

Miller WC, DeLong MR (1988) Parkinsonian symptomatology. An anatomical and physiological analysis. Ann NY Acad Sci 515: 287–302

Morihisa JM, Nakamura RK, Freed WJ, Mishkin M, Wyatt RJ (1984): Adrenal medulla grafts survive and exhibit catecholamine-specific fluorescence in the primate brain. Exp Neurol 84: 643–653

Müller TH, Unsicker K (1986) Nerve growth factor and dexamethasone modulate synthesis and storage of catecholamines in cultured rat adrenal medullary cells: dependence on postnatal age. J Neurochem 46: 516–524

Nishino H, Ono T, Takahashi J, Kimura M, Shiosaka S, Tohyama M (1986) Transplants in the peri- and intraventricular region grow better than those in the central parenchyma of the caudate. Neurosci Lett 64: 184–190

Olanow CW, Cahill D, Cox C (1988): Autologous transplantation of adrenal medulla to caudate nucleus in Parkinson's disease. Neurol 38 [Suppl] 1: 142

Olson L, Backlund E-O, Gerhardt G, Hoffer B, Lindvall O, Rose G, Seiger A, Strömberg I (1986) Nigral and adrenal grafts in Parkinsonism: recent basic and clinical studies. In: Yahr MD, Bergmann KJ (eds) Parkinson's disease. Adv Neurol 45: 85–94

Patel-Vaidya U, Wells MR, Freed WJ (1985) Survival of dissociated adrenal chromaffin cells of rat and monkey transplanted into rat brain. Cell Tiss Res 240: 281–285

Peterson DI, Price ML, Small CS, Linda L (1988) Autopsy findings in a patient that had an adrenal-to-brain transplant for Parkinson's disease. Neurol 38 [Suppl] 1: 144

Pezzolli G, Goodman R, Ferrante C, Silani V, Yebenes J, Truong D, Jackson-Lewis V, Fahn S (1988) Human fetal adrenal medullary cells reduce experimental parkinsonism in monkey. In: Gash DM, Sladek JR Jr (eds) Transplantation into the mammalian CNS: preclinical and clinical studies. Prog Brain Res 74 (in press)

Prochiantz A, Di Porzio U, Dato A, Berger B, Glowinski J (1979 a) In vitro maturation of mesencephalic dopaminergic neurons from embryos is enhanced in presence of their striatal target cells. Proc Nat Acad Sci (USA) 76: 5387–5391

Prochiantz A, Daguet MC, Herbet A, Glowinski J (1979 b) Specific stimulation of in vitro maturation of mesencephalic dopaminergic neurons by striatal membranes. Nature 293: 570–572

Redmond E, Sladek JR Jr, Roth RH, Collier TJ, Elsworth JD, Deutch AY, Haber S (1986) Fetal neuronal grafts in monkeys given methylphenyltetrahydropyridine. Lancet i: 1125–1127

Riederer P, Rausch WD, Birkmayer W, Jellinger K, Seemann D (1978) CNS modulation of adrenal tyrosine hydroxylase in Parkinson's disease and metabolic encephalopathies. J Neural Transm [Suppl] 14: 121

Rydel RE, Greene LA (1987) Acidic and basic fibroblast growth factors promote stable neurite outgrowth and neuronal differentiation in cultures of PC12 cells. J Neurosci 7: 3639–3653

Schmidt RH, Björklund A, Stenevi U, Dunnett SB, Gage FH (1983) Activity of intrastriatal nigral suspension implants as assessed by measurements of dopamine synthesis and metabolism. Acta Physiol Scand [Suppl] 522: 19–28

Seeger RC, Rayner SA, Banerjee A, Chung H, Lang WE, Neustein HB, Benedict WF (1977) Morphology, growth, chromosomal pattern and fibrinolytic activity of two new human neuroblastoma lines. Cancer Res 37: 1364–1367

Silani V, Pezzolli G, Motti E, Ferrante C, Falini A, Pizzuti A, Zecchinelli A, Moggio M, Buscaglia M, Scarlato G (1988) Human fetal chromaffin cells for neurotransplantation in parkinsonian patients. In: Gash DM, Sladek JR Jr (eds) Transplantation into the mammalian CNS: preclinical and clinical studies. Prog Brain Res 74 (in press)

Sladek JR Jr, Redmond DE Jr, Roth RH, Elsworth JD, Deutch AY, Haber SN, Blount J, Collier TJ (1988) Long term reversal of MPTP-induced parkinsonism in primates with fetal dopamine nerve cell transplants. In: Gash DM, Sladek JR Jr (eds) Transplantation into the mammalian CNS: preclinical and clinical studies. Prog Brain Res 74 (in press)

Sladek JR Jr, Gash DM (1988) Nerve-cell grafting in Parkinson's disease. Review article. J Neurosurg 68: 337–351

Strecker RE, Sharp T, Brundin P, Zetterström T, Ungerstedt U, Björklund A (1987) Autoregulation of dopamine release and metabolism by intrastriatal nigral grafts as revealed by intracerebral dialysis. Neuroscience 22: 169–178

Strömberg I, Herrera-Marschitz M, Ungerstedt U, Ebendal T, Olson L (1985) Chronic implants of chromaffin tissue into the dopamine-denervated striatum. Effects of NGF on graft survival, fiber growth and rotational behavior. Exp Brain Res 60: 335–349

Strömberg I, Bygdeman M, Goldstein M, Seiger A, Olson L (1986) Human fetal substantia nigra grafted to the dopamine-denervated striatum of immunosuppressed rats: evidence for functional reinnervation. Neurosci Lett 71: 271–276

Tanner CM, Goetz CG, Gilley DW, Shannon KM, Stebbins GT, Klawans HL, Wilson RS, Penn RM (1988) Behavioral aspects of intrastriatal adrenal medulla transplant surgery in Parkinson's disease (PD). Neurology 38 [Suppl] 1: 143

Ungerstedt U, Arbuthnott GW (1970) Quantitative recording of rotational behavior in rats after 6-hydroxy-dopamine lesions of the nigrostriatal dopamine system. Brain Res 24: 485–493

Vines G (1988) First British implant of fetal tissue. New Scientist: 22

Widner H, Brundin P, Björklund A, Möller E (1987) Immunological aspects of neural grafting in the mammalian central nervous system. In: Gash DM, Sladek JR Jr (eds) Transplantation into the mammalian CNS: preclinical and clinical studies. Prog Brain Res 74 (in press)

Correspondence: Dr. W. H. Oertel, Department of Neurology, Klinikum Großhadern, University of Munich, Marchioninistrasse 15, D-8000 München 70, Federal Republic of Germany.

L-dopa in Parkinson's disease

E. Schneider

Department of Neurology, General Hospital Hamburg-Harburg,
Federal Republic of Germany

Summary

L-dopa at present is the most potent drug in the treatment of Parkinson's disease (PD). The initiation of L-dopa has brought about an improvement in the quality of life and an increase in life expectancy. After long-term administration a declined efficacy, fluctuations in mobility and dyskinesias are to be seen.

It was supposed that L-dopa by producing toxic radicals may be the cause not only for the long-term treatment failure but also for the progression of the disease. Arguments in favor or against an early use of L-dopa are presented.

Introduction

At present L-dopa, in combination with a peripheral decarboxylase inhibitor, is generally accepted as the most potent drug in the treatment in Parkinson's disease (PD). Nevertheless, the necessary long term administration does have its short-comings such as declining therapeutic response, fluctuations in mobility and dyskinesias which can be observed after 3–5 years of treatment. Soon it became evident that the progressive disease could not be halted. In addition, there were arguments against an early use of the drug, assuming that levodopa itself rather than the disease process may be the cause of the long term problems (Fahn and Calne, 1978). Early treatment was defined as the initiation of L-dopa as soon as the diagnosis was made whereas late treatment means starting dopa when the symptomatology is beginning to interfere with patient's psychological well-being, social life or occupational activities and other antiparkinsonian drugs have lost their effectiveness.

Evidence against early treatment

The arguments for delaying L-dopa therapy recently have been discussed by Melamed (1986): From a clinical point of view, it could be shown that the declining response as well as side-effects, e.g. fluctuations, seems to be more related to the duration of treatment than to disease duration (Lesser et. al., 1979). Likewise, since the same magnitude of improvement was to be seen irrespective of severity and stage of PD makes the assumption improbable that disease progression is the main cause of deteriorating responsiveness. The beneficial effects of drug holidays may be interpreted as the result of detoxication with e.g. increased receptor sensitivity, improved L-dopa absorption and enhancement of conversion to dopamine in the CNS. Futhermore, the occasionally good response to L-dopa, the positive effects of dopamine agonists and also the results of parenteral L-dopa administration hardly explain levodopa tolerance as being mainly based on disease progression. Experimental data have shown that the effectiveness of L-dopa is not bound to the striatal dopaminergic neurons, as the substance is also decarboxylized outside the dopaminergic terminals, e.g. in serotonergic neurons where exogenous L-dopa is converted into functional and receptor-available dopamine molecules. Chronic L-dopa administration causes a down-regulation of dopamine receptors. In rats chronically pretreated with L-dopa there is a smaller and shorter striatal dopamine elevation after L-dopa administration. The possibility is also discussed that an accelerated progression of the neuronal degeneration by the autoxidation of dopamine may be producing cytoxic free radicals and quinones. Therefore, increased L-dopa and dopamine levels might prove to be dangerous.

Evidence for early treatment

Evidence for not delaying L-dopa treatment mainly came from clinical studies. At first, L-dopa has brought about not only improvement in the quality of life but also a considerable increase in life expectancy of the parkinsonian patients. This has been shown by numerous investigators. Markham and Diamond (1981, 1986) demonstrated that L-dopa helps patients at all stages of the disease to a comparable degree and does not lose effectiveness over time. Hoehn (1983) showed that patients receiving L-dopa immediately after establishing their diagnosis were better off after 10 years with regard to disability scores and survival rates than those in whom treatment was started at some later time. Also Guillard et al. (1986) could find a more pronounced deterioration of the akinetic symptomatology in relation to disease duration and L-dopa administration in patients receiving L-dopa 7 years after diagnosis has been made, who showed a faster decline in L-dopa effectiveness than a group in which L-dopa was administered after about 1.8 years.

Whereas wearing-off phenomena showed an increase with duration of therapy, in some studies dyskinesias and so-called true on-off phenomena did not. Bonnet et al. (1987) concluded from their data that long term aggravation in PD might result from non-dopaminergic lesions by differentiating between parkinsonian score with and without L-dopa. The residual score at the moment of maximal L-dopa effect was considered as reflexion of non-dopaminergic neuronal lesions. Whereas akinesia, rigidity, and tremor showed a rather constant response to L-dopa over time, at least during the first 9 years of treatment, gait, stability, and dysarthria were prone to a more pronounced deterioration and increasingly less favourable response to L-dopa, thus contributing disproportionately to the worsening of the global score. On the other hand this explanation does not take into consideration the problem of wearing-off phenomena. These seem to be best explained by the progressing degeneration of the dopaminergic neurons resulting in a decreased storage capacity for dopamine. This idea, discussed for many years, has experienced confirmation by the results of positron emission tomography studies with L-[F^{18}]-fluorodopa. Leenders et al. (1986) showed a decreased striatal storage of dopamine, more pronounced in parkinsonian patients with fluctuations than in the beginning stages of PD. Despite the reduction in storage capacity the response to L-dopa is preserved when the drug is administered intravenously as was shown by several investigators. Contrary to expectations, a continued good clinical response during levodopa long-term infusion therapy could be observed, even when the dosage of levodopa was gradually reduced (Sage et al., 1987). From this it can be concluded that the receptor sensitivity is maintained. Furthermore, the results of absorption studies show that an alteration in the peripherial pharmakokinetics of L- dopa does not explain the different response to L- dopa with progressing disease. Gancher et al. (1987) found a virtually identical absorption, clearance, distribution volumes, and distribution and elimination half-lives in de novo, stable, and fluctuating patients.

Progression of Parkinson's disease without L-dopa

The question when to start L-dopa may be additionally illustrated by the natural course of PD without L-dopa. In this field, there are only very limited and reliable data. Hoehn and Yahr (1967) who analyzed retrospectively 866 patients hospitalized between 1949–1964 found that in the pre-L-dopa era about a quarter with a disease duration of less than 5 years were severely disabled or dead. After a disease duration of 5 to 9 years two thirds were dead and after 10 to 14 years of illness 80% had died. Stern (1987), summarizing the survival statistics of different studies, points to the fact of an improvement of survival rates only during the initial years of L-dopa treatment, nearly reaching the mortality rates for the times before introduction of L-dopa. As these results, presumably without exception, are based on patient samples which were treated in the years when L-dopa was first marketed, one has to assume that in most cases there had been late rather than early treatment.

But how long can L-dopa treatment actually be delayed? This question was followed by Goetz et al. (1987) in 100 patients who had never received L-dopa before. Of these patients only 17 never had to receive L-dopa during a maximal follow-up time of 96 months. In 55 patients L-dopa seemed to be necessary immediately after referral to the special ward and in 28 patients after 6 months. Treatment was initiated when job security was threatened or postural instability became obvious. 50% of the patients received levodopa within 3 years of the onset of symptoms. For the whole group the mean disease duration was 48 months.

Effects of low-dose L-dopa therapy

Beside the recommendation to delay L-dopa treatment, a low-dose regime has also been advocated. Poewe et al. (1986) analyzed the course of 35 patients with idiopathic PD who received only small doses of L-dopa e.g. up 500 mg/day in combination with an inhibitor. Patients were followed up for 6.2 years. In 94% there was a marked to mild improvement after 1 year, although overall improvement was less compared to those with high dosage treatment. After 6 years, only one third derived worthwhile benefits, whereas in the majority there was a decline below baseline level. Concerning side-effects, it was found that 54% of the low-dose group experienced peak-dose dyskinesias compared to 88% in the high-dose group. The figures for off-period dystonia were 20% to 44%, for end-of-dose effects 52% to 80%, for on-off oscillations 6% to 20%. In this retrospective study no convincing effect in favour of a mere low-dose L-dopa treatment can be seen, as the low-dose regime, although reducing side-effects, is accompanied by less favourable therapeutic results.

Response fluctuations and pharmakokinetics of L-dopa

L-dopa is mainly absorbed in the proximal small bowel. Absorption takes place via a saturable, large neutral amino acid (LNAA) transport system. Retardation in gastric emptying brings about a delay in L-dopa resorption. Also for the brain, there is a saturable carrier system transporting other LNAAs too. Therefore, there will be competition for L-dopa absorption in the mucosa of the intestine and at the blood brain barrier. As was clearly demonstrated by Nutt et al. (1984) transport of L-dopa into the brain–indicated by its clinical effects–not only depends on the plasma concentration of L-dopa but also of the level of other LNAAs. There is a marked difference in the plasma concentration of L-dopa depending on whether L-dopa has been given in an empty stomach after 12 hours fasting or after breakfast. In addition, the L-dopa effect reached by a constant L-dopa infusion can be interrupted by intravenuously given LNAAs. The same can be expected from orally ingested amino acids, thus leading to on-off phenomena in an unpredictable manner. Recently, Leenders et al. (1986) have shown that even in healthy persons the entry of L-[F^{18}]-fluorodopa is markedly reduced when the plasma level of LNAAs is elevated two-to threefold. Why then do we have fluctuations only in advanced PD? First, one has to keep in mind that L-dopa has a very short plasma half life. Secondly, because of the reduced storage capacity of the dopaminergic neurons there is an increased dependency on the continuous delivery of L-dopa to the brain. The magnitude of response seems to follow the all or nothing rule, already triggering fluctuations then when there is only a small reduction in the levodopa level. Thus, predictable and unpredictable, paroxysmal on-off may be based on the same pharmakodynamic mechanism (Nutt, 1987).

Dietary therapy for motor fluctuations

The positive effects of a low-protein diet on the efficacy of levodopa were first observed many years ago. The study of these problems gained momentum when the relation between food intake and L-dopa plasma concentration was more clearly analyzed (see above). Up to now, the studies with a low-protein diet have brought contradictory results. A diet which restricted protein intake during daytime and balancing the necessary amount of protein to bedtime was followed by a reduction in parkinsonian disability and fluctuations (Pincus and Barry, 1987). In contrast, Juncos et al. (1987) found that only an excessive protein intake but not one meeting the recommended daily allowance had any significant effect on plasma concentration of L-dopa and LNAAs and parkinsonian scores, respectively.

References

Bonnet A-M, Loria Y, Saint-Hilaire M-H, Lhermitte F, Agid Y (1987) Does long-term aggravation of Parkinson's disease result from non-dopaminergic lesions? Neurology 37: 1539–1542

Fahn S, Calne DB (1978) Considerations in the management of parkinsonism. Neurology 28: 5–7

Gancher ST, Nutt IG, Woodward WR (1987) Peripheral pharmakokinetics of levodopa in untreated, stable, and fluctuating parkinsonian patients. Neurology 37: 940–944

Goetz CG, Tanner CM, Shannon KM (1987) Progression of Parkinson's disease without levodopa. Neurology 37: 695–698

Guillard A, Chastang C, Fenelon G (1986) Étude à long terme de 416 cas de maladie de parkinson. Facteurs de pronostic et implications thérapeutiques. Rev Neurol (Paris) 142: 207–214

Hoehn MM, Yahr M (1967) Parkinsonism: onset, progression, and mortality. Neurology 17: 427–442

Hoehn MM (1983) Parkinsonism treated with levodopa, progression and mortality. J Neural Transm [Suppl] 19: 253–264

Juncos JL, Fabbrini G, Mouradian MM, Serrati C, Chase TN (1987) Dietary influences on the antiparkinsonian response to levodopa. Arch Neurol 44: 1003–1005

Leenders KL, Palmer AJ, Quinn N, Clark J, Firnau G, Garnett ES, Nahmias C, Jones T, Marsden CD (1986) Brain dopamine metabolism in patients with Parkinson's disease measured with positron emission tomography. J Neurol Neurosurg Psychiatr 49: 853–860

Leenders KL, Poewe WH, Palmer AJ, Brenton DP, Frackowiak RSJ (1986) Inhibition of L-[^{18}F] Fluorodopa uptake into human brain by amino acids demonstrated by positron emission tomography. Ann Neurol 20: 258–262

Lesser RP, Fahn S, Snider SR, et al (1979) Analysis of the clinical problems in parkinsonism and the complications of long-term levodopa therapy. Neurology 29: 1253–1260

Markham CH, Diamond SG (1981) Evidence to support early levodopa therapy in Parkinson's disease. Neurology 31: 125–131

Markham CH, Diamond SG (1986) Long-term followup of early DOPA treatment in Parkinson's disease. Ann Neurol 19: 362–365

Melamed E (1986) Initiation of levodopa therapy in parkinsonian patients should be delayed until the advanced stages of the disease. Arch Neurol 43: 402–405

Nutt JG, Woodward WR, Hammerstad JP, Carter JH, Anderson JL (1984) The on-off phenomenon in Parkinson's disease. N Engl J Med 310: 483–488

Nutt JG (1987) On-off phenomenon: Relation to levodopa pharmakokinetics and pharmacodynamics. Ann Neurol 22: 535–540

Pincus JH, Barry KM (1987) Plasma levels of amino acids correlate with motor fluctuations in parkinsonism. Arch Neurol 44: 1006–1009

Poewe WH, Lees AJ, Stern GM (1986) Low-dose L-dopa therapy in Parkinson's disease: A 6-year follow-up study. Neurology 36: 1528–1530

Sage JI, Trooskin S, Heikkila R (1987) Long-term duodenal infusion of levodopa for on-off phenomena in parkinsonism: Continued good clinical response associated with gradually declining levodopa in take. Ann Neurol 22: 173

Stern G (1987) Prognosis in Parkinson's disease. In: Marsden CD, Fahn S (eds) Movement disorders 2. Butterworths, London, pp 91–98

Correspondence: Prof. Dr. E. Schneider, Department of Neurology, General Hospital Harburg, Eissendorfer Pferdeweg 52, D-2100 Hamburg 90, Federal Republic of Germany.

Pharmacokinetic investigations of various levodopa formulations

M. Gerlach[1], W. Kuhn[2], and H. Przuntek[3]

[1] Department of Neurology, University of Würzburg,
[2] Department of Psychiatry, University of Würzburg,
[3] Department of Neurology, Ruhr-Universität Bochum, Federal Republic of Germany

Summary

We investigated the single-dose pharmacokinetics of 5 different levodopa sustained-release (SR) formulations in healthy volunteers. All SR formulations showed a very different profile from that of a standard formulation. In all cases, the SR formulations prolonged the time to peak levodopa blood levels. The rate of absorption is reduced as evidenced by the peak levels being lower and smoother. In some cases there is a tendency to produce a plateau rather than a peak, which lasts for between 1 and 4 hours. On the basis of available pharmacokinetic parameters, we selected an SR formulation which would appear to be the most suitable candidate. A pilot trial with this formulation in 5 Parkinson patients with motor fluctuations confirmed the results of the volunteer studies.

Introduction

Despite the high incidence of side-effects such as dyskinesia, on-off phenomenon, and psychosis, particularly during long-term treatment, replacement therapy with L-3,4-dihydroxy-phenylalanine (levodopa) is still the first-line approach to Parkinson's disease. The molecular-biochemical processes that are involved in the pathogenesis of fluctuations in motor performance have not yet been fully elucidated. On the basis of numerous studies showing a correlation between the incidence of fluctuations in motor performance and the levodopa plasma concentration profile, it could be established that they were mainly caused by pronounced fluctuations in the levodopa plasma concentrations. To optimize antiparkinsonian therapy, levodopa was to be pharmaceutically formulated in such a way that more constant plasma levels could be ensured, especially since the chronic intravenous administration of levodopa is practicable neither in inpatients nor in outpatients with Parkinson's disease. The most promising candidates have so far been sustained-release levodopa formulations which release levodopa slowly and continuously over an extended period of time.

In this paper we shall present our pharmacokinetic studies on the new sustained-release (SR) levodopa formulations and, based on these results, give a detailed description of the complex processes involved in the absorption, distribution and elimination of levodopa.

Materials and methods

In a randomized cross-over study in six male volunteers, the kinetics of various SR levodopa formulations following single doses were compared with a standard preparation. Based on the pharmacokinetic data obtained, the formulation that proved most suitable for Parkinson patients was selected for a pilot trial.

Study design, subjects, test preparations, inclusion and exclusion criteria, parameters, and the statistical analysis of the study in the volunteers have already been extensively described in a previous publication (Gerlach et al., 1988).

Five patients with idiopathic Parkinson's disease, who had received previous levodopa therapy, took part in an uncontrolled clinical trial. The patient demographics are shown in Table 1.

The patients took 3 capsules of SR formulation 11/10 (300 mg levodopa SR, Asta Pharma AG, Federal Republic of Germany) and 75 mg benserazide (Pharmacie des Hôpitaux de Paris) before breakfast for 7 days; five hours after the first dose, another capsule of formulation 11/10 (100 mg levodopa SR) as well as 25 mg benserazide were administered. On days 1 and 7 of the

Table 1. Patient demographics

Pat.	Sex	Age	Weight in kg	Height in cm	Duration of the disease in years	Hoehn and Yahr stage
1	f	59	63	165	7	I
2	m	68	74	178	15	IV
3	m	56	99	186	10	III
4	m	48	82	198	12	IV
5	m	54	74	178	13	IV

trial, levodopa plasma levels were measured. On these two days the above dosage was mandatory; on the other days dose variations were permissible.

Blood samples were taken at fixed times (pre-dose; 10, 30, 45, 60 and 90 min; as well as 2, 2.5, 3, 3.5, 4, 5, 6, 7, 8, 9, 10 and 12 hours after the first dose) in the sitting position through an indwelling venous catheter into an EDTA tube filled with 100 μl 0.5% sodium bisulfite solution. The plasma obtained by rapid centrifugation was immediately frozen at −20° C for subsequent analysis. Analysis was by means of reversed-phase HPLC and electrochemical detection following separation of deproteinised plasma on C_{18} cartridges (Gerlach et al., 1986).

Results

Fig. 1 shows the levodopa plasma concentration profiles (medians) following single doses of 5 SR formulations (11/7, 11/10 and 12/2–4), the regular-release formulation 12/1 and the standard

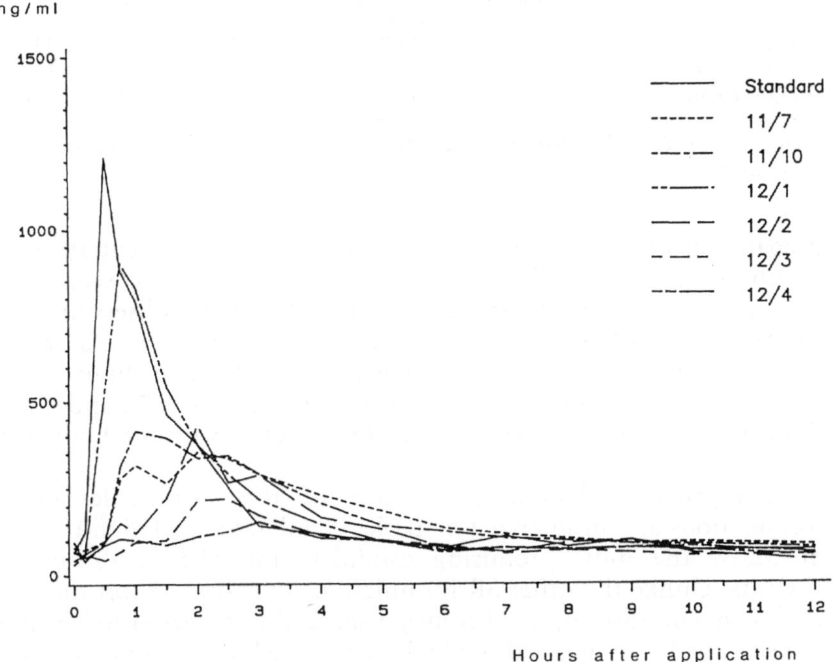

Fig. 1. Plasma levodopa concentrations (medians) versus time following a single oral administration of various SR formulations (11/7, 11/10, 12/2–4), a non-SR formulation 12/1 and a standard preparation in 6 healthy volunteers (in each case 100 mg levodopa + 25 mg benserazide were administered)

Table 2. Pharmacokinetic parameters (medians) of various levodopa formulations administered to 6 healthy volunteers

	Formulations						
	Standard	Non-SR	SR				
Lot no.	12/5	12/1	12/2	12/3	12/4	11/7	11/10
C_{max} (ng/ml)	1242	952	490	297	162	494	526
t_{max} (h)	0.50	0.88	2.00	2.50	4.50	2.00	1.00
AUC (ng × h/ml)	1945	1703	575	755	44	1136	1322
Rel. bioavailability (in %)	100	88	30	39	2	71	82

Table 3. Half-value duration (HVD) for various levodopa formulations in healthy volunteers (calculation of HVD was performed only in those subjects whose concentration-time curve showed exactly two intersections with the $C_{max}/2$ line)

	Formulations[a]				
	Standard	Non-SR	SR		
Lot no.	12/5	12/1	12/2	11/7	11/10
Number of subj.	6	5	5	5	6
HVD median (min)	58.0	71.5	101.8	164.8	196.4

[a] Due to the low C_{max} half-value duration was not calculated for formulation 12/3 and 12/4

preparation in male volunteers. The time to peak concentration is clearly prolonged for all SR formulations. The sharp peak seen with formulation 12/1 and the standard preparation is not observed with the SR formulations. A consistant increase in levodopa levels over a constant period of time is not obvious; in fact, it can be observed for varying periods of time only with formulations 11/7, 11/10 and 12/2. Return to endogenous levodopa baseline levels is within 8 hours for all preparations studied.

The pharmacokinetic parameters of the various levodopa formulations are summarized in Tables 2 and 3. Formulation 11/10 is obviously the most promising candidate for trial in Parkinson patients. Unlike the other SR formulations, it shows a short time to peak concentration ($t_{max} = 1$ hour), a markedly increased half-value duration of 3.3 hours, and the highest relative bioavailability (82%).

Fig. 2 shows the time course of the medians of the levodopa plasma levels on days 1 and 7 of the trial in 5 Parkinson patients following oral administration of SR formulation 11/10. The medians show that, on days 1 and 7, the time to peak levodopa concentration

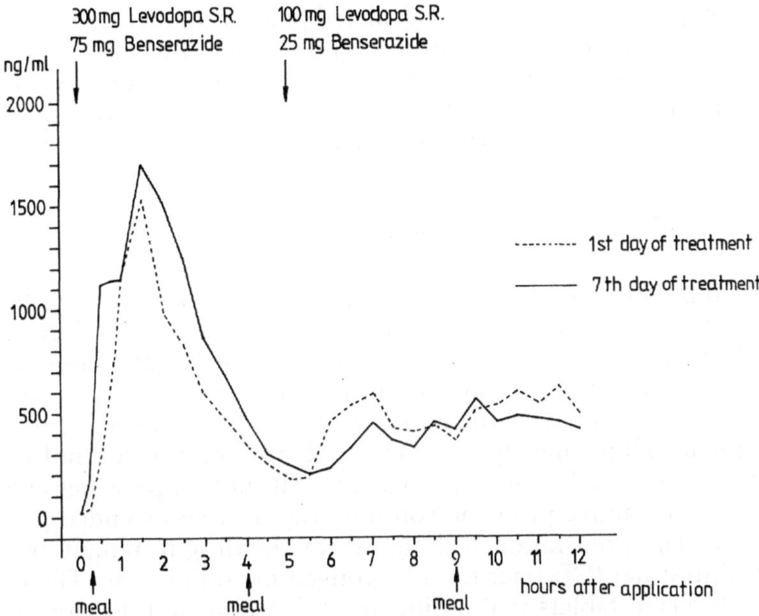

Fig. 2. Time course of the levodopa plasma concentration (medians) following oral administration of levodopa SR formulation 11/10 (300 mg levodopa SR and 75 mg benserazide mornings before breakfast and 100 mg levodopa SR and 25 mg benserazide 5 hours later) in 5 Parkinson patients

Table 4. Comparison of the pharmacokinetic parameters (medians) of the levodopa SR formulation 11/10 in healthy volunteers and patients

	Patients (n = 5; 300 mg levodopa SR)		Healthy volunteers (n = 6; 100 mg levodopa SR)
	day 1	day 7	
C_{max} (ng/ml)	1210	1699	526
t_{max} (h)	1.5	1.5	1.0
AUC_{0-5h} (h × ng/ml)	3000	4369	1111
HVD (min)	134.2	157.9	196.4

AUC Area under curve, *HVD* half-value duration

in plasma is somewhat prolonged while on the test compound (cf. t_{max} values in Table 4). Interesting findings are the smoother levodopa plasma concentration curve, the tendency toward a plateau, particularly after the second dose, and the similar profile of the curves on days 1 and 7.

Comparison between the calculated pharmacokinetic parameters of the study in healthy volunteers and the trial in Parkinson patients (see Table 4) shows a linear behavior of the area under the curve (AUC) and C_{max} values for the two doses (100 mg and 300 mg levodopa SR) despite the different sources of the data.

Discussion

A number of studies (Nutt et al., 1986; Cedarbaum et al., 1987; Goetz et al., 1987; Juncos et al., 1987; Marsden et al., 1987; Gerlach et al., 1988) have recently been published which aimed at an improved management of fluctuations in motor performance by a pharmacokinetic approach, a particularly severe problem in patients on chronic levodopa therapy. A variety of pharmaceutical-technological principles are used to achieve this goal. All these approaches aim at releasing the active principle continuously over an extended period of time. The preparation is designed for the drug to remain in the gastrointestinal (GI) tract for a prolonged period of time. The controlled-release tablets (CR-3-Sinemet, CR-4-Sinemet) developed by MSD Sharp & Dohme, in which levodopa is embedded in a synthetic polymer matrix, are based on the assumption that the slowly dissolving matrix does not immediately pass the pylorus and therefore slowly and continuously releases the active principle in the stomach. The hydrodynamically balanced system (HBS) developed by Hoffmann-La Roche is also based on the principle of altering drug release by a specific matrix composition. In this formulation, levodopa is embedded in a hydrocolloidal matrix, which swells as soon as the gelatin capsule has dissolved in the stomach and, as a consequence, the contents have come into contact with the gastric juice. This slimy, voluminous complex, whose specific gravity is less than 1, is now supposed to float on the gastric contents for several hours, with levodopa being continuously released from the hydrated matrix for a prolonged period of time. Another possible way of influencing the release of active principle is the use of coated pellets. This method—employed by Asta Pharma AG—uses enteric-coated pellets in which levodopa is embedded and which are enclosed by a gelatin capsule. An additional film on the pellets is to achieve continuous levodopa release in the intestinal tract.

Comparison of all pharmacokinetic results available so far shows that all levodopa SR formulations live up to their name in that they release the drug in a sustained fashion; however, this release, and thus the blood concentration profile, is not uniform over several hours. The levodopa concentration-time curves show a smoother pattern as compared to the standard preparations, without the sharp peaks typically seen with the standard preparations. The apparent plateau over 6 hours on the CR III levodopa formulation [published by the study group of Chase (Juncos et al., 1987)] cannot be confirmed statistically, since the times of measurement were spaced too far apart. The relevance of this observation, therefore, warrants a good deal of caution. The parameter half-value duration used for assessing SR formulations is prolonged up to 4 hours with these formulations compared to the standard preparations. Relative bioavailability is reduced compared to that of the standard preparations (by between approximately 20% and 80% depending on the preparation).

This pharmacokinetic behavior is due to the limited time the formulation remains in the GI tract on the one hand and to the rapid levodopa metabolism on the other (see Fig. 3). Given that modern levodopa therapy makes use of coadministration of the peripheral enzyme inhibitors benserazide and carbidopa, decarboxylation is of minor importance in the biotransformation of levodopa. However, levodopa is rapidly degraded and inactivated by other metabolic

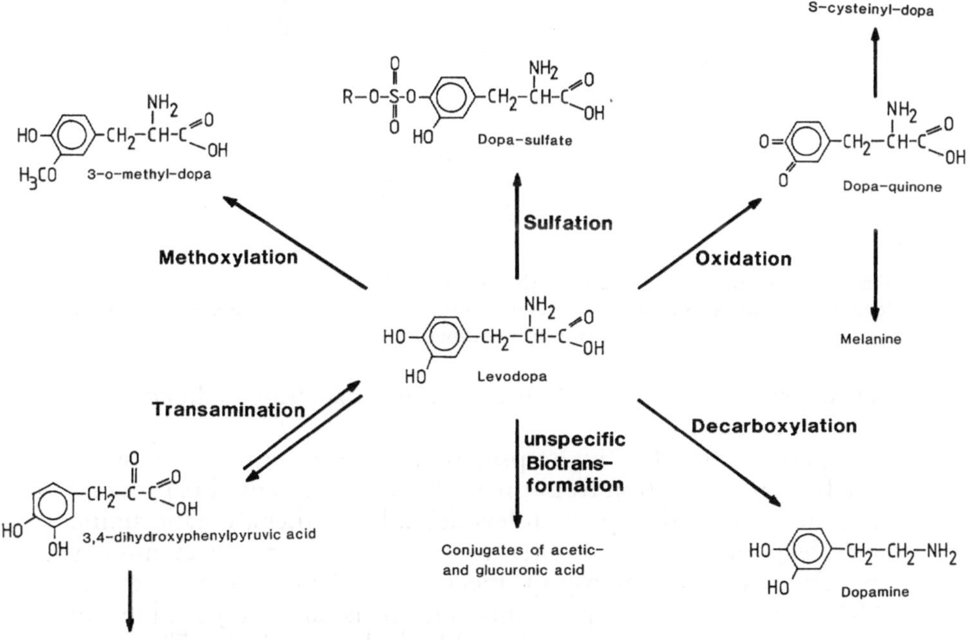

Fig. 3. Metabolism of levodopa

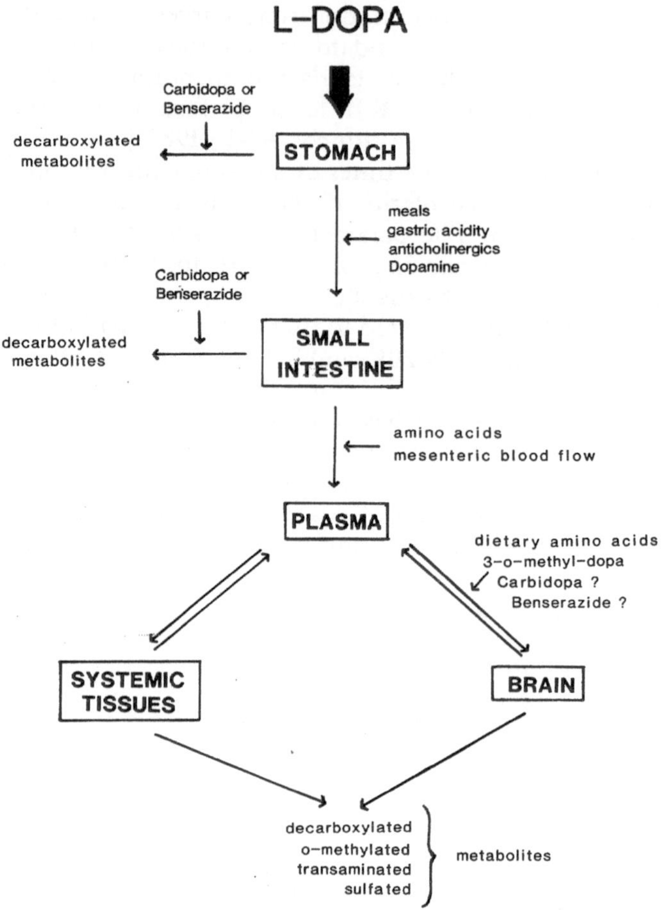

Fig. 4. Schematic representation of the absorption, distribution and metabolism of levodopa. Small arrows represent factors influencing these processes (according to Nutt and Fellman, 1984)

pathways: methoxylation, transamination, oxidation, sulfoconjugation, glucuronidation.

Apart from the high presystemic and systemic rate of metabolism, absorption from the GI tract plays a central part in oral antiparkinsonian therapy with levodopa. Like other aromatic amino acids, levodopa is absorbed only from the duodenum and jejunum by an active saturable transport mechanism (Gundert-Remy et al., 1983). An active transport mechanism is also responsible for levodopa passage across the blood-brain-barrier (BBB). The saturability of this transport system prompted studies in patients on a low-protein diet. It could be shown that Parkinson patients with fluctuations in motor performance experienced an improvement on a low-protein diet (Pincus et al., 1987).

The principal factors determining the parmacodynamic actions of levodopa include release from the stomach into the small intestine, duration and extent of enteral absorption, extent of presystemic and systemic metabolism, and passage across the BBB (Fig. 4).

Summing up, it can be said that the available pharmacokinetic data suggest that at least one of the levodopa SR formulations studied by us is a promising candidate for improving antiparkinsonian therapy (longer dosing intervals, reduction in the side-effects described above). Preliminary clinical experience with this new SR formulation 11/10 has shown that Parkinson patients with levodopa-dose-related fluctuations in motor performance experience a definite improvement. Given the complexity of the on-off phenomenon and of the pathogenesis, which has not yet been fully elucidated, further clinical trials will have to be done to show whether a pharmacokinetic approach alone is sufficient to achieve an improvement in these clinical symptoms.

References

Cedarbaum JM, Breck L, Kutt H, McDowell FH (1987) Controlled-release levodopa/carbidopa. II Sinemet CR 4 treatment of response fluctuations in Parkinson's disease. Neurology 37: 1607–1612

Gerlach M, Klaunzer N, Przuntek H (1986) Determination of L-Dopa and 3-O-methyl-Dopa in human plasma by extraction using C_{18} cartridges followed by high-performance liquid chromatographic analysis with electrochemical detection. J Chromatogr 380: 379–385

Gerlach M, Gebhardt B, Kuhn W, Przuntek H (1988) Pharmacokinetic studies with sustained-release formulation of levodopa in healthy volunteers. J Neural Transm [Suppl] 27: 211–218

Goetz CG, Tanner CM, Klawans HL, Shannon KM, Carroll VS (1987) Parkinson's disease and motor fluctuations: Long-acting carbidopa/levodopa (CR-4-Sinemet). Neurology 37: 875–878

Gundert-Remy UR, Hildebrandt R, Stiehl A, Weber E, Zürcher G, Da Prada M (1983) Intestinal absorption of levodopa in man. Eur J Clin Pharmacol 25: 69–72

Juncos IL, Fabbrini G, Mouvadian MM, Serrati C, Kask AM, Chase TN (1987) Controlled release levodopa treatment of motor fluctuations in Parkinson's disease. Neurol Neurosurg Psychiatr 50: 194–198

Marsden CD, Rinne UK, Koella WP, Dubius R (eds) ⟨Madopar⟩ HBS. International workshop on the "On-Off"-phenomenon in Parkinson's

disease: New possibilities for its management, Agno, 1985. Europ Neurol 27 [Suppl] 1
Nutt IG, Fellman IH (1984) Pharmocokinetics of levodopa. Clin Neuropharmacol 7: 35–49
Nutt IG, Woodward WR, Carter IH (1986) Clinical and biochemical studies with controlled-release levodopa/carbidopa. Neurology 36: 1206–1211
Pincus IH, Barry K (1987) Influence of dietary protein on motor fluctuations in Parkinson's disease. Arch Neurol 44: 270–272

Correspondence: Dr. M. Gerlach, Department of Neurology, University of Würzburg, Josef-Schneider-Strasse 11, D-8700 Würzburg, Federal Republic of Germany.

Iron therapy in Parkinson's disease

Stimulation of endogenous presynaptic L-DOPA biosynthesis by the iron compound oxyferriscorbone

W. Birkmayer and **J. G. D. Birkmayer**

Birkmayer-Institut für Parkinsontherapie, Wien, Austria

Summary

The iron compound oxyferriscorbone® has been used as novel medication in more than 100 Parkinson patients. In 66% of the patients a beneficial clinical effect was observed 22% of the patients showed a very good (better than 30%) improvement of disability, 44% a moderate (up to 30%) improvement. The effect of iron was dependent on the dosage and the severity of the case but independent of the mode of application (i.v. or i.m.). The action of the iron medication lasts from 24 hours up to 4 days depending on the severity of the case.

Introduction

The dopamine deficiency in the basal ganglia is the direct cause of the motoric impairment of Parkinson patients. This dopamine deficiency has been overcome by substituting L-DOPA. This first causative therapy of Parkinson's disease was introduced by Birkmayer and Hornykiewicz in 1961. Tyrosine, the immediate precursor of L-DOPA, does not show any dopaminergic effect, which indicates that the metabolic transformation of tyrosine to DOPA is not functioning in Parkinsonian patients. The enzyme catalyzing the production of L-DOPA from tyrosine is the tyrosine-hydroxylase (Fig. 1). Therapeutic application of alphamethyl-p-tyrosine, an inhibitor of tyrosine hydroxylase, deteriorated the disability of Parkinson patients (Birkmayer, Mentasti, 1969), indicating that this enzyme plays a key role in this disease.

Compelling evidence for this assumption was provided by McGeer (1971) in showing that the activity of this enzyme is dimin-

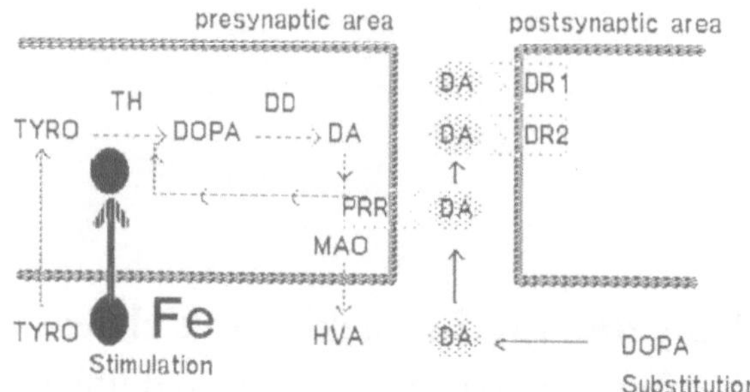

Fig. 1. Dopaminergic Neurotransmission. *TH* tyrosinehydroxylase, *DD* DOPA-decarboxylase, *MAO* monoaminoxidase (G. D. Birkmayer, 1988)

ished in the substantia nigra of Parkinson patients. It has been found that this enzyme is an iron containing protein with tetrahydrobiopterin as coenzyme (Nagatsu et al., 1981). Recently, it could be shown that iron can stimulate tyrosine hydroxylase in vitro up to 20 fold (Rausch et al., 1988). On the basis of these findings we started to treat Parkinson patients with iron. The crucial feature of our approach was the use of oxyferriscorbone® which is a very special compound in which iron is tightly bound to ascorbic acid in its $Fe^{(2+)}$ and its $Fe^{(3+)}$ form.

Methods

The disability of the patients was scored according to Birkmayer and Neumayer (1972). Patients were divided in 3 groups:

Group of slight cases: Disability up to 30%

Group of moderate cases: Disability between 30% and 60%

Group of severe cases: Disability greater than 60%

Administration of oxyferriscorbone

Patients in the slight case group received one ampule oxyferriscorbone containing 15 mg iron intramuscularly twice a week. After 10 injections the disability of the individual patients was recorded and the serum iron level determined. Patients in whom the serum iron was below the normal range of 80 mg/100 ml received then 2 ampule (30 mg iron) twice a week. The treatment was continued with one i.m. injection of oxyferriscorbone per week in those patients who exhibited a moderate to a very good clinical improvement. In non-responders the therapy was changed to i.v. infusions of one ampule per day. If this approach yielded a clinical benefit, infusion therapy was continued for 30 days. After this time the iron application was discontinued and the patients were followed up for several weeks. Those patients who deteriorated during the surveillance period received a minor dosage of oxyferriscorbone (1 ampule per week i.m.). This amount was sufficient to achieve a long-lasting effect. In some patients iron medication has to be restarted after a period of 3 months. In these cases oxyferriscorbone was administered as an infusion (100 ml) or as i.m. injection. The beneficial effect of the i.m. iron injections lasted four to seven days. It was determined by examining the disability before and 1 hour as well as 6 hours after the administration of oxyferriscorbone according to the disability scale of Birkmayer and Neumayer (1972).

Results

Patients receiving the the usual Parkinson-therapy Sinemet® or Madopar®, with additives such as deprenyl or dopamine agonists, were examined before and after receiving oxyferriscorbone. The following results were obtained (Fig. 2).

22% of the patients exhibited an improvement in their disability by 40% to 50% which, from the clinical point of view, represented a very good response. From the 100 patients evaluated 21 showed a 30% and 23 a 20% improvement of their disability. This change can be regarded as a moderate response. In 20 patients the improvement was not better than 10%. We have to admit that a 10% change can only be measured reliably after frequent examination at various intervals. For this reason we have included these patients in the group of non-responders. 14 patients were real non-responders. They did not show a benefit in any of their symptoms, regardless of the dosage or the mode of administration of oxyferriscorbone.

In order to gain information about the mechanism of the action of oxyferriscorbone we measured the dopamine metabolites homovanilinic acid (HVA) and vanil-mandelic acid (VMA) in the urine

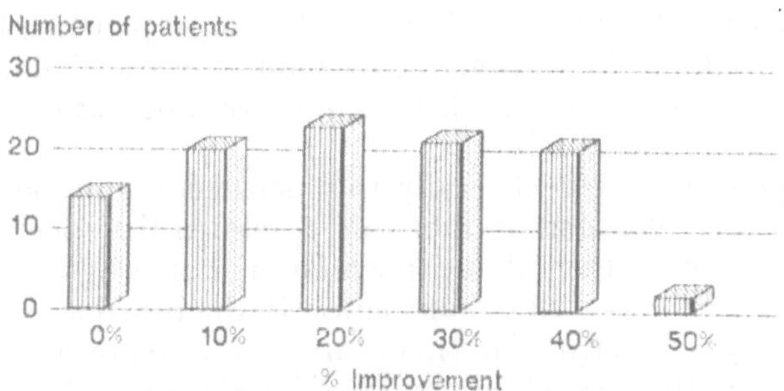

Fig. 2. Improvement of disability by oxyferriscorbone

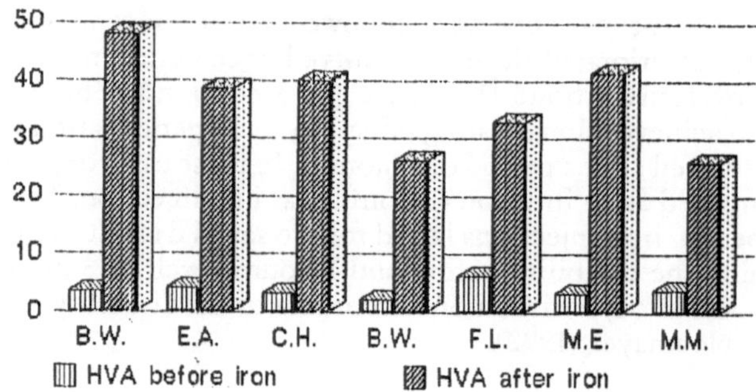

Fig. 3. Increase of urinary HVA level by oxyferriscorbone: before iron, after iron

before and after iron treatment. Treatment with L-DOPA leads to a significant increase in the urine level of HVA (Birkmayer and Riederer, 1985). If we assume that the applied iron stimulates the tyrosine hydroxylase and thus increases the endogenous L-DOPA concentration, we would expect an increase in the HVA or VMA urine level. This turned out to be the case. In all the patients in which oxyferriscorbone led to an improvement of disability an increase in the dopamine metabolite HVA could be detected (Fig. 3).

Discussion

This study confirms and extends our previous reports on the clinical benefit of the iron oxyferriscorbone for Parkinson patients (Birkmayer and Birkmayer, 1986; Birkmayer and Birkmayer, 1987). In the meantime our findings have been confirmed by Iliceto et al. (1987) and Ott (1988). These investigators observed a similar improvement of the disability in about the same percentage of patients. According to their observation there is no doubt about the clinical benefit of oxyferriscorbone to Parkinson patients. It should be added that of the various other iron drugs available on the European pharmaceutical market oxyferriscorbone was the only one exhibiting a beneficial effect.

The question arises why only this particular compound works and others not. We assume that this is due to the special galenic formulation of oxyferriscorbone in which the iron is tightly bound to ascorbic acid, thus avoiding dissociation in the blood and capturing of the iron by transferrin. Another point is whether or not oxyferriscorbone passes the blood brain barrier. The elevation of the concentration of the dopamine metabolite HVA in the urine provides only indirect evidence. A direct proof has been attempted, using magnetic resonance imaging (MRI) of the particular brain area of Parkinson patients before and 30 minutes after oxyferriscorbone administration. No conclusive results have been obtained thus far. As we applied only 15 mg of iron the amount of iron reaching the distinct brain areas might be below the limit of detection. On the other hand it has been found by MRI that a high percentage of Parkinson patients do have a reduced iron content in certain basal ganglia (Rutledge et al., 1987). Perhaps it needs more than 30 minutes to fill up the low iron pools in the basal ganglia.

References

Birkmayer W, Hornykiewicz O (1961) Der L-Dioxyphenylalanin-L-DOPA-Effekt bei der Parkinson-Akinese. Wien Klin Wochenschr 73: 787

Birkmayer W (1969) Der Alpha-methyl-p-Tyrosin-Effekt bei extrapyramidalen Erkrankungen. Wien Klin Wochenschr 81: 10

Birkmayer W, Neumayer E (1972) Die moderne medikamentöse Behandlung des Parkinsonismus. Z Neurol 202: 257

Birkmayer W, Riederer P (1985) Die Parkinson-Krankheit. Springer, Wien New York

Birkmayer W, Birkmayer JD (1986) Iron—a new aid in the treatment of Parkinson patients. J Neural Transm 67: 287

Birkmayer JD, Birkmayer W (1987) Improvement of disability and akinesia of patients with Parkinson's disease by intravenous iron substitution. Ann Clin Sci 17: 32

Iliceto G, DeMari M, Federico F, et al (1987) Iron in Parkinson's disease: preliminary report. Ital J Neurol Sci [Suppl] 7: 62

McGeer PL, McGeer EG, Wada JA (1971) Distribution of tyrosine hydroxylase in human and animal brain. J Neurochem 18: 1647

Nagatsu T, Oka K, Yamamoto T et al (1981) Catecholaminergic enzymes in Parkinson's disease and related extrapyramidal diseases. In: Riederer P, Usdin E (eds) Transmitter biochemistry of human brain tissue. Macmillan, London, p 291

Ott E, Birkmayer W, Birkmayer GD (1988) Oxyferriscorbone in der ambulanten Behandlung von Parkinson-Patienten. Tagungsberichte, Österr Parkinson-Gesellschaft (in press)

Rausch WD, Hirata Y, Nagatsu T, et al (1988) Tyrosine hydroxylase activity in caudate nucleus from Parkinson's disease: effect of iron and phosphorylating agents. J Neurochem 50: 202—208

Rutledge JN, Hilal SK et al (1987) Study of movement disorders and brain iron by MR. AJNR 8: 397—411

Correspondence: Prof. Dr. W. Birkmayer, Schwarzspanierstrasse 15, A-1090 Wien, Austria.

indirect evidence, a direct proof has been attempted, using magnetic resonance relaxation (MRI) of the particular brain area of Parkinson patients before and 30 minutes after oral ferrous iron administration. No conclusive results have been obtained thus far. As we applied only 15 mg of iron the amount of iron reaching the brain might be below the limit of detection. On the other hand, it has been found for [illegible] a high percentage of Parkinson patients [illegible] a reduced iron content in [illegible] (Riederer et al., 19[illegible]). [illegible] more than 40 minutes to fill up [illegible] in the basal ganglia.

References

[illegible]

Provisional experiences with the combination of L-dopa and L-deprenyl

G. Ulm[1] and F. Fornadi[2]

[1]Paracelsus-Elena-Klinik, Kassel,
[2]Parcelsus-Nordseeklinik, Helgoland, Federal Republic of Germany

Summary

In a retrospective study three different groups of patients are compared: patients on L-dopa on its own, patients on L-dopa plus L-deprenyl and patients on L-dopa plus Ergot-derivatives. The aim of the study was to find out if there was any advantage gained from one of these strategies. The study showed that the combination-therapy was superior to L-dopa alone; it did not matter whether L-dopa was combined with L-deprenyl or with an Ergot-derivative.

Introduction

Despite great progress in the research on Parkinson's disease some basic problems remain unresolved. One central question is by which factors the disease is triggered, i. e. the question of etiology.

The MP TP-model has advanced our understanding to a certain extent (Marsden, 1986): it appears that certain toxic radicals have a specific affinity to nigrostriatal cells and can trigger a progressive degenerative process in that area (Glover, 1986). Apart from a search for further toxic substances and the discovery of their active mechanisms, the emphasis on the therapeutic side is on the neutralisation of these radicals. The first substance to be credited with such a neutralising effect is L-deprenyl (Knoll, 1987). It is not absolutely clear, however, to what extent L-deprenyl has this protective effect in humans and can thus improve the clinical picture and the quality and expectancy of life. As it is, existing evidence from the few studies that have so far been made in this field points towards a positive effect (Birkmayer et al., 1985; Sandler et al., 1987; Tetrud et al., 1987).

Clinical management

The permission to use L-deprenyl (brand name in the Federal Republic "Movergan®") was granted sixteen months ago by the Federal Health Authority in Berlin coupled with the recommendation to usw L-deprenyl in combination with L-Dopa and for patients suffering from fluctuations. Studies relating to this problem had already been made and these showed above all an improvement of end-of-dose-akinesia, making a small reduction of L-dopa possible (Czanda et. al., 1980; Ulm et al., 1987).

This recommendation was aimed at the advanced stage of the disease with complications already in evidence. Thus the drug is a complement to Bromocriptine, which has so far been recommended in combination with L-dopa for treatment at the earlier stages or as monotherapy at the very beginning of the disease.

Apart from its easy application and good tolerance, L-deprenyl has become an important option because of its neutralisation of toxic radicals. If a degeneration of nigrostriatal cells could in fact be slowed down it would have to be used as soon as the diagnosis is confirmed.

We make use of this potentially causal therapy by beginning treatment with a combination of L-deprenyl and a small dose of L-dopa. (There are of course exceptions.) Ergot-preparations are only applied at a later stage. We have thus in effect reversed the guidelines issued by the Federal Health Authority at the time of admission.

What follows is a discussion of the question of whether there are already any indications that such a combination might be successful and more effective than a combination of L-dopa and Ergot-derivatives or L-dopa-therapy on its own.

Material and method

We have carried out a retrospective analysis of our patient sample. We have studied the cases of patients from Paracelsus-Elena-Klinik, Kassel (120 beds) and Paracelsus Nordseeklinik, Helgoland (30 beds). Both hospitals treat about 1.500 patients a year. As a rule these patients stay with us for several weeks. The relevant medical data of all patients have been recorded since 1986. In each case we put 150 items on record. These include case history, admission and release reports, concomitant diseases, therapeutic measures and side-effects of treatment.

We devided patients into three categories that were subsequently compared:

MAIN CHARACTERISTICS OF PATIENTS	Group I levodopa	Group II + deprenyl	Group III + ergots
Number of patients	133	113	33
Male/female %	56.7/43.3	60.9/39.1	52.8/47.2
Age (y)	68.4	64.1	66.8
Age at onset (y)	59.8	53.7	54.7
Duration/disease (y)	8.5	10.3	11.7
Duration/levodopa(y)	5.4	7.1	8.5
Dur./depr./ergot (y)	-	2.3	3.69

Fig. 1

Group I: patients treated with L-dopa for at least two years,

Group II: patients treated with a combination of L-dopa and L-deprenyl for at least two years,

Group III: patients treated with L-dopa and Ergot-derivatives for at least two years.

Motor movements were assessed both according to the Webster scale and CURS. The sum score was used in either case and with Webster and ankinesia a subscore was formed by adding up items I, IV and V.

The demographic data are schown in Fig. 1.

The groups are homogenous in respect of age, sex and age of manifestation of disease. Group II has the longest duration of the disease and the longest time of L-dopa-therapy. This is due to the fact that we use Ergot-derivatives later, i. e. in more advanced cases.

We compared these three groups in respect of the actual symptoms expressed by CURS-sum score and the progression of the prevailing symptoms during the time in question.

Results

The actual situation is shown in Fig. 2. Group II (L-dopa plus L-deprenyl-group) has the least pronounced state of disease. This group is significantly better ($p < 0.005$) than group I (L-dopa only). Group III (L-dopa plus Ergot-derivatives) seems to be the worst. However, this is the group with the longest duration of disease, which means that the figures were already below those of the other group when treatment was begun.

The progression of symptoms is shown in Fig. 3.

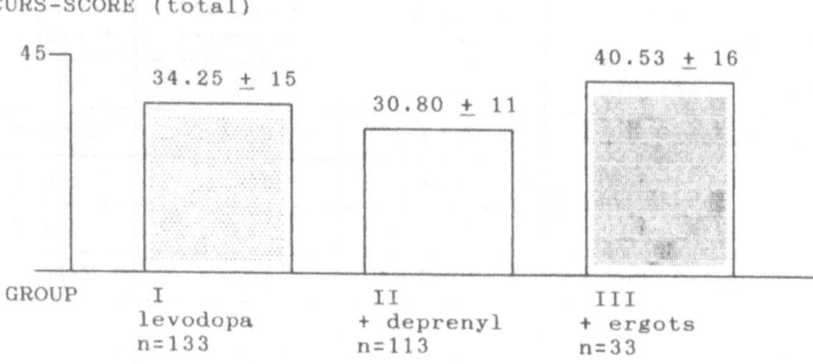

Fig. 2. Difference sign. $p < 0.05$

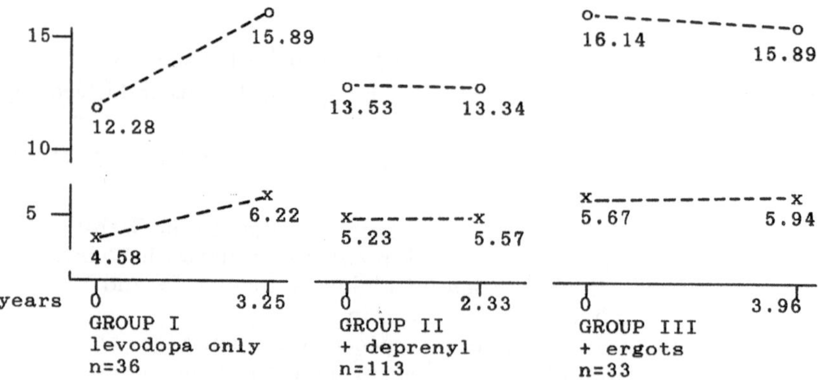

Fig. 3. Worsening in group I is significant. $p < 0.01$, o total, x akinesia-score (sum 1, 4, 5)

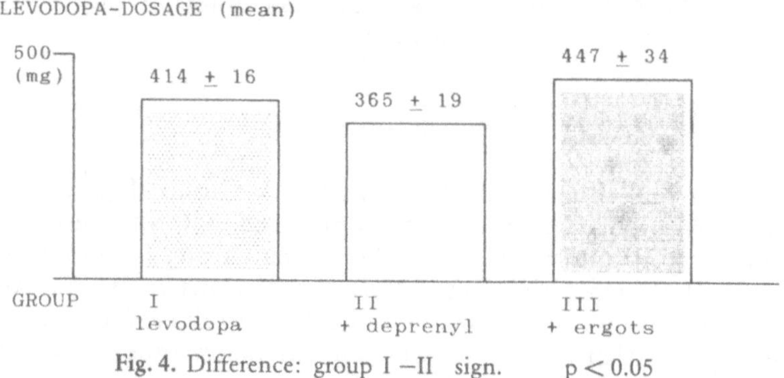

Fig. 4. Difference: group I–II sign. $p < 0.05$
group I–III not sign. $p = 0.30$
group II–III sign. $p < 0.05$

The results are insignificant deterioration in group II and III, while group I deteriorated significantly ($p < 0.001$). This applies to the overall score for motor movements (total score) and also to the subscore for akinesia.

The dosages of L-dopa used in each group are shown in Fig. 4.

Discussion

The study shows a slight deterioration of patients treated with a combination of L-dopa and L-deprenyl or L-dopa and Ergot-derivatives. There is no significant difference between the two groups. Patients who were given L-dopa only showed a significant deterioration of all symptoms. This group compares even more unfavourably with group II when we consider that their CURS sum score is worse than with the combination group, despite a shorter duration of disease and of L-dopa-treatment. At first glance the combination of L-dopa plus L-deprenyl seems superior to the combination of L-dopa plus Ergot-derivatives, but we have to take into account that for group III the initial figures were worse in any case.

The study shows a clear advantage of a combination of two different compounds over L-dopa on its own. (L-dopa "on its own" means always L-dopa + Decarboxylase-inhibitor, of course.)

There is a variety of possible reasons for this. In general the combination of two different compounds can help to keep the L-dopa-dosage down an this probably means a relief of the presynaptic neuron.

Probably the followings effects can be achieved simultaneously:

a) stimulation of pre- and postsynaptic receptors by Ergot-derivatives keep the receptors stabilised
b) blocking L-dopa-metabolism by L-deprenyl and
c) perhaps neutralisation of toxic radicals by L-deprenyl

The study will be repeated after a period of three or five years, respectively. It may be possible in this way to form a more accurate picture of the tendency which we have so far outlined. Furthermore a triple combination (L-dopa plus Ergot-derivatives) will be compared to the other groups.

References

Birkmayer W, Knoll J, Riederer P, Youdim MBH, Hars V, Marton J (1985) Increased life expectancy resulting from addition of L-deprenyl to madopar treatment in Parkinson's disease: a long term study. J Neural Transm 644: 113–127

Csanda E, Antal J, Fornadi F (1980) Clinical experience in extrapyramidal diseases with selective MAO-B-inhibitor, L-deprenyl. In: Magyar K (ed) Monoamine exidase and their selective inhibitors. Akadémiai Kiadó, Budapest/Pergamon Press, Oxford, pp 127–132

Glover V, Gibb C, Sandler M (1986) Monoamine oxidase B (MAO-B) is the major catalyst of l-methyl-4-phenyl-1, 2, 3, 6-tetrahydropyridine (MPTP) oxidation in human brain and other tissues. Neurosci Lett 64: 216–220

Knoll J (1987) R-(-)-deprenyl (selegeline, movergan®) facilitates the activity of the nigrostriatal dopaminergic neuron. J Neural Transm [Suppl] 25: 45–66

Lees AJ (1987) Surrent controversies in the use of selegiline hydrochloride. J Neural Transm [Suppl] 25: 157–162

Marsden CD, Sandler M (1986) The MPTP story: an introduction. J Neural Transm [Suppl] 20: 1–3

Sandler M, Willoughby J, Glover V, Gibb C (1987) Selegiline and the prophylaxis of Parkinson's disease. J Neural Transm [Suppl] 25: 35–43

Tetrud JW, Langston JW (1987) R-(-)-deprenyl as a possible protective agent in Parkinson's disease. J Neural Transm [Suppl] 25: 69–79

Ulm G, Fornadi F (1987) R-(-)-deprenyl in the treatment of end-of-dose akinesia. J Neural Transm [Suppl] 25: 163–172

Correspondence: Dr. G. Ulm, Paracelsus-Elena-Klinik, D-3500 Kassel, Federal Republic of Germany.

Clinical pharmacology of amantadine and derivatives

J. D. Parkes

Department of Neurology, King's College School of Medicine and Dentistry and Institute of Psychiatry, London, U.K.

Summary

Amantadine has a definite although minor place in the clinical management of parkinsonism. It is simple to administer, well tolerated by patients, with rapid if usually minor to moderate benefit rather than dramatic response, and with few serious adverse effects.

Introduction

Amantadine is a stable, crystalline, water-soluble amine salt, with a structure unrelated to that of any other antimicrobial or antiparkinsonian drug. The amine penetrates all cell membranes, including those of the nervous system.

Amantadine has antiviral activity only against certain RNA viruses. It specifically blocks the penetration of sensitive strains into host cells, and may also prevent viral uncoating after cellular entry. It is of proven value in viral diseases in man only in prophylaxis and treatment of influenza A. Its effect here is incomplete, but may be sufficient to reduce the serological response to infection, due to diminution of virus numbers. Unlike many other antiviral drugs, amantadine is almost non-toxic. Amantadine is not of proven clinical value in the management of herpes zoster, sub-acute sclerosing leucoencephalitis, or Jakob-Creutzfeldt disease.

Amantadine in Parkinson's disease

The activity of amantadine in Parkinson's disease is unrelated to the antiviral effect, and was found by chance by Schwab and his colleagues (1969), when the drug was given for influenza prophylaxis to a woman with Parkinson's disease. The clinical profile of amantadine was established in this first study, with in most patients modest response, but in a few subjects dramatic inprovement. The degree of response may be greater than with anticholinergics, less than with levodopa or dopamine agonists, but only a few direct comparisons have been done. In any case, in parkinsonism, the therapeutic index depends upon side effects as well as benefit, and these are different in nature with amantadine from other antiparkinsonian drugs (McEvoy, 1987). An additive effect of amantadine with anticholinergics and levodopa has been demonstrated (Webster and Sawjer, 1984).

The mode of action of amantadine in parkinsonism remains unclear. At least some of the benefit is likely to be due to augmentation of catecholamine metabolism and increased dopamine release to nerve stimuli. Amantadine probably relieves only the signs and symptoms of the illness, without changing the course of disease, but like other catecholamine-reuptake blocking drugs, amantadine may prevent neuronal uptake of endogenous or exogenous toxins such as MPP.

Amantadine derivatives

The properties of adamantane, the parent hydrocarbon, and its derivatives, have been explored in detail by Stetter (1954). Many adamantyl analogues of drugs have been prepared, but of these only amantadine and its salts have been widely investigated for their antiviral and antiparkinsonian effects. The most interesting of these compounds include amantadine spiro derivatives and the optical isomers of alpha-methyl-1-adamantine methyl amine, rimantadine, which is more active than amantadine against some influenza viruses. Amantadine and rimantadine have different pharmacokinetic profiles, which probably largely determine the different side effects (Hayden et al., 1983).

Pharmacokinetic studies

Simple, sensitive and specific methods have been developed to determine amantadine concentration in body fluids and tissues. Plasma values in patients during chronic treatment with amantadine hydrochloride 200–300 mg daily vary from 0.27 μgm/ml^{-1}– 0.77 μgm/ml^{-1} after one week of treatment, slightly higher after three weeks (0.68 μgm/ml^{-1}—1.01 μgm/ml^{-1}). The plasma concentration bears some relation to both motor effects in parkinsonians, and antiviral prophylactic effects in healthy adults (Aoki et al., 1985).

Amantadine is totally absorbed from the gut. The drug does not seem to be metabolished in man, although converted to an N-methyl derivative in some species. The major elimination is urinary, the rate of excretion depending on the amount present in the body, so the plasma levels rise until the rate of elimination approaches the rate of administration. There is considerable variation in excretion rate in different individuals, and this rate is influenced considerably by urinary pH. An initial dose three times the maintenance dose will build up to a stable body level in one day.

Effect of amantadine on neurotransmitter systems

A number of different effects of amantadine on neurotransmitter systems have been described, but the mechanism of antiparkinsonian effect remains elusive.

Acetylcholine

Amantadine has little if any anticholinergic effect, and does not act like scopolamine, atropine, benztropine, or other well-known, centrally-acting anticholinergics. In clinical practice, amantadine ist stimulant, and does not possess the sedative- or memory-impairing effects of some of these anticholinergic drugs.

5-HT

Amantadine in 10^{-4} M concentration will shift the dose-response curve for 5-HT uptake into smooth muscle or human blood platelets. However, amphetamine is 5—10 times more potent than amantadine in this respect.

Catecholamines

Amantadine affects the synthesis, accumulation, release and reuptake of catecholamines in the central and peripheral nervous system. These effects are slightly different dependent on the tissue studied, heart or brain region, but all will result in an increased availability of transmitter. As judged from animal studies, these effects may occur in man only with higher concentrations of amantadine than are likely to be achieved in conventional therapy. Amantadine has no effect on the de-amination of ^{14}C-5-HT or ^{3}HNA at high concentrations either in vitro or in vivo.

Amantadine in Parkinson's disease

Schwab et al. (1969) reported that 66 per cent of a group of 163 patients with Parkinson's disease improved after taking amantadine hydrochloride. Amantadine had been given to prevent influenza to a 58-year-old woman with moderately severe bilateral Parkinson's disease, who then had a partial remission of her symptoms, which returned on stopping the drug after six weeks. One hundred and seven parkinsonian patients were then given amantadine 200 mg daily and improved, 48 were unchanged, and 3 were worse. In 5 patients the results were uncertain. Of those who improved, one third showed a decline in response after 4—8 weeks, whilst in others, benefit remained for 3—8 months. On amantadine withdrawal, there was a prompt exacerbation of symptoms. Five patients who did respond initially did so when amantadine was resumed 1—3 months later, whilst in other patients treatment was a failure from the onset.

A quarter of patients described some side effects whilst taking amantadine, but these were usually mild. Side effects included jitteriness, insomnia, abdominal uneasiness, loss of appetite, slight feelings of dizziness, and potentiation of the side effects of mental confusion and hallucination as a result of benzhexol and benztropine. These features were reversible within 36 hours of stopping amantadine.

Subsequent studies have confirmed these initial results. Thirteen patients with idiopathic Parkinson's disease given amantadine 100 mg twice daily as a single treatment by Mann et al. (1971) showed a moderate improvement in bradykinesia, tremor and rigidity scores commencing on the second day. Fasano et al. (1970) described 16 patients with Parkinson's disease given amantadine 100 mg intravenously, 2 patients with an excellent, 6 with a good, and 2 with a slight response, the effect of amantadine persisting for 24—48 hours, but, since this preparation may cause venous thrombosis, further investigation has been limited. Most trials have shown improvement of around 15—30 per cent in total disability scores, with amantadine 200 mg daily (see Parkes et al., 1974). In contrast, however, Hunter et al. (1970) found amantadine gave little or no therapeutic benefit, and Greenburg (1971) reported that only 3 of 26 patients showed any improvement. The response to amantadine is not dependent on the presence of additional anticholinergic drugs, and patients with Parkinson's disease and post-encephalitic parkinsonism respond, at least initially, in a similar fashion.

Amantadine dosage

Amantadine 100, 300 and 500 mg daily was given by Parkes et al. (1970) to patients with Parkinson's disease, each dosage for a 2-week period. The patients described the preferred dose as 300 mg daily, and this resulted in a 26 per cent reduction in the initial disability score. Nine of 43 patients did not respond to any amantadine dosage.

Amantadine sulphate, citrate and hydrochloride

Amantadine hydrochloride is more rapidly absorbed than either the sulphate or the citrate (Völler, 1972), and produces a more rapid onset of effect, with a comparatively high concentration of this form in the serum. There may be no distinct clinical advantage in using one particular preparation, although Fünfgeld (1970) considered the clinical response to amantadine sulphate superior to that of hydrochloride. The excretory conditions for the sulphate and the hydrochloride are essentially similar.

Long-term treatment with amantadine

Several investigators have studied the long-term use of amantadine. Some patients show a decline in initial response to amantadine after 4—12 weeks, but this is not the majority, and in a group of 26 patients given amantadine 200—600 mg daily, alone or combined with anticholinergic drugs over a period of one year, improvement at this time was rated at 29 per cent (Parkes et al., 1971). Other studies have shown efficacy at 4 years (Timberlake and Vance, 1978), and a single 75-year-old female attending King's College Hospital has now taken amantadine 200 mg daily for 18 years, with continued benefit. Of considerable importance is the observation by Völler (1970) that amantadine gives considerable additional clinical effects in patients taking sub-optimal dosages of levodopa, or where levodopa treatment is limited by side effects.

Amantadine improves the major symptoms of parkinsonism including tremor, rigidity and akinesia, though to a much lesser extent than levodopa. Dallos et al. (1972) compared one month's treatment with levodopa 4 g daily with one month of amantadine 400 mg daily in two separate patient groups. In 34 patients with levodopa, and 21 with amantadine, there was a 33 per cent improvement in tremor, rigidity and akinesia on levodopa, as compared with 23 per cent on amantadine. Considerable further improvement with levodopa would be expected after one month of treatment, but no further improvement with amantadine.

Effects of amantadine in drug-induced parkinsonism

Kelly and Abazzahab (1971) found that rigidity and tremor, as well as tardive dystonia, caused by haloperidol and phenothiazines, were reduced in certain patients by amantadine 100–200 mg daily. Improvement in these symptoms was rapid, and in addition, amantadine improved akathisia. Most important, the degree of psychosis of the patients was not altered. Several subsequent studies have shown a similar effect of amantadine in drug-induced parkinsonism without any alteration in tardive dyskinesia (see, e.g. Borison, 1983). The degree of symptomatic improvement caused by amantadine in these patients is generally slight to moderate, with some improvement in gait, rigidity and facial expression. However, Mindham et al. (1972) found that amantadine was no more effective than placebo in the treatment of fluphenazine-induced motor disorders. Overall, these results show a greater disparity. Amantadine is probably of slight value in some patients with drug-induced extrapyramidal disorders, but further studies are needed to identify which patients respond, and why.

Side effects of amantadine treatment

The incidence of unwanted drug effects with amantadine is low. Nausea, vomiting and involuntary movements, which may result from levodopa, are very infrequent with amantadine. Addiction and tolerance to amantadine have not been reported. Amantadine does however cause the unique side effect of livedo reticularis of the skin, with leg oedema in some patients. In elderly subjects with parkinsonism, amantadine may have a useful alerting effect.

References

Aoki FY, Stiver HG, Sitar DS, Boudreault A, Ogilvie R (1985) Prophylactic amantadine dose and plasma concentrations—effect relationships in healthy adults. Clin Pharmacol Ther 128: 128–136

Borison RL (1983) Amantadine in the management of extrapyramidal side effects. Clin Neuropharmacol 6 [Suppl] 1: S 57–63

Dallos V, Heathfield K, Stone P (1972) Use of amantadine in Parkinson's disease. Result of a double-blind trial. Br Med J 4: 24–26

Fasano VA, Urciuoli R, Broggi G (1970) Primo osservazioni sull'uso dell'amantadina per via endovenosa e dulla associazione dell'amantadina per via orale con farmaci anticolinergici e con levo-dopa nel trattamento della malattia di Parkinson. Minerva Med 61: 2895

Fünfgeld EW (1970) Amantadin-Wirkung bei Parkinsonismus. Dtsch Med Wochenschr 95: 1834–1836

Greenburg J (1971) Treatment of parkinsonism with amantadine hydrochloride. Pennsylv Med 74: 54

Hayden FG, Hoffman HE, Spyker DA (1983) Differences in side effects of amantadine hydrochloride and rimantadine hydrochloride relative to differences in pharmacokinetics. Antimicrob Agents Chemother 23: 458–464

Hunter KR, Stern GM, Laurence DR, Armitage P (1970) Amantadine in parkinsonism. Lancet i: 1127–1129

Kelly JT, Abazzahab FS (1971) The antiparkinsonian properties of amantadine in drug-induced parkinsonism. J Clin Pharmacol 11: 211

McEvoy JP (1987) A double-blind crossover comparison of antiparkinsonian drug therapy: amantadine versus anticholinergics in 90 normal volunteers, with an emphasis on differential effects on memory function. J Clin Psychiatr 48 [Suppl] 20–23

Mann DC, Pearce LA, Waterbury LD (1971) Amantadine for Parkinson's disease. Neurology 21: 958–962

Mindham RHS, Gaind R, Anstree BH, Rimmer L (1972) Comparison of amantadine, orphenadrine and placebo in the control of phenothiazine-induced parkinsonism. Psychol Med 2: 406

Parkes JD, Baxter RCH, Curzon G, Knill-Jones RP, Knott PJ, Marsden CD, Tattersall R, Vollum B (1971) Treatment of Parkinson's disease with amantadine and levodopa. Lancet i: 1083–1085

Parkes JD, Baxter RB, Marsden CD, Rees JE (1974) Comparative study of amantadine, benzhexol and levodopa in the treatment of Parkinson's disease. J Neurol Neurosurg Psychiatr 37: 422–426

Parkes JD, Zilkha KJ, Calver DM, Knill-Jones RP (1970) Controlled trial of amantadine hydrochloride in Parkinson's disease. Lancet i: 259–262

Schwab RS, England AC, Poskanzer DC, Young RR (1969) Amantadine in the treatment of Parkinson's disease. J Am Med Ass 208: 1168

Stetter H (1954) Die Chemie der organischen Ringsbetreme mit Utotropin-Struktur. Angew Chem 66: 217

Timberlake WH, Vance MA (1978) Four-year treatment of patients with parkinsonism using amantadine alone or with levodopa. Ann Neurol 3: 119–128

Völler G (1972) Zur Therapie des Parkinson-Syndroms mit Amantadin. Münch Med Wochenschr 14: 1066

Webster DD, Sawjer GT (1984) The combined use of amantadine HCI and levodopa/carbidopa in Parkinson's disease. Curr Ther Res 35: 1010–1013

Correspondence: Dr. J. D. Parkes, Department of Neurology, Institute of Psychiatry, De Crespigny Park, London SE 5 8 AF, United Kingdom.

Dopamine agonist treatment in early Parkinson's disease

U. K. Rinne

Department of Neurology, University of Turku, Turku, Finland

Summary

Due to insufficient antiparkinsonian efficacy, dopamine agonists alone do not seem to be useful as the primary treatment for the majority of parkinsonian patients, despite the fact that they produce only few fluctuations in disability. Early combination of dopamine agonist and a low dose of levodopa seems not only to improve parkinsonian disability, but also to inhibit the development of fluctuations in disability, especially end-of-dose disturbances and dyskinesias. Thus it appears advisable that the management of parkinsonian patients should begin in the early phase of the disease with this kind of combination.

Introduction

By correction of the deficiency in striatal dopamine, levodopa treatment considerably alleviates the symptoms of Parkinson's disease and substantially improves the quality of life of parkinsonian patients. Levodopa has not, however, resolved the long-term management of Parkinson's disease. The beneficial effect is not permanent and there are many late problems in long-term levodopa therapy, especially different kinds of fluctuations in disability (Rinne et al., 1976; Yahr, 1976; Rinne, 1981). New therapeutic strategies are therefore needed in the treatment of patients. For this purpose, extensive studies have been carried out over the past few years on the antiparkinsonian efficacy of dopamine agonists that stimulate the dopamine receptors directly and thus do not necessarily depend on the nigral dopamine neurons, which are progressively failing.

Initially it was shown that dopamine agonists are beneficial in advanced parkinsonian patients with a deteriorating levodopa response and fluctuations in disability (Calne et al., 1974; Rinne, 1983). Furthermore, during the past few years I have been studying the therapeutic role of dopamine agonists alone or in early combination with levodopa as initial treatment of early Parkinson's disease (Rinne, 1983, 1985, 1987).

Dopamine agonist alone

Each of the dopamine agonists available for clinical trials—bromocriptine, mesulergine, lisuride, pergolide, terguride, CQ 32-084, CQA 206-291 and CM 29-712—has a different neuropharmacological profile, and has distinctive effects both on D-1 and D-2 dopamine receptors and also on other neurotransmitter receptors. This would suggest that they might also elicit different clinical responses in the treatment of parkinsonian patients. However, our trials with previously non-levodopa-treated patients have shown that the degree of therapeutic responses to the various dopamine agonists studied was only moderate, and roughly similar in each case. Furthermore, the therapeutic efficacy of these compounds seems to be equal to that of levodopa in regard to tremor, although rigidity and hypokinesia were better alleviated by levodopa (Rinne, 1983).

Long-term follow-up studies have shown that only a small proportion of the parkinsonian patients seem to have really long-term benefit and tolerance during treatment with a dopamine agonist alone. Thus, after three years of treatment, only roughly one fourth of the patients were still undergoing the treatment, and even less thereafter (Table 1). Furthermore, the mean improvement in parkinsonian disability seems to be only moderate, and clearly decreased over the period of treatment. There were some distressing side effects, but the main reason for withdrawal of dopamine agonist treatment was insufficient therapeutic response. Consequently,

Table 1. Number of de novo parkinsonian patients under long-term treatment with bromocriptine, pergolide or lisuride

		Duration of treatment (years)									
	Initially	1		2		3		4		5	
Drug	N	N	(%)	N	(%)	N	(%)	N	(%)	N	(%)
Bromocriptine	76	32	(42)	22	(29)	21	(28)	13	(17)	5	(7)
Pergolide	20	12	(60)	7	(35)	5	(25)	2	(10)	1	(5)
Lisuride	30	16	(53)	10	(33)	6	(20)	5	(17)	4	(13)

levodopa had to be added to the regimen in most of the patients. However, the therapeutic responses seem to be very good in regard with the development of fluctuations in disability. There did not develop any end-of-dose disturbances and there were only a few cases with peak-dose dyskinesias or randomly-occurring freezing episodes during periods of three to five years' treatment (Rinne, 1985, 1986, 1987).

Thus, all in all, dopamine agonists do not seem to be useful anti-parkinsonian agents alone as the primary treatment for the majority of parkinsonian patients, despite the fact that they produce only few fluctuations in disability.

Early combination of dopamine agonist and levodopa

Because the dopamine agonists available so far are not good enough to be a primary treatment in Parkinson's disease, the important question arises whether early addition of levodopa to the regimen could offer any further therapeutic benefit and result in a better therapeutic index than with either drug alone. Indeed, own recent findings obtained from a 5-year follow-up with bromocriptine (Rinne, 1985, 1987) or with pergolide (Rinne, 1986), and the three years' results of a randomized trial with lisuride (Rinne, 1986), do in fact indicate that dopamine agonists in early combination with levodopa might be effective for these purposes. Briefly, during combined treatment the daily dosage of levodopa needed for optimal therapeutic response was significantly lower than when using levodopa alone. Furthermore, this kind of treatment regimen resulted in a therapeutic response equal to that achieved with high-dose levodopa alone but with significantly fewer fluctuations in disability, especially end-of-dose disturbances and dyskinesias.

Thus these long-term follow-up studies give strong support for the early combination of a dopamine agonist and levodopa as a new strategy in the primary treatment of patients with Parkinson's disease.

Discussion

According to the present results the dopamine agonists studied are not good enough to improve parkinsonian disability alone. What is the reason for this? These dopamine agonists stimulate only or predominantly D-2 dopamine receptors, whereas good therapeutic response seems to need simultaneous stimulation of both D-2 and

D-1 receptors, as has also been demonstrated in experimental animals (Robertson and Robertson, 1986; Starr and Starr, 1987). Dopamine created from exogenous levodopa is therefore needed together with the dopamine agonist. Theoretical support for this approach derives from the hypothesis that conformational changes in dopamine receptors induced by dopamine could explain the synergistic effect of levodopa and dopamine agonists (Goldstein et al., 1985). On the other hand, according to experimental studies, dopamine-induced stimulation of D-1 receptors potentiates the action of dopamine agonists on D-2 receptors (Waddington, 1986).

Why are there less end-of-dose disturbances and dyskinesias during treatment with a dopamine agonist alone or in early combination with levodopa than with high-dose levodopa alone? Combined treatment with a dopamine agonist and levodopa seems to give therapeutic responses with a lower levodopa dosage than with the drug used alone. This as such might be an important factor for decreasing the risk of long-term complications with levodopa treatment (Birkmayer, 1976; Rajput et al., 1984; Poewe et al., 1986). Prevention of fluctuations in disability, especially end-of-dose failure, may result not simply from the lower dose of levodopa, but also from other factors. Indeed, the longer pharmacokinetic action of dopamine agonist on the striatal receptor level than with levodopa alone might be affecting the therapeutic responses obtained.

On the other hand, dopamine receptor mechanisms and storage failure of dopamine in the striatum might be involved. It has been suggested that end-of-dose disturbances may be due to a failure in the synthesis, storage and release of dopamine in the nigrostriatal dopamine neurons because of faulty feedback control via presynaptic D-2 autoreceptors (Carlsson, 1983). Indeed, recent PET studies have demonstrated decreased storage capacity of striatal dopamine in end-of-dose failure (Leenders et al., 1985). In order to prevent this pathophysiological process, it is possible that early combined treatment with levodopa and a dopamine agonist like bromocriptine and lisuride, having a great affinity to presynaptic D-2 autoreceptors, might maintain the normal functioning of the striatal dopamine neurotransmission for a longer period than does levodopa alone. Indeed, according to experimental studies, bromocriptine is able to block levodopa-induced changes in the striatal dopamine receptors (Ogawa et al., 1984). On the other hand, the effects of microiontophoretically applied lisuride on the firing rate of dopamine neurons suggest a long-lasting effect of lisuride on D-2 autoreceptors (Gessa, 1988).

Thus, early combination of a dopamine agonist and levodopa seems to sustain a better balance between postsynaptic D-2 and D-1 receptors and presynaptic D-2 autoreceptors than high-dose levodopa alone. In this way it is possible to maintain the normal functioning of striatal dopamine neurotransmission for a longer period than with levodopa alone, resulting in better management of Parkinson's disease with fewer fluctuations in disability. Therefore it appears advisable that the treatment of parkinsonian patients should begin in the early phase of the disease with a dopamine agonist combined with a low dose of levodopa.

References

Birkmayer W (1976) Medical treatment of Parkinson's disease: General review, past and present. In: Birkmayer W, Hornykiewicz O (eds) Advances in parkinsonism. Editiones Roche, Basle, pp 407–423

Calne DB, Teychenne PF, Claveria LE, Eastman R, Greenacre JK, Petrie A (1974) Bromocriptine in parkinsonism. Br Med J 4: 442–444

Carlsson A (1983) Are "on-off" effects during chronic L-Dopa treatment due to faulty feedback control of the nigrostriatal dopamine pathway? J Neural Transm 19: 153–161

Gessa GL (1988) Agonist and antagonist actions of lisuride on dopamine neurons: electrophysiological evidence. J Neural Transm [Suppl] 27: 201–210

Goldstein M, Lieberman A, Meller E (1985) A possible molecular mechanism for the antiparkinsonian action of bromocriptine in combination with levodopa. Trends Pharmacol Sci 6: 436–437

Leenders KL, Herold S, Palmer AJ, Turton D, Quinn N, Jones T, Frackowiak RSJ, Marsden CD (1985) Human cerebral dopamine system measured in vivo using PET. J Cerebral Blood Flow Metab 5/1: 157–158

Ogawa N, Yamawaki Y, Kuroda H, Takayama H, Ota Z (1984) Differences in the effects of levodopa and bromocriptine on rat striatal dopamine receptors. Neurosciences 10: 259–266

Poewe WH, Lees AJ, Stern GM (1986) Low-dose L-dopa therapy in Parkinson's disease: A 6-year follow-up study. Neurology 36: 1528–1530

Rajput AH, Stern W, Laverty WH (1984) Chronic low-dose levodopa therapy in Parkinson's disease: an argument for delaying levodopa therapy. Neurology 34: 991–996

Rinne UK (1981) Treatment of Parkinson's disease: problems with a progressing disease. J Neural Transm 51: 161–174

Rinne UK (1983) New ergot derivatives in the treatment of Parkinson's disease. In: Calne DB, Horowski R, McDonald RJ, Wuttke W (eds) Lisuride and other dopamine agonists. Raven Press, New York, pp 431–442

Rinne UK (1985) Combined bromocriptine-levodopa therapy early in Parkinson's disease. Neurology 35: 1196–1198

Rinne UK (1986) Dopamine agonists as primary treatment in Parkinson's disease. In: Yahr MD, Bergmann KJ (eds) Advances in neurology. Raven Press, New York, pp 519–523

Rinne UK (1987) Early combination of bromocriptine and levodopa in the treatment of Parkinson's disease: A 5-year follow-up. Neurology 37: 826–828

Rinne UK, Sonninen V, Siirtola T (1976) Long-term treatment of parkinsonism with L-Dopa and decarboxylase inhibitor: A clinical and biochemical approach. In: Birkmayer W, Hornykiewicz O (eds) Advances in parkinsonism. Editiones Roche, Basle, pp 555–565

Robertson GS, Robertson HA (1986) Synergistic effects of D-1 and D-2 dopamine agonists on turning behaviour in rats. Brain Res 284: 387–390

Starr BS, Starr MS (1987) Behavioural interactions involving D_1 and D_2 dopamine receptors in on-habituated mice. Neuropharmacology 26: 613–619

Waddington JL (1986) Behavioural correlates of the action of selective D-1 dopamine receptor antagonists. Impact of SCH 23390 and SKF 83566, and functionally interactive D-1: D-2 receptor system. Biochem Pharmacol 35: 3661–3667

Yahr MD (1976) Evaluation of long-term therapy in Parkinson's disease: Mortality and therapeutic efficacy. In: Birkmayer W, Hornykiewicz O (eds) Advances in parkinsonism. Editiones Roche, Basle, pp 435–443

Correspondence: Prof. Dr. U. K. Rinne, Department of Neurology, University of Turku, SF-20520 Turku, Finland.

Subcutaneous apomorphine in Parkinson's disease

C. M. H. Stibe, P. A. Kempster, A. J. Lees, and G. M. Stern

Department of Neurology, Faculty of Clinical Sciences,
Middlesex and University College Hospitals, London, U.K.

Summary

Motor function in patients who are disabled by parkinsonian on-off oscillations can be significantly improved by administration of subcutaneous apomorphine by diurnal infusion or intermittent injection. In conjunction with use of the peripherally acting dopamine receptor antagonist domperidone, this treatment is well tolerated and has produced prolonged benefit in patients suffering from the most intractable complication of prolonged L-dopa therapy.

Introduction

Fluctuating and unpredictable response to orally administered L-dopa medication often prevents effective control of motor disability in patients with Parkinson's disease. Oscillation of motor function eventually develops in most patients who are initially responsive to L-dopa. Once established, severe motor oscillations are difficult to manage with conventional treatment. Constant infusion techniques to maintain a constant level of central dopamine receptor stimulation, and use of dopamine receptor agonists which do not share the transport and metabolic pathways of L-dopa are possible therapeutic approaches to this problem. However, parenteral treatment with ergolide dopamine receptor agonists may be complicated by neuropsychiatric side effects at doses which produce equivalent motor responses to L-dopa medication (Critchley et al., 1986).

We have administered the non-ergolide dopamine receptor agonist apomorphine by subcutaneous infusion and multiple injection with benefit in patients disabled by fluctuating motor response to oral L-dopa (Stibe et al., 1988). Side effects which in the past had limited the oral use of this agent such as vomiting, postural hypotension and pre-renal uraemia (Cotzias et al., 1970) are preventable with simultaneous oral administration of domperidone, a peripherally acting dopamine antagonist (Corsini et al., 1979). An initial comparative study of the therapeutic effects of subcutaneous apomorphine and lisuride in several patients showed that apomorphine provided equivalent or better motor responses without the neuropsychiatric side-effects seen with lisuride (Stibe et al., 1987). Our experience with subcutaneous apomorphine treatment of patients with severe motor oscillations and also of patients with other forms of parkinsonian motor disturbances is described.

Apomorphine treatment for motor oscillations

Patients and methods

19 non-demented patients with disabling motor oscillations were treated with subcutaneous apomorphine. This patient group (12 male, 7 female, mean age 56 years, mean duration of disease 14 years) all showed clear periodic responsiveness to L-dopa medication with improved motor function and dyskinetic movements during on phases; 10 of these patients had previously experienced hallucinations or confusion due to antiparkinsonian medication. Initial assessment by patient-diary recording of motor function established the pattern of motor fluctuations on conventional treatment. On this basis, the decision to administer apomorphine by infusion pump or by intermittent injection was made by considering the number of off phases experienced per day. Patients with in excess of five off periods per waking day were generally treated by infusion pump.

Infusion technique

11 patients were treated by constant diurnal subcutaneous infusion of apomorphine by a small portable infusion pump. A delivery needle was inserted into the anterior abdominal wall and changed daily by patients. Infusion treatment was commenced while patients were in hospital and apomorphine and L-dopa doses were adjusted to achieve optimum control of motor function. Infusions were begun soon after waking and discontinued before bedtime. Domperidone 20 mg three times a day was administered orally. In addition, patients were educated in the use of a pulse function of the pumps, enabling additional apomorphine boluses to be self-administered at times of waning therapeutic effects.

Multiple injections

8 less severely affected patients received intermittent self administered subcutaneous apomorphine injections. Apomorphine was delivered by an insulin syringe mounted in a "Penject" (Hypoguard Ltd, Melton, Suffolk, U.K.) which allowed preset doses to be given. Optimum dose for effective motor response was established under observation in hospital and patients were educated to inject themselves in anticipation of impending off phases. Patients also took oral domperidone 20 mg thrice daily.

Results

Infusion patients

The 11 patients using pumps all gained worthwhile benefit and many patients showed striking improvement, regaining considerable independence. Soon after commencing apomorphine infusions, the mean number of hours off per day was reduced from 10.1 to 3.8 (mean 6.3 hours, $p < 0.05$ by Student's t test). Mean disability when off was also significantly reduced. Although the overall severity of dyskinetic movements was only mildly reduced, 4 patients showed striking improvement of involuntary movements. Mean infusion rate of apomorphine was 0.04 mg/kg/hour. Improved motor performance was maintained in all patients after a mean treatment period of fourteen months. Several patients were able to further reduce the amount of daily off time once expertise was gained with anticipatory bolus doses. A mean reduction in daily L-dopa dose of 209 mg was possible. In no case could L-dopa medication be completely discontinued, although apomorphine infusion alone sustained motor function after morning L-dopa "start-up" in two patients.

Multiple injections

Similarly good results were obtained in all 8 patients treated with intermittent injections. The mean daily number of hours off fell from 6 to 2.7 (mean reduction 3.3 hours, $p < 0.05$). Injected doses of apomorphine ranged from 0.8 mg to 6 mg. After several months of treatment, one patient chose to switch to infusion pump treatment to avoid multiple injections.

Objective therapeutic responses to apomorphine doses

Responses to single apomorphine doses in individual patients in general mirrored their usual response to L-dopa, with a similar pat-

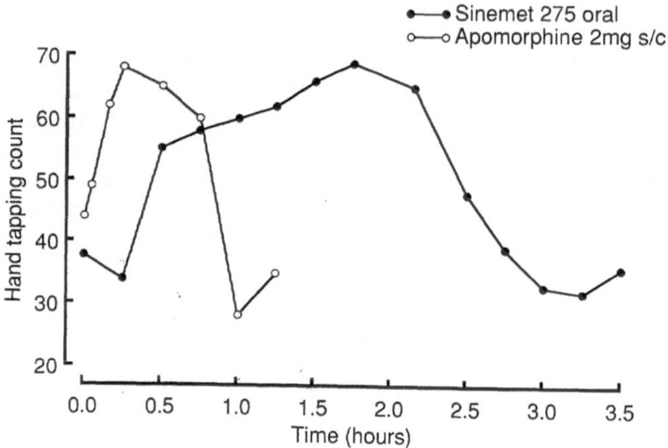

Fig. 1. Comparison of motor response to oral L-dopa/carbidopa and subcutaneous apomorphine doses by serial hand tapping count estimations

tern and degree of motor improvement and involuntary movements. Motor response was apparent within 5—15 minutes of injections and lasted for 40—90 minutes. Comparison of quality and duration of motor response to apomorphine and L-dopa in a typical patient is shown in Fig. 1. Although repeat doses were sometimes required, off phases could virtually always be reversed with subcutaneous apomorphine boluses. Reassessment of response to L-dopa doses was carried out in four patients after twelve months treatment with diurnal pump infusion. Each showed that large amplitude oscillations remained.

Treatment of other parkinsonian motor deficits

Three patients with severe, non-oscillating parkinsonian motor disabilities which showed weak responsiveness to L-dopa were treated with apomorphine. Motor response to apomorphine was not superior to that of L-dopa and these patients derived relatively little benefit.

Painful off period dystonias in two of the infusion patients and one multiple injection patient were effectively controlled by apomorphine. One of the severely disabled, non-oscillating patients, despite generally modest benefits from apomorphine treatment, showed consistent reversal of periodic painful dystonic spasms by injections. A further two patients, with only mild fluctuation of motor function in relation to L-dopa doses, had distressing episodes of belching during off periods. Investigations suggested that this phenomenon was caused by off period oesophageal dysmotility (Kempster et al., 1988). Apomorphine injections suppressed belching in both patients.

One patient with disabling diphasic dyskinesias was treated with apomorphine pump infusion. In keeping with previous studies of treatment of this complication by intravenous L-dopa infusion (Quinn et al., 1984), diphasic involuntary movements could not be prevented by apomorphine.

Adverse reactions to apomorphine

All patients using pumps developed itchy red nodules at needle sites which persisted for 1–2 weeks. Peripheral blood eosinophilia was also present in all of these patients. Side-effects such as nausea, vomiting and postural hypotension, which might have been expected to limit the practicability of this form of treatment, were contained by concurrent administration of domperidone. Biochemical tests of renal function remained normal in all patients. Several patients experienced mild drowsiness when first exposed to apomorphine but this subsided with continued treatment. One patient had transient visual hallucinations without other evidence of cognitive disturbance. However, no patient experienced neuropsychiatric side-effects which limited apomorphine treatment, although three patients had previously experienced psychotic reactions to L-dopa or lisuride medication.

Discussion

Subcutaneous apomorphine led to sustained improvement and increased independence in daily living in all patients whose major cause of disability was inconstant response to oral L-dopa medication. Apomorphine has reliable and effective dopamine agonist activity when administered subcutaneously. Several patients commented that a major benefit of this treatment was a degree of control over what had hitherto been erratic and unpredictable fluctuations of motor function. In conjunction with domperidone administration, the use of apomorphine was not limited by peripheral dopamine receptor stimulatory side-effects or by neuro-psychiatric toxicity. Motor response to apomorphine was generally equivalent to that of L-dopa in individual patients. Predictably therefore, patients with parkinsonian motor disabilities which are weakly influenced by L-dopa medication are unlikely to be significantly improved by this treatment.

Although subcutaneous apomorphine appreciably reduced the number and severity of off phases in patients with motor oscillations, some periods of incapacitating immobility still occurred despite constant-rate infusions. These were usually overcome by additional bolus doses of the drug. Most patients continued to require L-dopa for optimum response. Objective assessment of motor responses to L-dopa doses after one year of treatment showed no evidence that the underlying nature of the motor fluctuations was ameliorated by long term diurnal constant rate apomorphine infusion. In this respect, it may be relevant that, all of the infusion patients still received significant phasic central dopamine receptor stimulation, both due to residual oral L-dopa medication and to bolus apomorphine doses. Infusion pumps were generally well tolerated, but the complexity of this technique may limit its use in some patients. The intermittent injection "Penject" device was easier to use and is probably the more practical treatment for oscillating patients with less frequent off periods.

Subcutaneous apomorphine produced benefit in patients complaining of unpleasant positive motor phenomena in off periods, even if these were not associated with marked oscillation of motor performance. Off period dystonic episodes, which were prolonged and distressing before commencing apomorphine, were quickly curtailed by injections. Similarly, the unusual complication of off period belching was successfully treated in two patients. The prompt and reliable onset of response to subcutaneous apomorphine was the most important factor in the control of positive motor disturbances in off phases.

It is our experience that subcutaneous administration of apomorphine can effectively stabilise motor function in the subgroup of parkinsonian patients with fluctuating and erratic response to oral medication. It is suggested that patients disabled by this complication should be offered a trial of subcutaneous apomorphine treatment before consideration of more hazardous therapeutic measures such as brain implantation of dopamine secreting tissue.

References

Corsini GU, Del Zompo M, Gessa GL, Mangoni A (1979) Therapeutic efficacy of apomorphine combined with an extracerebral inhibitor of dopamine receptors in Parkinson's disease. Lancet i: 954–956

Cotzias GC, Papavassilou PS, Fehling C, Kaufman B, Mena I (1970) Similarities between neurological effects of L-dopa and apomorphine. N Engl J Med 282: 31–33

Critchley P, Perez F, Quinn N, Coleman R, Parkes D, Marsden CD (1986). Psychosis and the lisuride pump. Lancet ii: 349

Kempster PA, Lees AJ, Crichton P, Frankel JP, Shorvon P (1988) Off period belching due to a reversible disturbance of oesophageal motility in Parkinson's disease and its treatment with apomorphine. Movement Disorders (in press)

Quinn N, Parkes JD, Marsden CD (1984). Control of the on/off phenomenon by intravenous infusion of L-dopa. Neurology 34: 1131–1136

Stibe CMH, Lees AJ, Stern GM (1987) Subcutaneous infusion of apomorphine and lisuride in the treatment of parkinsonian on-off oscillations. Lancet i: 871

Stibe CMH, Lees AJ, Kempster PA, Stern GM (1988) Subcutaneous apomorphine in parkinsonian on-off oscillations. Lancet i: 403–406

Correspondence: Dr. P. A. Kempster, Neurology Department, The Middlesex Hospital, Mortimer Street, London, W 1 N 8 AA, United Kingdom.

Continuous dopaminergic stimulation with parenteral lisuride in complicated Parkinson's disease

S. Bittkau[1], **I. R. Klewin**[2], **I. Suchy**[3], and **H. Przuntek**[2]

[1] Department of Neurology, University of Würzburg, Federal Republic of Germany
[2] Department of Neurology, St. Josef University-Hospital, Bochum, Federal Republic of Germany
[3] Department of Neuroendocrinology, Schering AG, Berlin

Summary

The parenteral infusion of lisuride in the treatment of fluctuations in complicated Parkinson's disease is highly effective. We report about 18 patients, who were treated in our institutions. In our patient-group we observed a significant ($p < 0.01$) drop in the UPDRS—scores for rigor, akinesia and speech and for OFF-hours (− 30%), tremor remained nearly uneffected. Two deaths following myocardial infarction and pneumonia were not attributable to therapy. 6 patients experienced hallucinations, in 2 of these infusion had to be stopped. In 2 patients with drug-induced mental alterations these phenomena seemed to depend on the ratio of lisuride to levodopa as on the infusion-profile.

This new therapy is time-consuming and needs a highly individual approach, but remains most effective in the treatment of motor fluctuations and of dyskinesias, offering possibly even further insight into the pathogenesis of these phenomena.

Introduction

The most important therapeutic problems of advanced stages of Parkinson's disease are the changing response to levodopa-therapy, the fluctuations of motor performance, and the mental side effects of antiparkinsonian drugs. Since it has been shown that fluctuations of mobility in Parkinsonian patients are correlated with fluctuating plasma dopamine levels after oral intake of levodopa, pharmaceutical research is strongly oriented towards developing methods which produce a constant plasma level of the effective substance or continuous stimulation of dopaminergic striatal receptors. Though constant levodopa infusions produce constant dopamine levels and abolish fluctuations in mobility (Shoulson et al., 1975), intravenous levodopa is not feasible for long-term treatment for practical reasons.

In the search for new dopaminergic substances, it was lisuride which seemed best suited to the continuous s. c. application (Obeso et al., 1983) because the hydrogen maleate of lisuride is water soluble and effective in the microgramm range. Lisuride is lipophilic and passes quickly through the blood brain barrier. This has been shown in Parkinsonian patients by Parkes et al., 1981, who observed a clear cut improvement of akinesia, rigidity and tremor within 10 minutes of an intravenous bolus of 150 mcg of lisuride.

Lisuride is well absorbed from the intestines, but it undergoes a first pass metabolism in the liver resulting in a bioavailability of 10—22%. The elimination of lisuride from the blood is quite rapid with a half-life of 1.7 hour in healthy young volunteers and a half-life of 2.2 hours in Parkinsonian patients (Hümpel et al., 1981). Lisuride is a semisynthetic ergot derivative. It shows a high affinity for D 2-receptors (Horowski and Wachtel; Riederer et al., 1983). Lisuride is mainly a D 2-agonist. Therapeutic doses of lisuride strongly stimulate postsynaptic D 2-receptors. Presynaptic inhibition which occurs after low doses has been demonstrated biochemically (Kehr, 1977) and may be seen occasionally in patients at the beginning of treatment as a transient worsening of tremor or akinesia. Lisuride is also a D 1-agonist, but as such is much less potent than at the D 2-receptor. Lisuride also exhibits partial serotonin agonistic properties at low doses and it is adrenolytic at high doses. Lisuride has successfully been used for the oral treatment of Parkinson's disease (McDonald and Horowski, 1983).

Subcutaneous infusion of lisuride has been performed in more than 100 patients with extrapyramidal disorders for up to 2½ years, until now mainly at four centres (Bittkau and Przuntek, 1986; Critchley et al., 1986; Obeso et al., 1986; Ruggieri et al., 1986). Only patients with severe motor fluctuations whose response to combined, fractionated oral treatment had become unsatisfactory, were included.

It was obvious rather soon that the continuous infusion of lisuride resulted in an improvement of motor performance which ranged from moderate to striking, but that the use of the new treatment method was complicated by mental side effects. These side effects might be some mild euphoria, agitation. They might appear as nightmares or unthreatening visual hallucinations, but they might also result in threatening paranoic delusions and in severe confusion.

The reports from the four centres differed insofar as in one most of the patients interrupted the new therapy during the first weeks because of psychotic side effects (Critchley et al., 1988) while the others were either much less troubled by such effects (Obeso et al., 1988; Stocchi et al., 1988) or kept the patient under treatment by use of various comedications (Bittkau et al., 1988).

A comparative evalutation (Suchy and Krause, 1988) pointed towards the importance of preceding and concomittant medications but could not fully elucidate the causes of the differing results.

Therefore we (IRK, HP) undertook another longterm trial in which we intended to show the effect of the subcutaneous infusion of lisuride on the parameters rigidity, tremor, akinesia and speech in patients with distinct fluctuations of motor performance or wearing-off of the duration of response to levodopa doses.

Furthermore, we (SB) analysed the course of two patients treated with levodopa and lisuride more than two years; we particularly analysed the improvement of mobility and the mental side-effects in correlation with the dosage of levodopa and lisuride.

Patients and methods

In the interval from November 1986 until January 1988 we treated 16 patients suffering from parkinsonism with subcutaneous infusion of lisuride.

The criteria for treatment consisted of distinct fluctuations of motor performance and wearing-off of the duration of response to levodopa doses.

Patients: 11 male, 5 female with a mean age of 66.3 ± 5.5 years and a mean age at onset of parkinsonism of 57.4 ± 7.4 years with a mean duration of Parkinsonism of 8.8 ± 5.0 years. After a clinical examination the patients were scored by the Unified-Parkinson-Disease-Rating-Scale (UPDRS).

Before starting the subcutaneous infusion of lisuride, and every two days subsequently, we carried out the motor performance test according to Schoppe. The patients were also asked to fill in an hourly self-evaluation of their mobility (totally blocked–moderately impaired mobility–freely mobile) and of side-effects. We then adapted the lisuride infusion-rate inversely to the individual 24 hour mobility profile. The computerized minipump we used (Disetronic MRS 2) allows an hourly variation of the infusion rate to be programmed for 24 hours. The infusion system itself has to be changed every 48 hours.

The finally resulting dosage of lisuride ranged from 0.25 up to 2.0 mg per day. The observed parameters during therapy were tremor, akinesis, rigidity and speech. Their scores were examined previous to, during therapy and at reaching the optimal dosage of lisuride and explored, by the Wilcoxon test for matched samples, for statistically significant differences.

Results

11 patients showed an akinetic-rigid type, 4 an equivalent type and 1 a type with tremor dominance.

The effect on the parameters tremor, rigidity, akinesia, and speech measured by the UPDRS (0—4) before therapy and at reaching the optimal dosage is shown in Figs. 1—4. The mean score for tremor was reduced from 1.4 ± 1.4 to 1.3 ± 1.1. The difference was not statistically significant. The mean score for rigor went from 3.0 ± 1.0 to 2.2 ± 0.8 ($p < 0.01$). Akinesia improved from 3.0 ± 0.9 to 2.2 ± 1.1 ($p < 0.01$).

Speech also improved significantly under therapy from 3.1 ± 1.1 to 2.4 ± 1.1 ($p < 0.01$). The off-phases could be significantly reduced by 30%.

Two multimorbid patients died from intercurrent diseases during therapy (cardiac infarct and pneumonia).

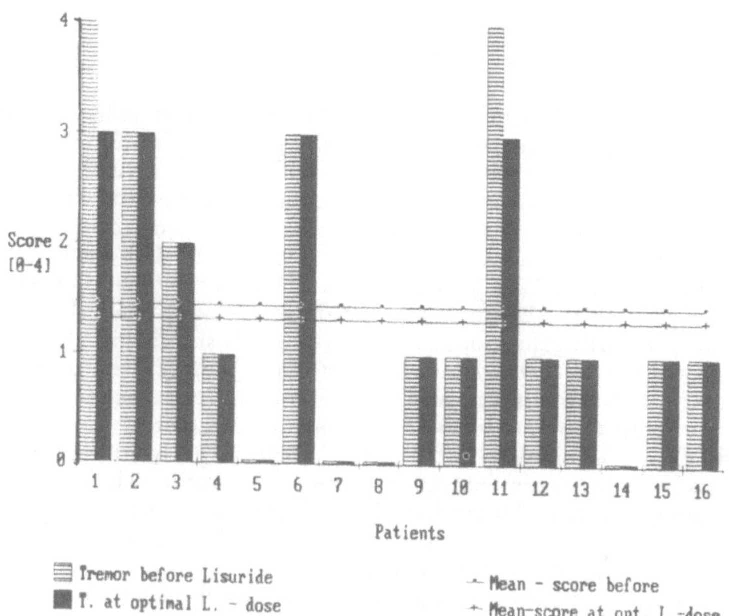

Fig. 1. Tremor — scores (UPDRS) of 16 patients before and at optimal lisuride dose

In two patients we had to prematurely stop the subcutaneous application of lisuride because of a developing psychosis getting out of control.

In two patients we could change from subcutaneous application of lisuride to oral after 1 and 5 months. The side-effects were gastrointestinal symptoms in 5 patients reaching from slight nausea to severe constipation. In one patient a carcinoma of the sigmoid was masked by the symptoms of an ileus. In 6 patients psychotic reactions such as visual and acoustical hallucinations could be observed. Sometimes very vivid and colourful dreams went ahead of psychotic symptoms. 4 patients showed orthostatic dysregulation. Only 5 were free of any side-effects at all.

Beside the two psychotic reactions that went out of control there was no severe side-effect that could not be treated symptomatically to prevent breaking off the treatment with lisuride.

Long-term results

Furthermore we report on 2 patients, who were treated with the lisuride-pump for 28 months and experienced most severe psychiatric side-effects within the first year of therapy, which could be improved by changing the correlation of levodopa and lisuride.

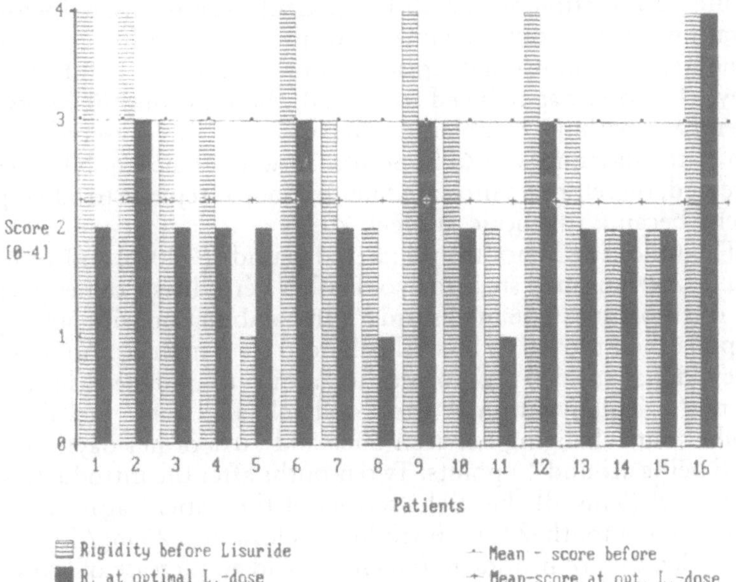

Fig. 2. Rigidity – scores (UPDRS) of 16 patients before and at optimal lisuride doses

Pat. C. S., female, at start of trial 55 years, idiopathic M. Parkinson known for 12 years, levodopa treatment for 10 years, Hoehn + Yahr stage V, levodopa intake 1.000 mg (plus 250 mg Benserazid), no other antiparkinsonian drug tolerated, severe depression, since 3–4 years frequent episodes of drug-induced confusion, hallucinations and psychotic events, bound to wheelchair, OFF-periods over 90% of the daytime. Columbia University Rating Scale (CURS): 87 points. [For the method of the lisuride infusion see (Obeso et al., 1986; Bittkau and Przuntek, 1988).]

In December 1985 we started the trial and increased lisuride within 9 days to a maximum dose of 2.1 mg/d and concommitantly reduced levodopa to 450 mg. Except for some nausea she tolerated the dose well for one week, her mobile hours increased from 10 to 80% of daytime, her CURS dropped to 40 points. After 10 days with 2.1 mg of Lisuride/d she developed oral dyskinesias of increasing intensity. We reduced lisuride to 1.7 mg/d and levodopa to 350 mg/d. 5 days later a short-lasting paranoid-hallucinatory psychosis on a still clear sensorium could be observed. Lisuride was reduced to 1.3 mg/d. The following days her mobility worsened while her dyskinesias persisted. We elevated levodopa to 400 mg (day 37). Till day 50 she had frequent episodes of non-threatening hallucinations (familiar persons, cats) without confusion. On a hospital stay for routine examination at day 60 the patient decompensated without alteration of her drug regime and presented with a paranoid-hallucinatory confusional episode more severe at night, lasting 4 days. Lisuride was reduced to 1 mg/d, the following 10 weeks the patient was without complaints. Her mobility was preserved about 70% of the daytime, CURS 51 points. At week 18 her mobility again decreased, therefore we increased her levodopa stepwise to 500 mg/d, patient became psychotic at week 20.

The following 5 months we altered lisuride (– 10%) and/or levodopa (+ 25% in small steps to avoid either her akinesia or her psychiatric disturbances. 0.86 mg lisuride/d in combination with 500 mg of levopoda then proved to be well balanced for 3 months. An increase of levodopa to 600 mg forced us again to reduce lisuride to 0.72 mg/d. This regime was then well tolerated for 7 months, her mobile hours changing between 60% and 70% of her daytime, her CURS being around 45 points. Two months after the introduction of L-Deprenyl (5 mg/d) the night terrors of the patient again became worse, so at month 22 we introduced clozapine 25 mg/d. After a further decrease of lisuride to 0.64 mg/d and an increase of levodopa to 700 mg/d the patient seemed to have even fewer short hallucinations and vivid dreams while her mobility was about the same as before.

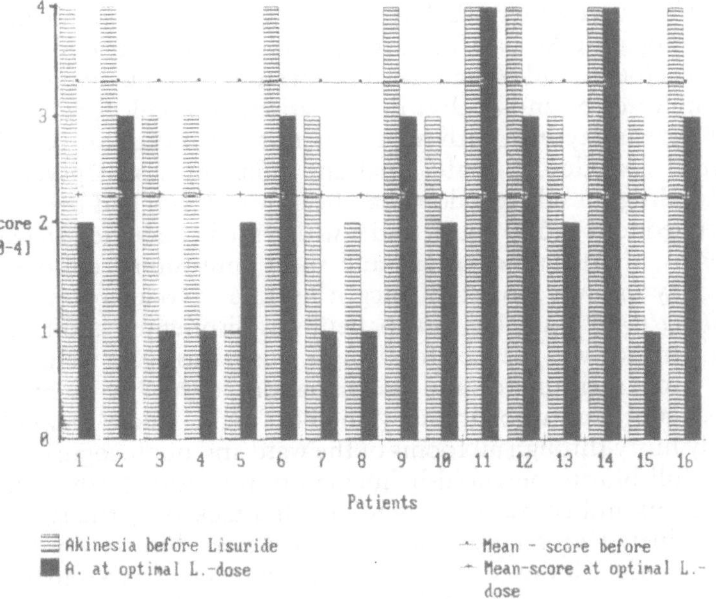

Fig. 3. Akinesia – scores (UPDRS) of 16 patients before and at optimal lisuride doses

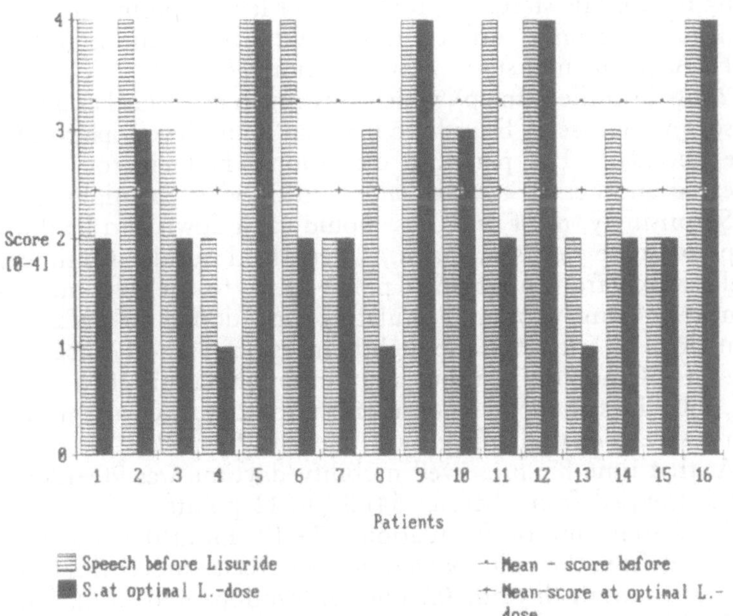

Fig. 4. Speech – scores (UPDRS) of 16 patients before and at optimal lisuride doses

The other patient (L. N.) was a 66 year old retired medical doctor, idiopathic M. Parkinson for 11 years, 10 years treated with levodopa, Hoehn and Yahr stage IV, severe on-off-fluctuations, partly complicated by end-of-dose deterioration, immobile for about 50%—60% of his daytime, with mild dementia. Over the past years he had several episodes of confusional and nonconfusional psychosis, all other drugs beside levodopa had been withdrawn and levodopa had been reduced to 500 mg/d, under which medication he still had vivid dreams, short hallucinations and short confusional episodes.

Day one we infused 80 mcg/h lisuride, levodopa was halved to 250 mg/d, 10 hours later the patient was disoriented and hallucinating. Lisuride was reduced to 1.8 mg/d, his status again would deteriorate from nightmares and hallucinations (5th—8th night), over a moderately confusional state (9th—14th night, where he would restlessly hurry through all rooms of the ward and not recognize his wife) to a full blown, peranoid-hallucinatory psychosis (15th night) with delusions and confusion, where he felt attacked by the staff and by hallucinated prosecutors, which even forced him to try to escape through the window of a 6th floor. Lisuride was reduced to 0.6 mg/d, mild neuroleptics tranquilizers were tried (melperon 50—0—75 mg, triflupromazin 50 mg, thioridazin-HCL 30 mg, promethazin HCL 25 mg b.i.d.). His status improved under levomepromazin 12.5—0—25 mg and he only suffered from mild hallucinations (6th—10th week). As his mental status again deteriorated, his local GP injected 2 × 2 mg fluspirilen (Imap), which very much made his Parkinsonism worse, an increase of lisuride to 0.85 mg/d and levodopa to 350 mg were necessary. His paranoid confusional hallucinatory state reappeared another 3—4 weeks later, mostly worse at night time.

Surprisingly, most nights he would calm down and lie down to sleep for some hours at 4 a.m., when the Lisuride infusionpump would switch from 6 mcg/h (8 p.m.—4 a.m. to the daytime rate of 57 mcg/h (4 a.m.—8 p.m.). Postulating some dopamine-deficiency at night-time, we increased the nightly infusion-rate to 19 mcg/h (daytime-rate 50 mcg/h). Without any further alteration of the medication, with the same total amount of lisuride/day, the patient was free of any severe mental side effects the next 10 weeks.

At that time his preserved mobility/daytime was 90—100%, his CURS dropped from (before: 54) 32 to 12 points.

After some minor fluctuations (7.—10. month) his status was stable and, from the point of view of side-effects, uneventful from the 11th—17th month (lisuride 0.74 mg/d, levodopa 450 mg/d, L-deprenyl 5—7.5 mg/d). At that time we used two different application-schedules (hourly changing "saw—tooth" profile and the above described continuous profile under "blind" study-conditions). Four times his status deteriorated very greatly as far as side-effects were concerned at each change of the infusion profile. We then introduced clozapin 25 mg, and later 37.5 mg/d. The patient is now on 0.67 mg lisuride, 550 mg levodopa and 37 mg clozapin/d. His preserved mobility varies between 70%—90%, his CURS is between 20—25. He still experiences hallucinations (1—3 ×/week) and quite often they may be called "pseudohallucinations", as the patient controlls the reality-aspect and then neglects his phantasies. As for mnestic disturbances, both patients were unable to remember more complex hallucinations and in any case did not remember confusional episodes.

As to the character of the hallucinations, we were mostly informed of visual hallucinations of familiar persons, sometimes of animals, but also the deformation of a spot to a trait, the coloration of the trait in reddish colours etc. Rarely visual hallucinations were accompanied by auditory and in another case by olfactory hallucinations.

Mostly the hallucinations were non-threatening in character. In two further patients we were also able to observe relatively good to excellent mobility at the time of their psychic alterations, but we would also observe overt psychosis and persistent hypo- to akinesia.

Mobility-wise the therapeutic effect of Lisuride was good, the s. c. infusion helped to reduce dyskinesias as well as off-periods very much. Patient C. S. still suffered from wearing-off-phenomena, but had stable times with preserved mobility. Early morning dystonias were nearly abolished, turning around at night was mostly possible. Intellectually patients seemed more "awake", they reported better concentration-capacities and seemed less depressed (data not shown).

Discussion

In advanced stages of parkinsonism with oscillations of motor performance, which means distinct on-off-phases, it is necessary to apply dopamine agonists. In order to attain a continuous stimulation of the receptors the dopamine agonist lisuride is subcutaneously applied by microcomputer-pump according to individual needs. The advantage of parenteral as distinct from oral application is the possibility of programming the lisuride dosage per hour according to the individual mobility profile. In two cases the subcutaneous could be changed to oral application without loss of mobility. This might be due to a normalisation of the receptor sensitivity on one hand or to a supersensitivity of the D 2-receptors in the striatal regions.

The observed side-effects (gastrointestinal-like nausea, vomiting and subileus, orthostatic dysregulation) are tolerable and can be symptomatically treated. We could demonstrate an antiparkinson effect of combined levodopa and s. c. continuous lisuride infusion therapy in patients with severe Parkinson's disease comparable to the group C described by Suchy and Krause, 1988. The mobility was improved and the off-phases were significantly reduced.

When reducing the levodopa doses and slowly increasing the lisuride-dose, in contrast to group C described by Suchy and Krause, we observed less psychic side-effects.

The acute psychosis is however a difficult therapeutic problem and may lead to the interruption of lisuride therapy. To get more knowledge about the reasons that patients become more or less psychotic, we analyzed the data of two patients, who were treted with levodopa and s. c. continuous lisuride therapy for more than 28 months. From these data, presented here, it seems that a too high ratio of lisuride to levodopa seems to be less well tolerated as far as side-effects are concerned than a lower ratio, but this problem can be overcome, even in patients, who do not tolerate oral DA-agonists in low doses, with the continuous s. c. application by a mini-infusion-pump.

In patient L. N. we could see, that a too low (10% of the daytime) infusion-rate at night-time was accompanied by more mental side-effects than a higher (30%) infusion rate, the most severe mental side-effects occurred in these patients (like in some further patients treated by us, data not shown) within the first three months. Therefore the dose-adaptation in patients with psychotic side effects should be done with the utmost caution—we would suggest, that within the first 2—3 months one should start with even lower doses of lisuride, to prevent possible side-effects.

a) Ratio lisuride: levodopa

If one calculates from approximately equipotent antiparkinsonian doses (levodopa 500 mg p. o., bromocriptin 50 mg p. o., lisuride 0.15 mg i. v. (Quinn et al., 1983), one might estimate, that 0.1 mg lisuride are equipotent with 100 mg levodopa + decarboxylase-inhibitor. From that estimation, the ratio in the first patient decreased from 2.5 : 1 to 1.2 : 1, in the second patient from 3 : 1 to 1.2 : 1.

In both patients it seemed, that the lower ratio was better tolerated as far as mental side effects are concerned than the initial higher ratio.

Beside its D 2-agonistic properties levodopa has strong D 1-agonistic properties, whereas lisuride has only partial agonistic properties at the D 1-receptor (Fuxe et al., 1983). Chronic levodopa treatment leads to a down-regulation of dopamine receptors, shown in the striatum (Ponzio et al., 1984; Reches et al., 1983; M. B. Schneider et al., 1984). Whether a reduced intake of levodopa after addition of lisuride helps to correct this imbalance towards less hyposensitive DA-receptors and whether areas with high D 1-receptor density are included in that process, should be the subject of further research.

At least this hypothesized mechanism might play a role in the clinically seeming up-regulation observed in some of our patients. The higher sensitivity of DA-neurons projecting to the mesolimbic system might account for the higher prevalence of psychiatric disturbances, which could be reversed after re-introduction of higher levodopa intake (at 2 years of the observation-period).

b) Tolerance to parenteral lisuride

By measuring the plasma levels of lisuride after s. c. infusion and comparing these with the respective levels after oral intake, we have shown that we can easily achieve 10 times higher values by the parenteral route (Bittkau and Przuntek, 1988). These constant levels seem to be better tolerated as far as side-effects are concerned.

c) Level of the night-time infusion rate

Switching off the pump or lowering the lisuride-infusion rate at night-time very much causes the plasma level to drop down because of the short plasma half-life of the substance of 2—3 hours (Burns and Calne, 1983). The more the plasma-levels fluctuate to a large extent, the more the dopaminergic stimulation will fluctuate at the receptor-site. There are even reports in the literature (Lippem, 1976; Shukla et al., 1985), where a persistent paranoid psychosis began in close time-relationship to the abrupt discontinuation of a constant long-term application of bromocriptine, so that not only an exceeding dopaminergic-stimulation, but the imbalance per se may account for the induction of these phenomena.

Furthermore, there seems to be a balance within the neurotransmitter system between the dopamine-, serotonin-, noradrenaline- and the cholinergic-system. In 1972 Birkmayer raised the idea, that the imbalance of these systems might favour or produce the drug-induced psychoses and even some of the confusional states. These counter-regulation-mechanisms seem to get less effective with time and quantity of exposure to dopaminergic drugs, mainly of levodopa (Celesia and Barr, 1970; Klawans, 1982; Moskovitz et al., 1978; Sweet et al., 1976) and to a lesser degree with age, other illnesses, etc. as well. The patients reported here are just very responsive to fluctuations in their receptor-stimulation. This may also be the reason, that patient L. N. got better on a more continuous lisuride-infusion profile. Further theories to explain the generation of psychosis include dopamine as a false neurotransmitter in serotonergic neurons (Birkmayer et al., 1972; Birkmayer and Riederer, 1975), an increased turnover of serotonin, predominantly in the nucleus ruber, for which hypothesis the reported efficacy of methysergide (Klawans, 1974) might be indicative, in contrast to the use of L-tryptophan in managing (less severe) dopaminergic-drug-induced psychoses (Birkmayer and Neumayer, 1972; Gehlen and Müller, 1974).

d) Changes of receptor-sensitivity within the first three months

We had to reduce the lisuride dose significantly within the first 3 months for developing mental side effects not only in the patients reported here. Even after a significant dose-reduction the most severe side-effects, such as paranoid (confusional) psychosis developed not until after some weeks of a constant lisuride infusion. In other patients (data not shown) we could see, at a steady state, that as we lowered the infusion rate of lisuride (0.6—0.7 mg/d) there was an increasing efficacy in mobility over the same period. Although these are only preliminary data, it is worth examining the idea of developing "hypersensitivity" under a constant dopaminergic stimulation (Klawans, 1975)—or—alternatively—whether hyposensitive receptors change towards "normal" with either the dopamine-agonist or the amount of levodopa intake being responsible for this process (Ponzio, 1984; Reches, 1984; M. B. Schneider, 1984). Moskovitz (1978) explained the levodopa-induced psychosis by a "kindling phenomenon" and paralleled this effect with hypothesized hypersensitive dopamine pathways in Schizophrenia, Chorea-Huntington and Wilson's disease.

The questions remain, whether we are allowed to compare on-off-fluctuations of mobility with "psychic fluctuations" (Przuntek, 1988) the more so because these phenomena both occur only after some years of dopaminergic therapy, and even more, as these phenomena often occur concomitantly.

This would mean that an optimized continuous stimulation of D 1- and D 2-receptors should have the best results in substitution therapy. If this is true, it should be tested whether also in early stages of parkinsonism the continuous stimulation of D1- and D2-receptors shows better results than the discontinuous application of antiparkinsonian drugs.

References

Birkmayer W, Neumayer W (1972) Die Behandlung der DOPA-Psychosen mit L-Tryptophan. Der Nervenarzt 43/2: 76–78

Birkmayer W, Riederer P (1975) Responsibility of extrastriatal areas for the appearance of psychotic symptoms (clinical and biochemical human post mortem findings). J Neural Transm 37: 175–182

Bittkau S, Przuntek W (1986) Psychosis and the lisuride pump. Lancet ii: 349 (letter)

Bittkau S, Przuntek H (1988) Chronic s. c. lisuride in parkinson's disease—motor performance and avoidance of psychiatric side effects. J Neural Transm [Suppl] 27: 35–54

Burns RS, Calne DB (1983) Disposition of dopaminergic ergot compounds following oral administration. In: Calne DB, et al (eds) Lisuride and other dopamine agonists. Raven Press, New York, pp 153–160

Celesia GG, Barr AN (1979) Psychosis and other psychiatric manifestations of levodopa therapy. Arch Neurol 23: 193–200

Critchley P, Grandas Perez F, Quinn N, Coleman R, Parkes D, Marsden CD (1986) Psychosis and the lisuride pump. Lancet ii: 349 (letter)

Critchley PHS, Grandas Perez F, Quinn N, Coleman R, Parkes D, Marsden CD (1988) Continuous subcutaneous lisuride infusions in Parkinson's disease. J Neural Transm [Suppl] 27: 55–60

Fuxe K, Agnati LE, Köhler C, Andersson K, Eneroth P, Calza L, Ögren SO (1983) Heterogenity of brain dopamine systems: Possible discrimination of different types of dopamine systems and receptors by ergot drugs. In: Calne DB, Horowski R, Mc Donald RJ, Wittke W (eds) Lisuride and other dopamine-agonists. Raven Press, New York, pp 11–32

Gehlen W, Müller J (1974) Zur Therapie der Dopa-Psychosen mit L-Tryptophan. Dtsch Med Wochenschr 99: 457–463

Horowski R, Wachtel H (1976) Direct dopaminergic action of lisuride hydrogen maleate, an ergot derivative, in mice. Eur J Pharmacol 36: 373–383

Hümpel M, Nieuweboer B, Wendt H, Hasan SH (1981) Radioimmunoassay of plasma lisuride in man following intravenous and oral administration of lisuride hydrogen maleate: effects on plasma prolactin level. Eur J Clin Pharmacol 20: 47–51

Kehr W (1977) Effect of lisuride and other ergot derivatives on mono-aminergic mechanisms in rat brain. Eur J Pharmacol 41: 261–273

Klawans HL, Crosset P, Dana N (1975) Effect of chronic amphetamine exposure on stereotyped behavior: implications for pathogenesis of L-dopa-induced dyskinesias. In: Calne DB, Chase TN, Barbeau A (eds) Advances in Neurology, vol 9. Raven Press, New York

Lippem S (1976) Psychosis in patient on bromocriptine and levodopa with carbidopa. Lancet ii: 571

McDonald RJ, Horowski R (1983) Lisuride in the treatment of parkinsonism. Eur Neurol 240–255

Moskovitz C, Moses H, Klawans HL (1978) Levodopa-induced psychosis: a kindling phenomenon. Am J Psychiatry 135: 669–675

Obeso JA, Luquin MR, Martinez Lage JM (1983) Lisuride infusion for Parkinson's disease. Ann Neurol 14: 134

Obeso JA, Luquin MR, Martinez-Lage JM (1986) Lisuride infusion pump: a device for the treatment of motor fluctuations in Parkinson's disease. Lancet i: 467–470

Obeso JA, Luquin MR, Vaamonde J, Martinez Lage JM (1988) Subcutaneous administration of lisuride in the treatment of complex motor fluctuations in Parkinson's disease. J Neural Transm [Suppl] 27: 17–25

Parkes JD, Schachter M, Marsden CD, Smith B, Wilson A (1981) Lisuride in parkinsonism. Ann Neurol 9: 48–52

Ponzio F, Cimino M, Achilli G, Lipartiti M, Perego C, Vantini G, Algeri S (1984) In vivo and in vitro evidence of dopaminergic system down regulation induced by chronic L-dopa. Life Sci 34: 2107–2116

Przuntek H (1988) Personal communication

Quinn N, Marsden CD, Schachter M, Thompson C, Lang AE, Parkes JD (1983) Intravenous lisuride in extrapyramidal disorders. Raven Press, pp 383–393

Reches A, Wagner HR, Jackson-Lewis V, Yablonskaya-Alter E, Fahn S (1984) Chronic levodopa or pergolide administration induces down-regulation of dopamine receptors in denervated striatum. Neurology 34: 1208–1212

Riederer P, Reanolds GP, Danielczyk W, Jellinger K, Seeman D (1983) Desensitization of striatal spiperone-binding sites by dopaminergic agonists in Parkinson's disease. Raven Press, pp 375–381

Ruggieri S, Stocchi F, Agnoli A (1986) Lisuride infusion pump for Parkinson's disease. Lancet ii: 348–349 (letter)

Schneider MB, Murrin LC, Pfeiffer RF, Deupree JD (1984) Dopamine receptors: effects of chronic L-dopa and bromocriptine treatment in an animal model of Parkinson's disease. Clin Neuropharm 7/3: 247–257

Shoulson I, Glaubiger GA, Chase TN (1975) On-off response: clinical and biochemical correlations during oral and intravenous levodopa administration in parkinsonian patients. Neurology 25: 1144–1148

Shukla S, Turner WJ, Newman G (1978) Bromocriptin-related psychosis and treatment. Biol Psychiat 20: 305–310

Stocchi F, Ruggieri S, Antonini A, Baronti F, Brughitta G, Bellantuono P, Bravi D, Agnoli A (1988) Subcutaneous lisuride infusion in Parkinson's disease: clinical results using different modes of administration. J Neural Transm [Suppl] 27: 27–33

Suchy I, Krause W (1988) Continuous subcutaneous infusion of lisuride in parkinsonian patients with fluctuations in mobility. De Gruytes (in press)

Sweet RD, Mc Dowell FH, Feigenson JS, Loranger AW, Goodell H (1976) Mental symptoms in Parkinson's disease during chronic treatment with levodopa. Neurology 26: 30–310

Correspondence: Dr. S. Bittkau, Department of Neurology, University of Würzburg, Josef-Schneider-Strasse 11, D-8700 Würzburg, Federal Republic of Germany.

Discussion*

Discussions on Thursday, May 5th 1988

Discussion following the paper by Poewe and Gerstenbrand

Streifler: I was wondering whether you tried to identify certain features which have a similarity to the clinical appearance of the disease itself, where one can ask oneself "to what extent are such features 'prodromal' traits"? Apart from the basic personality traits which such people may have, and which we certainly found in a group of European patients, it occurred to me to wonder whether it might not be worth trying to study a group of Central Africans parkinsonians and to establish *their* pre-morbid personality traits. Might we not discover some differences, in spite of the same underlying structural catecholamine deficits?

Poewe: We have not divided our findings according to the stages in the patients' lives, so we were unable to do what Ward and coworkers did in their twin study, which was to look at various stages of life during the patients' pre-morbid life and to see at what stage the differences appeared. I did not comment on cultural differences: we might even have a job to find enough parkinsonian patients in Central Africa!

Korczyn: I have two questions: first, you have done multivariate analysis to discover which of these traits are interrelated and which are independent. I would like to know whether smoking is actually linked to a particular personality trait or is it an independent protective factor? I ask because we have examined psychiatric patients with drug-induced Parkinson's disease and found that the appearance of the drug-induced Parkinsonism was inversely related to smoking habits in these patients, and that there was a dose-effect, in other

* For reasons of space the discussions following all presentations have been edited to avoid cluttering them up with well-established facts available in the literature and with trivialities. Remarks originally made in German were translated, and the overall English text was checked for consistency of style by Dr. Karl Blau, London.

words that the more they smoked the more immune they were to the extrapyramidal effects produced by phenothiazines or butyrophenones. My second question, which you briefly touched on but did not have time to go into in detail, is whether the appearance of these personality traits is due to the dopamine deficiency. Clearly, as you mentioned, these traits do not progress during the development of the disease, and do not get better when you administer levodopa or dopaminergic drugs, so I suppose this is an unlikely hypothesis.

Poewe: I agree it is only a hypothesis, but what you say is true, and certainly we, and others, have not seen any progression of these signs with the progression of the disease. Coming to your first point about smoking and multivariate analysis, this is still part of an ongoing study.

Danielczyk: May I make a marginal comment on smokers: I used to feel that a smokers use cigarettes as a means of suppressing irritation or frustration, and that they tend to be introverts. Have there been any studies on this?

Poewe: I only know of one study on the personality of smokers: there was a report by the Royal College of Physicians in Britain, and among other tests they administered Eysenck's Personality-inventory to smokers and non-smokers. They found that the smokers tended to be more aggressive.

Brune: Did you find similar behavioural traits among other brain-diseases as you found in Parkinson patients?

Poewe: I only know of one such disease, which I really ought not to reveal, because the studies are still in progress. They are patients with essential tremor.

Discussion following the paper by Danielczyk and Fischer

Korczyn: We have just completed a case-control study in which we compared a group of patients with Parkinson's disease who had dementia with those who did not have dementia. We excluded patients with obvious other reasons for dementia, such as multi-infarcts on their CT-scan, or other diseases. We found that the patients with dementia had significant pre-morbid differences: they came from a lower social class and they had a lower educational level than their non-demented counterparts. Also they were more likely to have experienced the holocaust and they also exhibited more vascular risk factors such as hypertension or diabetes mellitus, although they showed no obvious changes on their CT. So these findings are in line with your conclusion that there may be vascular factors responsible for the development of dementia.

Discussion following the paper by Steinberg

Kuhn: I should like to ask how you define bradyphrenia.
Steinberg: Bradyphrenia is distinguished by the fact that the psychomotor sub-tests that are measured against time show below average or poor performance as compared with those that are measured independently of time which might give average or in some cases even above average results.
Streifler: I find it interesting that this figure of 25% of depression in parkinsonian patients seems to have become some sort of "holy" number. We have heard that you found it in your patient material, it is a figure that is also found in Mayeux's papers. But whatever the figure may be, and I am not trying to cast doubt on your figures, it is conceivable that it might be related to the serotonin deficiency and the raphe-system which is also affected in Parkinson's disease. Why this figure should be so uniform is hard to understand, it cannot be due to only one factor, or to only one neurotransmitter. I can recall the relationship which Professor Birkmayer constructed on a theoretical basis in his paper on depression in Parkinson's disease, and especially the pharmacological factors which they have in common, so we are attacking the problem from different directions—and coming up with the same figures!

Discussion following the paper by Streifler

Anonymous: Could you distinguish precisely between the pain of coxarthrosis and night cramps? And what are you doing against them?
Streifler: Coxarthrosis is a problem of differential diagnosis, but it should be possible to diagnose by rotation of the leg in *articulatio coxae* and by radiological investigation. Therapeutically I have had success with non-steroid anti-inflammatory agents such as indomethacin.
Poewe: I should like to come back to the problem of painful dystonic feet. I have difficulty in classifying pain as primary or secondary, or as painful levodopa-induced foot dystonia.
Streifler: Well, of course I cannot be sure, but the evidence seems to me to point to central factors, but with additional factors such as muscular spasms, cramps and persistent deformity contributing to the pain as well. It is amazing how little the patients complain about the pain.
Streifler: The pain is mostly in the limbs, and more in the lower than in the upper extremities. Interestingly, it does not appear so much in the face.

Korczyn: The pain of paraesthesia is of course very stressful for the patients, but unfortunately this is subjective, and not something that can be objectively validated by any findings during examination of the patient. I wondered why you did not mention depression as a possible contributory factor.
Streifler: It is certainly very hard to make any objective assessment of pain. You are quite right, we should take depression into consideration.

Discussion of the paper by Korczyn

Poewe: Could I ask you about any abnormalities there may be in temperature regulation and in sweating in patients with Parkinson's disease? There are scattered reports of both hyperhydrosis and anhydrosis, and one occasionally sees a patient with a one-sided facial sweating deficit. Would you like to comment on these phenomena?
Korczyn: I think that this is an area that has not been sufficiently studied. If one sees a patient like the one you mentioned with defective hydrosis, I would be inclined to say that he is more likely to suffer from Shy-Drager syndrome than from Parkinson's disease.
Bogdahn: Paralytic ileus is a very frequent symptom of parkinsonian patients in the intensive-care unit. Do you have any recommendations for treatment of this problem specifically directed at the patients with Parkinson's disease?
Korczyn: We have indeed studied this quite extensively. The best way of managing the problem is to use betanechol (50—70 mg) in a small volume of saline, administered rectally, and this gave the expected results.
Streifler: I should have mentioned that many years ago we looked at the problem of dolichocolon or Hirschsprung's disease in parkinsonian patients, and we discovered a statistically increased incidence of dolichocolon in these patients, but we could not be certain: was this a primary or a secondary phenomenon? Could it possibly have been due to the anticholinergic drugs that we were giving the patients in those days? Another thing we studied in these patients was sebum secretion: it was raised. It was found that levodopa decreased sebum secretion; amantadine did the same thing.
Korczyn: Let me come back to the dolichocolon: is it a typical manifestation of Parkinson's disease or does it parallel the constipation from which the patients suffered? We tried to investigate the patients more closely, and in one case where we were able to do a post mortem we were struck by the appearance of the mesenteric plexus: it was really shrunken. This led us to review the literature, and as many of you know, there are a few papers about that describe abnormalities in

peripheral autonomic ganglia. So it could be that it is not only the brain which is affected but also the autonomic ganglia, and Lewy bodies have been described there. Since we know that dopamine occurs in peripheral autonomic ganglia, perhaps we need to re-investigate them.
Birkmayer: How do you explain the intolerance to hot environments?
Korczyn: We are of course aware of it, the patients do report it, but we have no explanation, we have not properly investigated it.
Birkmayer: I shall send you one of our papers on the subject: The intolerance to a hot environment is not dependent on dopamine, it depends on serotonin. Nearly all our patients receive L-tryptophan during the hot summer months, and it makes them feel much better.

Discussion of the paper by Kuhn and co-workers

Korczyn: Dr. Kuhn, let me ask you—and perhaps the previous speakers too—about the diagnosis of depression. You said that you used the Hamilton scale, and this is of course very commonly done. This scale may be very effective in picking up patients with primary dementia, but I do not know that it has been validated for patients with Parkinson's disease. There are reasons why it might not be valid as it stands: its questionnaire contains many items which are clearly abnormal in patients with Parkinson's disease: inability to perform daily activities is one such item. The same goes for many other rating scales that like this one has not been validated and yet is used in patients with Parkinson's disease. I think we should discover and validate tests for depression that are designed specifically for Parkinsonian patients.
Kuhn: We have used the Hamilton Rating Scale, as well as the Zung Scale and we have found a very high correlation between these two scales. In addition we have also measured depressive symptoms by DSM III.

Discussion of the paper by Laux

Steinberg: I have a comment and a question. My comment is that you said that psychic akinesia was a recently-introduced term, but in fact psychic akinesia was Naville's term for bradyphrenia, which he originally coined in 1922. My question is about the slide you showed of certain motor deficits which developed to an even greater extent in depressive patients than in patients with Parkinson's disease. How would you explain this?
Laux: It is not correct that this slide showed that depressives had the poorer motor functions, it is the other way round, they had better

motor functions than the parkinsonians: the "Motorische Leistungsserie" data in depression were clearly better than in Parkinson's disease.

Kraus: If I have correctly interpreted the table you presented, then in the test where they have to re-position pegs the Parkinson patients with depression obtained a better score than those without, and both these groups were inferior to the depressive patients, and I do not find this all that easy to explain. If we look at the movements involved, they seem to require relatively little in the way of automatic activity and to require a fresh impetus for each individual action involved. The test therefore appears to depend a great deal on the drive of the patient and should therefore be a good reflection of the patient's motivation.

Laux: That is quite right, although in this sub-test we have no significant difference between Parkinson patients with or without depression. However it is true that the purely depressed patients do remarkably well in this sub-test.

Discussion of the paper by Freund, Hefter, and Hömberg

Korczyn: I was not absolutely clear how you defined the sensory difficulty. These findings are of course of practical importance. We have studied patients with Parkinson's disease and tried to analyse their performance with visual feedback and during darkness in much the same way as you, and we too were struck by the enormous difference in performance in the dark, which was much more severely impaired compared to where they had visual feedback of what they were doing. In the dark they were slower and made more mistakes, the errors were larger and they were also more tremulous and the amplitude of the tremor was greater.

Hefter: That is what I would expect. The problem is that you find an increase that is out of proportion to what you would expect with normals. Of course you would expect some increase, even with normals, but in patients with Parkinson's disease the increase is out of all proportion greater. I was not prepared to say that this is due to the sensory input, but I would say that it has to do with the incorporation of the sensory information into the performance of the motor system. Obviously the targets had a severe effect on the motor task.

Streifler: The sensory input is also necessary for the anticipatory readiness to react, as has been established in the case of postural stability and other parkinsonian symptoms. This factor certainly influences the performance of the motor task.

Hefter: You are probably aware too of Professor Beneke's study of sequential movements, and of choice reaction-times. All of these factors affect the severity of the response. We are trying to puzzle out the distinct parts of the response, which is not easy: there is still much left to be done in the future.

Deuschl: My first question refers to your last point, concerning the sensory component: could it not be that the reason that more complex movements are more disturbed than simple ones is that you are asking the parkinsonian patient to aim at a broader or more narrow target? My second question concerns the first statement you made on the interaction between involuntary and voluntary movements: you demonstrated that both are disturbed. Did you mean that the involuntary movements, i. e. the tremor, was caused by the akinesia? Or would you say that these are two separate phenomena which of course interact when the patient performs voluntary actions?

Hefter: Involuntary movements are certainly a pathological sign: you do not normally see any interaction between physiological tremor and a voluntary task. What I wanted to show was that because of the involuntary activity, the slower the intended movement, the slower the voluntary activity will be. I think it is possible to separate the spline function from the contraction time. You saw how it was possible to construct the trajectory of the voluntary movement in the third example. Thus you can use this type of analysis you can isolate and compare the relative amounts of voluntary and involuntary activity. But so far I have not yet told you anything about what causes which: I only wanted to describe the interaction.

Lücking: Could you clarify the nature of the underlying tremor? You showed accelerometer tracings; did you do any EMG studies? Is it more in the nature of reciprocal alternation or what? I assume perhaps it may be more or less like an enhanced physiological synchronisation activity of the EMG. Or what do you think it is?

Hefter: It is complex: during an isometric task of the fingers we find a co-activation between agonist and antagonist in normal subjects. There is a slight time-lag, about 13 msec, between agonist and antagonist. In patients with a resting tremor the alternating pattern is transformed into a co-activation pattern. While in normal subjects you see one burst in the agonist and one in the antagonist, in the parkinsonian patients we see repetitive bursting during the task, synchronised in agonist and antagonist, and this may explain why the frequency of action tremor is much higher than that of resting tremor, at least double. It is not exactly double the frequency of the resting tremor for the simple reason that we see shifts of frequency in parkinsonian patients performing different tasks. So parkinsonian tremor has a certain range of freedom especially when different tasks are analysed.

Poewe: You have shown us examples of how increasing complexity, with the addition of sensory information processing, can slow down movements in the upper extremities of patients with Parkinson's disease. Now it is a common clinical finding to see Parkinsonians walk much faster and better when they are climbing stairs or when they have a pattern of bars on the floor to follow. How would you explain that?
Hefter: It depends a great deal on the frequency you are producing. When I walk across the floor I have a spatial frequency, and how well I walk depends on the adjustment of the internal frequency to this incoming spatial frequency. What we showed was that when we have an acoustic signal not matching the internal clock you will end up with a mismatch. In such circumstances some Parkinsonian patients will come to a halt and others will fall over. This strikes me as an interesting problem, and it seems to depend on the frequencies you are feeding into the system.

Discussion of the paper by Kraus, Tielsch, and Przuntek

Korczyn: In one of the slides you showed, where you tried to compare the results of the tests with the patients' clinical state as evaluated by the Hoehn and Yahr scale, I had the impression that you assumed that scale to be linear. I think that was why you were surprised when the test results were not linear, but in fact the Hoehn and Yahr scale is of course not linear. In fact, perhaps if you look at tapping-speed you can get a better idea of the severity of the disease, and using this as a criterion you can then perhaps see what the real differences are on the Hoehn and Yahr scale between stages one and two, and stages two and three and so on.
Kraus: The slide only dealt with the instrumental test. However, I agree with you that the Hoehn and Yahr score, as well as other types of clinical score are not linear.

Discussion of the paper by Steg

Lücking: Thank you Dr. Steg, this all fits in very well with the results of Beneke, who was mentioned earlier, who chose a simple task: squeeze and flex. When he measured the single simple movement squeeze or flex he did not find a great deal of disturbance, but when he combined the two into a complex movement then there was definite deterioration.
Gibb: This is certainly a potentially useful technique for making objective measurements. I should like to ask whether you have also looked at patients with primary endogenous depression.

Steg: Well we have not had time to do that yet. This is the "second generation" of instrumentation, the first was not suitable for routine work, only for establishing the validity of the technique. At present we are working on the development of the tests. I have just tried out the ordinary routine neurological assessment by this technique, and with it you can demonstrate—quantitatively and objectively—exactly what you see. So it is capable of very wide application. As yet we have not had time to do much more than this.

Da Prada: I have a question about the drug effects: can you discriminate between the quality of the effect of levodopa therapy and of levodopa combined with deprenyl or with something else?

Steg: We have recorded the effects of levodopa and have obtained dose-related curves. We have given bromocriptine too, but the observation of that takes more time than we have the patients for investigation. We take three of four hours establishing the effects of L-dopa. The bromocriptine effect appears later than that, it is more difficult to study. However, we see some indications that it has a different action-profile, but we are not yet certain about that. It seems as if L-dopa works more on kinetic functions, while bromocriptine works more on postural functions.

Korczyn: Another thing that emerged very nicely from your slides was the fact that normal patients start their M-phase, the manual phase, very briefly after first starting the P-phase.

Steg: Parkinsonian patients do not carry out their movements simultaneously. This has not been described previously, but the optoelectronic eye spots it immediately.

Przuntek: You were able to observe different effects with different drugs. Do these depend on the pharmacokinetics of the drugs or on their specific action?

Steg: At the moment we are just realising that we have to check the blood levels of the drugs in order to be certain of what we are observing: to correlate the level of L-dopa for instance, with the observed phenomena. We believe it is quite essential to do that kind of work in order to establish the correlations.

Przuntek: But you were able to observe a special effect on postural time with bromocriptine, and that differed between L-dopa and bromocriptine. Were you also able to observe different times, or different effects, for example, between lisuride and bromocriptine?

Steg: We have yet to do this.

Przuntek: Do you mean that the different antiparkinsonian drugs are effective in different regions of the brain?

Steg: Yes: this has been shown in animal experiments on rats, and I intend to show that it is equally true in humans.

Discussion of the paper by Scholz, Bacher, Bellenberg, Diener, and Dichgans

Przuntek: Dr. Scholz, you were able to improve the tremor with anticholinergic drugs. Can you also provoke the tremor with cholinergic drugs, for example with physostygmine?

Scholz: We have not tried this, but it is not necessary to do it directly, we can do it with the stress experiments. But yes, we could try physostigmine and see if it provokes parkinsonian tremor, that should not be any problem.

Hefter: I have some questions about pathological problems. You recorded from the forearm: now you know that the forearm has a natural frequency of about 4 Hertz. On the other hand, parkinsonian patients can easily produce voluntary alternating movements with a frequency of 4 Hertz, this is their internal resonance frequency. Might it not be that your results with the recording of tremor could be contaminated with passive or voluntary alternating movements?

Scholz: We cannot exclude this possibility, but can you imagine any patient making these alternating movements 50% of the time during the day? Who would do such a thing voluntarily? The other thing is that voluntary EMG-activity in parkinsonian patients is tremor modulated, but we can exclude non-modulated EMG-activity from voluntary movement by this method of analysis, because the computer only takes the modulated EMG-activity for the analysis.

Poewe: My question is probably a related one: when you use this method to monitor the therapeutic effects by long-term recording of tremor. Let us suppose that you do this in patients on, say, levodopa, who have levodopa-induced involuntary movements in their good or "on-periods": most of the time clinging, jerking, dystonic movements. What is the effect of these for your recording?

Scholz: These involuntary movements have a quite different frequency, they have a frequency of about 1—2 Hertz or even slower. So the computer looks for periods of five seconds to see whether there is a tremor present, and accepts this period if no tremor is present. Although voluntary movements may produce a shift in the curve, this is eliminated by the computer-programme.

Leenders: You showed a slide with several frequency histograms of essential tremor together with one of Parkinson's disease. Some of the patients with essential tremor looked like the patients with Parkinson's disease. We know that quite a lot of people with essential tremor go on to develop Parkinson's disease. Is there any predictive value in the histogram of essential tremor patients who look like parkinsonians? Do these patients in particular develop Parkinson's disease? Do you have any data on that?

Scholz: No, I am afraid we do not.

Discussion on the paper by Jörg (read by Prof. Maurer)

Przuntek: Are there any specific results that we could use for the early detection of Parkinson's disease?
Maurer: In my opinion, as a general observation, the neurophysiological methods give quite normal results in Parkinson's disease. I think this is the essential message of this paper. Dr. Jörg found out that the VEP's are prolonged, and that is the only evoked potential, in my own experience too, which is prolonged in this disease. A possible explanation, which Dr. Jörg mentioned, is that it might be due to a biochemical and electrophysiological disorder in the retina, so I think we shall have to leave this point open. Apart from this, I have some experience with auditory evoked potentials, and all that I can tell you about those is that the early components, which come from the brainstem, are also normal. So we have normal values which can be expected from neurophysiological methods concerned with SEP's and AEP's.

Discussion of the paper by Maurer

Kraus: Are there any differences between ON and OFF phases?
Maurer: We did not look for those.
Korczyn: We have done this in a similar way, in a case control study of patients with Parkinson's disease with and without dementia. Overall we found an increase in the delta range. In fact what we saw was that the major increase was in the 0–1 Hertz band, rather than later on, and we thought that the increase which we observed at the higher frequencies, say at 2, 3 or 4 Hertz, and even in the theta range, might well be a spill-over of the increased activity in the 0–1 Hertz band. Since you mentioned the increase in the theta range, I wonder what your opinion is: could this be a spill-over effect?
Maurer: In my opinion the delta range of 0–1 Hertz is very complicated in these patients, because I suspect that what you are getting on your charts is mostly artefacts, so I think that we should probably disregard this frequency range.
Horowski: Are your Alzheimer patients on neuroleptic drugs? Because some of the changes rather remind me of what you can sometimes see with neuroleptic drugs: the theta increase, for example.
Maurer: No, they were without medication.
Leenders: You have shown very nicely that patients with dementia have alterations in frequencies, while parkinsonian patients without dementia are virtually normal. Now we have this rather artificial distinction between cortical and subcortical dementias, and—though

I do not think it is necessarily quite true—people like patients with Huntington's chorea, progressive supranuclear palsy and Parkinson's disease were classified as being subcortical, while patients with Alzheimer's disease as having cortical dementia. But have you any experience with other types of dementia? And further, your data show that actually the cortex needs to be involved before dementia is really clinically present, don't they?

Maurer: Both are involved: I was able to show this with the P-300, which goes deeper into the allocortex. So I would say that it is a combination of damage to the allocortex or a lesion in those hippocampal formations together with a lesion in the association cortex, because P-300 initially goes down, and then comes up again to the association cortex. So I would say that it is both, as could be shown by the P-300, but the EEG pattern is only useful for evaluating the cortical damage or lesions.

Leenders: So without cortical involvement, no dementia?

Maurer: That's difficult to say in this case.

Leenders: But if the cortex is intact, can you still have dementia?

Maurer: As a neurophysiologist I would say that since the P-300 has initially to go down, and then comes up again, you know, the definitive answer cannot really be given, and so the answer has to be "well, there is a disturbance of the P-300 anyway".

Leenders: And what about my first question, about the P-300 and the Huntington's chorea dementias: have you had any results with those? Because they may reinforce your point.

Maurer: What we have is experience with Wilson's disease, and actually we have the same pattern. In Wilson's disease we have this frontal increase with a lesion pattern in the parietal and temporal area. We had the opportunity to treat such a patient, and in fact during the treatment the P-300 showed a tendency to revert to normal. So this is in fact a model for us: dementia, if our therapy is successful, should show a changed pattern, but actually, in our Alzheimer patients it did not change.

Korczyn: May I just comment on Dr. Leenders comment? I think the distinction should be made on the basis of the anatomical involvement. The distinction between the cortical and subcortical dementia is based on the anatomical structures that are primarily involved in the pathological process. I do not think that anyone has ever claimed, whatever people may think, that a patient can be demented without functional disturbances of the cortex, and of course all that the electrophysiological tests or the brain-mapping are telling us is what the cortex is doing, not whether it is directly involved in whatever other processes.

Scholz: As a clinician who is still using ordinary EEG's I am rather puzzled by your parkinsonian results. I remember a paper by Mrs. Berger of Freiburg who reported that 40% of her patients had slowing of the rhythm. How do you reconcile these facts?
Maurer: Throughout the literature most papers report normal EEG's, so in this case the question we must ask is, what stage of the disease are we talking about? Dr. Kuhn actually measured our collective of parkinsonian patients, and I think the crucial point is the exact stage of the disease: did Mrs. Berger also have demented patients in her group?
Scholz: She did not mention their clinical state.
Maurer: Because you know in Parkinson's disease with dementia there will be an enormous alteration in the EEG. This could well be the answer: it is the particular collective of patients with whom we happen to be dealing, not the methodology.
Lücking: With regard to the publication by Mrs. Berger of our institute, we did discuss this, and in those patients there was enhancement of theta and delta rhythms, as a sign of the involvement of cerebrovascular problems, and I think that this is true in a lot of patients. We have evidence from their CT-scans and from their histories that in these Parkinsonian patients there were additionally deteriorations as a sign of cerebrovascular disease.

Discussions on Friday, May 6th, 1988

Discussion of the paper by Leenders

Korczyn: I think that it is also important to realise that the diagnosis of Parkinson's disease varies in accuracy. When you see patients in the early stages, at best 80% of the diagnoses are correct and 20% of the patients are misdiagnosed. My question is whether it would not be possible to use this method to show the focal metabolic high-energy activity you saw in other diseases?
Leenders: If you take just one functional parameter in isolation that is not very powerful: combining several traces with several functions on them is much more powerful.
Horowski: You have shown increased cerebral blood flow with L-dopa. What was the situation with domperidone?
Leeenders: Domperidone prevented the increase.

Horowski: That is most likely the peripheral pharmacological effect via the dopamine receptors.

Leenders: McCullough and other people have shown in animal experiments that when they gave dopamine agonist stimulation they saw an increase in deoxyglucose. That has been challenged by others, and we could not find it, so we wanted to see which parts of the brain are activated in Parkinson's disease and which parts benefit from levodopa therapy, but it was not very obvious.

Poewe: There is an effect with levodopa which I see only in untreated as distinct from treated parkinsonians, but only on a chronic basis.

Leenders: I think that this is purely a dose-effect situation. We gave very large amounts in order to be surve that there was a clinical effect at the same time as we saw the vascular effect. The St Louis group gave much smaller doses: 250 mg, and that is not enough to show the response. But the same group has reported that there was a marked increase in blood flow in a larger group of treated parkinsonian patients when they were given moderate or low doses where one would hardly observe any dose effect.

Dichgans: Can you explain why the limbic system does not show up the injections of madopar?

Leenders: To be honest, I can only speculate, because at the moment I am not informed what the differences in concentration between striatum versus limbic cortex are. The images are scaled to the maximum activity, and if the concentration in the limbic cortex is very low then we don't see it on the converted colour scale.

Dichgans: In those cases where you have seen degeneration of the dopamine pathways in Parkinson's disease I assume that you scale the whole system up. So in those cases you should see the mesolimbic system if it were active.

Leenders: Right, but it depends on the relationships of the concentrations. To begin with that might be the case, but in healthy people I didn't particularly notice any activity there. But, as I showed you, I *did* see activity in the middle front of the cortex, so I should imagine that the limbic cortex has the same sort of level of activity as the middle front of the cortex.

Dichgans: Did you try to draw any conclusions from these observations?

Leenders: No, it's a big hassle to get it evaluated so that one manages to accumulate actual biological data. One is most of the time trying to overcome the difficulties of the technique, and that, at the moment, is the major drawback to analysing focal emissions.

Anonymous: I have a question related to the previous one: can you separate out the nucleus accumbens?

Leenders: Well, not by using fluorodopa, because the nucleus accumbens should be more or less similar to the rest of the striatum. So if one has no contrast one cannot make an image and we cannot see it, though of course it's there. So what one has to do is to compare the anatomical scans with the functional scans, and more work is now being done in getting high quality MI-scans matched by PET-scans. In MI-scans one has a pretty good idea where one is, anatomically, and if one matches such parameters one then also knows where to make the functional measurements.

Da Prada: Do you think an inhibitor of the COMT enzyme might be helpful with the resolution?

Leenders: That is a methodological aspect, but I think it is quite right. I didn't go into the problems of tracer kinetics, but while fluorodopa is methylated particularly in the nondopaminergic brain regions it is also methylated in the peripheral tissues. However, it is apparently not methylated so much in the striatum: the work done on monkeys by the MacMaster group in Canada has shown that the fluorodopa injected as a tracer is present as fluorodopamine in the striatum only for the first one and a half hours, while at the same time outside the striatum the tracer is present as methylated fluorodopa. To give a COMT-inhibitor together with the tracer would clean up the tracer kinetic problems, but we are awaiting the availability of a suitable COMT-inhibitor.

Riederer: Can you distinguish between intraneuronal and extraneuronal "staining"? What is the background staining in glia for example? If you used rather high concentrations of fluorodopa one would expect that fluorodopamine would be synthesised to some extent, and would accumulate in glial tissues.

Leenders: We always used trace amounts to begin with, and what we would have to explain is why so much tracer accumulated in the striatum and not in the rest of the brain. If you give dopa and it is concentrated in the striatum then you have to draw conclusions from the fact that you gave a particular compound. It's going into the brain and coming out of the brain in a certain dynamic way, and the differentiation of the kinetics in one region compared to another gives you functional amplification as well. The point is that you *can* do it, you can make actual measurements, as opposed to conventional nuclear magnetic resonance procedures, where the problem is that you cannot quantify. But the biological information you have to get from other considerations.

Discussion of the paper by Riederer, Sofic and Rausch

Wesemann: With regard to the D 1-receptor, you showed that receptor binding is not decreased, but the adenylate cyclase stimulated by D 1 is well decreased. Would you conclude that in Parkinson's disease it is not sufficient to stimulate the D 1-receptor, but that you must find a way of improving the coupling between the receptor and the adenylate cyclase system.

Riederer: We shall have to see which biochemical finding is of greater significance, the antagonist binding to the D 1-receptor or the stimulation of the adenylate cyclase system, the latter being a functional test. The clinicians using selective drugs either acting on the D 1- or D 2-receptors (or on both) will probably give us the final answer. I would say that the stimulatory properties are decreased. The D 1-receptor certainly loses its functional activity to some extent, as does the D 2-receptor in the very late stages of the disease, where the D 2-receptor does not respond at all. It shows normal receptor density and affinity, but the patient does not respond to any drug treatment, and so one can assume that in that system too, the receptor activity has a functional deficit, but at present we do not know at what step of the amplifier or effector system the deficit may be.

Herz: I would like to draw your attention to the distribution of neuropeptides in the striatum, which is not homogeneous: they are found at higher concentrations in particular sub-areas. Did you find anything analogous for the distribution of tyrosine hydroxylase?

Riederer: We did not study that in particular. We did study the caudal and rostral distribution densities of the D 2-receptors, assuming the degree of degeneration in these sub-areas was different. However, there were no differences in the receptor density and also not in the receptor affinity.

Jellinger: We tried to study the distribution of tyrosine hydroxylase using immunocytochemistry, and found no such differences. On the other hand it is extremely difficult to do this in the human *post mortem* brain, and in particular to localise the nerve endings.

Korczyn: You paid some attention to compensatory mechanisms, particularly spiroperidol-binding. My own feeling is that we do not in fact really know whether the number of dopamine-receptors in the striatum or in any other tissue is really at a minimum or a maximum or what it is. Neither do we know whether an increase in the number of receptors, if one occurs, is clinically relevant or physiologically important. I think the whole question of compensatory mechanisms has to be looked at more widely. Of course we do know that where there is damage to some dopaminergic neurons then others would increase their activity: you yourself have shown this in the ratio of HVA to the remaining dopamine. Moreover I think a distinction should be made, as far as supersensitivity goes, between the increase of activity towards endogenous dopamine and that towards an administered drug.

Discussion of the paper by Youdim

Jacob: Professor Youdim, my question is concerned with human neuropathology, and I want to consider the phenomena of marginal siderosis with marginal necrosis. Have you seen siderosis and parenchymal necrosis in your animals?
Youdim: We are quite sure that iron does get across the blood brain barrier into the brain. I want you to look at this slide: it represents the dentate gyrus. This intensity of iron at this site is equivalent to the intensity—maybe even greater—found in the globus pallidus: it is a very narrow band of iron, and one consistently sees this. What is interesting is that in Alzheimer's disease you get an accumulation of aluminium here, too. Halliwell and Gutteridge examined the lipid peroxidation activity of iron on its own and of aluminium on its own and putting aluminium together with iron. They found that aluminium alone had no lipid peroxidation effect, no neurotoxic effect. But if you put aluminium together with iron you get a 1500% increased synergistic action on lipid peroxidation cytotoxicity. We continually see the globus pallidus as the area that is implicated in dyskinesia. There is a recent paper showing that subjects on neuroleptic drugs have increased deposition of iron in the globus pallidus, exactly like you see in Hallervorden-Spatz disease which has very similar features to tardive dyskinesia. What we want to know is how the iron gets into this area, because transferrin, the iron-transporting protein, has a completely different distribution. So how is it transported? Is it just an effect of aging? I don't think it can be, because at least in rat brain iron reaches a maximum four weeks postnatally. So I think that in man, after the first decade, it remains constant throughout life.
Riederer: I would just like to remind you that in Parkinson's disease, as was shown by Kornhuber and Poewe, about 25% of the patients have a disturbance of the blood brain barrier, so it could well be that some iron is crossing the blood brain barrier in at least such a subgroup of parkinsonian patients.
Youdim: That may be so, but then why do the metals not increase non-specifically? The studies of Dexter and Marsden show that iron is specifically increased. They measured copper and zinc as well, and there was no change in those, so I don't think that the iron effect is related to the disturbance in the blood brain barrier.

Discussion of the paper by Höllt and Morris

Herz: You showed that lesioning of the substantia nigra has the same effect as haloperidol which mainly acts via the D 2-receptors, whereas inactivation of the D 1-receptors by the Schering compound does the opposite. Does that mean that the dopaminergic afferent pathways from the nigra to the striatum are mainly carrying information to the D 2-receptor?

Höllt: I cannot answer this question precisely: all that one can really identify is the overall effect. One knows, for example, that the D 1-receptor is coupled to adenylate cyclase. There was another system, like the leu-enkephalin system in the adrenal medulla, where you have a direct enhancing effect of the D 1-receptor on gene expression. On the other hand you have a D 2-receptor possibly coupled to the phosphoinositol system, and this might have a depressant effect, so you could even have things staying the same if both receptors are counteracting different post-receptor actions.

Herz: Okay, but only if the dominant effect seems to occur at the D 2-receptor and not at the D 1-receptor, that is the point.

Leenders: Are enkephalins decreased in Parkinson's disease?

Höllt: Only proenkephalins.

Leenders: But substantia nigra lesions or haloperidol increase the enkephalins. In Parkinson's disease we are dealing with lesions in the substantia nigra so what is the explanation for the decrease in enkephalins?

Höllt: There are no explanations for that. If in the rat a D 1-agonist can increase proenkephalin–and that might be independent of D 2-receptors and the enkephalins–we might get a counterbalance to the deficiency in Parkinson's disease.

Przuntek: We administered naloxone, an endorphin antagonist, to parkinson patients, and were able to observe an improvement, but only for fourteen days, and the same thing was also observed by Hefter and Freund of Düsseldorf. Can you explain this shortduration effect of naloxone?

Höllt: No, I cannot. We did a lot of experiments on rats which demonstrated that chronic haloperidol treatment actually had no major effects on our enkephalin in the striatum. It appears that there is no feedback system via the opioids. But the story is more complicated than that, because we do not have just one receptor, we have μ-, kappa- and δ-receptors, and naloxone has a different affinity for those. What we found by microinjection of specific labelling antagonists such as β-CNA or β-FNA was that specific effects could be obtained only with β-FNA, which antagonizes the kappa-receptor. If you usually give a large amount of naloxone maybe you can block this effect via the kappa-receptor, and then maybe after some time when you have cleared the naloxone by metabolism, there might be a tolerance-induction of enzyme, and then a smaller dose of naloxone may not have any affinity for the kappa-receptor, but this is pure speculation.

Discussion of the paper by Gibb and Lees

Korczyn: Was there any correlation between the prevalence of Lewy bodies and the dopamine loss or the loss of neurons in the substantia nigra? And was there a correlation between the extent of Lewy-body formation and the stages of the disease at which the patients were examined?

Gibb: What we believe happens during the course of Parkinson's disease is that Lewy-bodies occur at the earliest time that the cells begin to disappear from the substantia nigra. So Lewy-bodies appear to develop well before any symptoms become evident i. e. during the presymptomatic period. As the disease progresses the number of Lewy-bodies in the substantia nigra increases, and as the cells die, you then also lose Lewy-bodies. What happens to the Lewy-bodies when the cells die is not exactly clear, because you do not find an equivalent number of Lewy-bodies lying free in the neuropil. Presumably they are broken down. The other question you asked was whether the exclusion criteria are applicable to all stages of the disease. If you look at 300—400 cells, which is the practical limit, you can exclude Parkinson's disease at all stages. We think this is the case because the patients we selected covered all the different stages. The most important question perhaps concerns very long-standing Parkinson's disease—I mean of 20—30 years' duration: do the same criteria hold? We have done a more up-to-date study of about 30 such patients, and the answer was that they die. So I think the answer is that when you've looked at 300—400 cells and found no Lewy-bodies you can be fairly sure that it's not a case of Parkinson's disease. *This* is the important use of the Lewy-bodies in the diagnosis.

Jacob: My question concerns a very special type of Parkinson's disease described by Yahr and Mann, where there was hypermelanization, the melanification of neurons, signifying the total destruction of the neuron by melanin, not by Lewy-bodies, not by pale bodies. Your last slide showed that the Lewy-bodies produced a compression of melanin into the peripheral regions of the cells. In another type we see a dispersion of the melanin into fine granules. What I am suggesting is that we have to consider these neuronal changes.

Gibb: I should first of all like to question to what extent melanin has toxic effects within the substantia nigra, and this really relates to the extent to which cells are lost with aging. Because of course melanin accumulates from birth and throughout life: I mean it doesn't just stop at the age of 50, it goes on accumulating. Most estimates of the age-related cell loss and of the loss of striatal dopamine with age are probably overestimates, because such studies have probably not excluded patients with degenerative diseases, presymptomatic cases of Parkinson's disease, to take but one example. So the age-related contribution to cell loss in nigro-striatal dopamine depletion is probably not as great as it is thought to be, and certainly nowhere near the amount required to produce a parkinsonian disorder.

Poewe: It looks as if these incidental Lewy-bodies in healthy controls can be regarded as presymptomatic signs of incipient Parkinson's disease. How would you interpret the four-fold higher prevalence of Lewy-bodies in Alzheimer patients in the 60—70 year age-group, compared to age-matched healthy controls?

Gibb: The question is this: is it true that in Alzheimer's disease you really have a greater prevalence of Lewy-bodies than in healthy controls? The reason is this: when you do biochemical measurements too, of course you have Parkinson's disease on the one hand and you have Alzheimer's disease on the other, and probably in both disorders the nucleus basalis of Meynert is the principal area affected giving rise to dementia. If you have two pathologies together you have an overlap, so you need less of each pathology to produce a dementia. This is a rather complex concept: it is just that these patients have a dementia because they also have a Lewy-body disease affecting the nucleus basalis. It is this overlap which boosts the number of patients with Alzheimer's disease, but this gives you a false impression of the relationship between these disorders because you have no hard pathological evidence that there is this overlap.

Jellinger: I cannot confirm your studies, because in our large autopsy series of Alzheimer's disease—more than 100—with no clinical signs of Parkinson's disease, we have a 38—45% incidence of Lewy-bodies in the substantia nigra and locus coeruleus without involvement of the nucleus basalis of Meynert. On the other hand we also have a large autopsy series of parkinsonian patients, and one third of the aged parkinsonian patients with senile dementia have a true combination of both Alzheimer's and Parkinson's pathology. I think that this enables us to make some distinction: we do have a certain overlap, what we used to call "senile Parkinsonism", although these patients do not have a loss exceeding 40% of the substantia nigra. So we cannot say this as categorically as you did.

Gibb: There are many examples of the two disorders occurring together. The question—it is an epidemiological question—on that has to be solved at the pathological level by the application of biochemistry as well. It is no good just looking at the prevalence of Lewy-bodies and Alzheimer's disease, you have to know what the biochemical levels are as well. Why were the patients actually demented? Were they demented because of Alzheimer pathology or were they demented also because of the Lewy-body pathology, not only at the nucleus basalis but also at the cortex? Because of course Lewy-bodies occur in the cortex in about a third of patients.

Discussion of the paper by Jellinger and Lassmann

Gibb: I would like to ask whether in the brainstem, in the substantia nigra and in the locus coeruleus you see a different staining pattern with ubiquitin antibodies compared to antineurofilament antibodies, and whether there is any morphological difference in staining patterns.

Jellinger: The only difference is in the fact that the cortical Lewy-bodies showed much more intense immunoreactivity with both antibodies to ubiquitin and with polyclonal and one monoclonal antibody to paired helical filaments. I don't know the reason why, but I would suspect that probably both the immunoreactivity and the ultrastructure are more or less the structural correlates of some functional and biochemical changes that are progressively occurring.

Gibb: I think this is an important point, because there are many people who have been unable to conceive that Lewy-bodies in the cortex either occur sufficiently often or are sufficiently numerous to contribute to dementia. However, once can see them under the light microscope just with eosin staining in about one third of the patients. There are many cells that don't show classical Lewy-bodies but show pale bodies or other abnormalities. But from what I am getting to know it is becoming clear that ubiquitin antibodies, and other antibodies, show these abnormal cells much more clearly, and so I think one can be much more convinced by the often widespread cortical abnormalities in Parkinson's disease that may contribute to dementia in a small proportion of cases.

Jellinger: Unfortunately this is not the case, because with both antibodies against neurofilament protein and against paired helical filaments, in the *subcortical* areas you only see about two thirds of the Lewy-bodies. On the *cortical* you see a larger amount of Lewy-bodies than with the routine stains. That's completely the reverse of what happens with the neurofibrillary tangles, where you see many more with the immunocytochemical methods. So unfortunately we do not see all of the Lewy-bodies which are seen with routine stains with these immunocytochemical methods, and this may be due to some effect related to their functional stage or to some biochemical relationship.

Gibb: I agree with that comment, but I am in part reflecting on the recent results of a group in Nottingham, where ubiquitin antibodies clearly showed many cortical Lewy-bodies, pale bodies and inclusions. Whether this is more sensitive than silver stains or than the other antibodies is not yet clear to me.

Jellinger: That only applies to the cortical levels, not the subcortical. On the other hand I am conviced that the presence of Lewy-bodies in large numbers, even in the nucleus basalis, does not necessarily contribute to dementia.

Przuntek: I would like to ask whether the composition of the inclusion bodies in the MPTP-model and of the Lewy-bodies is the same.

Jellinger: Unfortunately we have no data on this. I don't know whether Lysia Forno has done any immunocytochemistry on it: all we know are her preliminary data at the electron microscopical level.

Gibb: On that point can I just say that up to now the inclusions found in the squirrel monkey MPTP-model have been quite different in every respect, indeed to the point of being found at different sites, for example the nucleus amygdalae in the rhesus monkey or in the squirrel monkey, and of course they never occur at that site in Parkinson's disease. At the light microscopical and ultrastructural levels they are quite different, but one might expect to stain them with ubiquitin antibodies: I wouldn't be surprised if that were the case.

Discussion of the paper by Da Prada

Przuntek: If you take the combination of the COMT-inhibitor and the MAO-inhibitor what happens to the catecholamines in the adrenal medulla?

Da Prada: I don't think you block their metabolism completely, even if you give the decarboxylase inhibitor: the inhibition is never 100%.

Youdim: What we have done is to combine the inhibitors with nomifensine: we pretreated rats with nomifensine and then gave them a COMT-inhibitor, or we gave them nomifensine followed by a selective MAO-B-inhibitor. In both of these animal models we saw a sustained "dopamine behaviour": typical hyperactivity, stereotypic behaviour and sniffing, suggesting that your COMT-inhibitor is in fact an active drug that works on the dopaminergic neuron and potentiates the dopamine that is formed post-synaptically.

Parkes: I don't know of any hard clinical data that, given the right dosage of each decarboxylase inhibitor, there is really any difference in the clinical response. At one stage the Canadians in particular were using progesterone in Parkinson's disease, and Paul Bedard and his group suggested that its action in potentiating the effect of levodopa might be via COMT-inhibition. Is that right?

Da Prada: I have no idea. In the rat at least, benserazide at low dosage is superior to carbidopa. In any case, because there is probably more 3-O-methyldopa after benserazide, this should be a good combination with the COMT-inhibitor. Only, to spare the amount of dopa I think it would be a good idea to combine both inhibitors.

Rinne: In Parkinson's disease it is not necessary to get high concentrations of dopamine in the brain. It is more important to maintain the levels of dopamine and so obtain long-lasting stimulation. So it may be possible, by using this combination, to get more prolonged dopaminergic stimulation in the striatum than by using levodopa alone.

Da Prada: I would agree with this hypothesis, because one can then improve the bioavailability of madopar: there is a small prolongation of the half-life and an increase in the area under the curve.

Korczyn: Is it possible to treat Parkinson's patients only with a combination of both inhibitors but without levodopa?

Da Prada: That is my question too: in the rat, even in the unilaterally-lesioned rat, I don't find rotation when I combine benserazide with the COMT-inhibitor. So I have the impression that the amount of endogenous dopa in the plasma is not enough to cross the blood brain barrier. Maybe it would be possible in humans.

Youdim: I would like to answer Dr Rinne's question. At least in the animal studies we have done with the COMT-inhibitor the inhibition produces the kind of overt animal behaviour which is definitely dopaminergic, because you can block it with haloperidol, and it seems to be a sustained release. You get exactly the same kind of behaviour if you combine an MAO-B-inhibitor with nomifensine.

Riederer: Dr Da Prada, have you combined deprenyl with the COMT-inhibitor? And what is the difference in the accumulation of amines in rat brain?

Da Prada: You get a potentiation of the effect, maybe of extraneuronal origin.

Youdim: That is an important point: why is it when you give deprenyl to the human and you measure dopamine, as we have done in collaboration with Riederer, that dopamine is increased by about 40—60%, while phenylethylamine is increased 3000%? It is because dopamine is still being metabolised by COMT, while the deprenyl only blocks the MAO-B, which is the sole metabolic pathway for phenylethylamine. This is why COMT-inhibitors are likely to play a valuable role as therapeutic drugs.

Discussion of the paper by Wisser

Anonymous: Each constituent of the cerebrospinal fluid is subject to various transport mechanisms. Maybe HVA predominantly originates in the central nervous system, but the CSF is in passive equilibrium with the serum compartment, and I wonder whether in Parkinson's disease the HVA level in the serum is decreased.
Wisser: I don't know, but there is certainly a high correlation between the concentrations of noradrenaline in the plasma and the CSF, with a correlation coefficient of 0.7.
Rinne: When we measure dopamine metabolites in urine, CSF and brain it is very important to consider whether these pathological effects on the dopamine system are localised in the substantia nigra neurons, or whether they occur everywhere in the body. I suppose that these findings demonstrate that in the early phase of the disease these changes are only found in the substantia nigra neurons, and maybe later on in a very final end-stage you can also find changes in other dopamine neurons. Therefore changes that occur in the early phase can only be seen in the brain, perhaps with PET-scanning.

Discussion of the paper by Ackenheil and Bondy

Parkes: What is your definition of "drug-free schizophrenia"?
Bondy: I told you that two thirds of the group were "drug-naïve", which means that they never previously received neuroleptics because it was their first admission to hospital. You could argue now that this is not schizophrenia: that may be true, but we have followed these patients for over six years, and most of these patients came back, and in the meantime the diagnosis of schizophrenia was confirmed. That is why I dare to call them schizophrenics. The other members of this drug-free group received no drugs for more than one year before they came back. Of course you can never be sure that they did not take any haloperidol 10 mg a day or not, but since we observed no changes in the binding site during treatment, since it always remained at a constant level, we have no anxiety on that score.
Parkes: Aren't you surprised that the treatment didn't alter that?
Bondy: Yes, we were surprised, and this is really also the main reason that makes us say that it's not a receptor. It does not resemble the brain dopamine-receptor, but it could be that it is a type of peripheral receptor which, unlike the central receptor, is not regulated.

Rinne: My question is related to the previous one, because I was wondering whether you have studied the binding more closely: are you really sure that this is a dopamine-receptor is it just some other kind of binding?

Bondy: I hope you didn't misunderstand me: I have always tried to say that this is a neuroleptic-binding, or spiperone-binding site, because nobody really knows whether it is a dopamine-receptor. There are some new reports which say that it is a kind of dopamine-receptor with stereoselectivity for agonists and antagonists, but up to now nobody has really come up with any function that looks very specifically like a dopamine-receptor or even specifically like a receptor with a stereoselective function.

Dichgans: Do you by chance have any data on relatives, especially identical twins?

Bondy: We do have both: we have investigated 50 relatives of 20 schizophrenics and found that within families this increase in spiperone-binding co-segregates with the disease. We found an increase in spiperone-binding in all relatives that were ever ill or were ever diagnosed as schizophrenics, but we also found increase spiperone-binding in some relatives that were not affected up to the time of the assay, which means that it could be a kind of marker for vulnerability to schizophrenia, and two of these relatives have the first manifestations of the disease. Furthermore, we have done twin-studies comprising 20 mono- and 20-dizygotic twins, and found that the spiperone-binding is genetically determined.

Parkes: Have you any data on circadian changes in lymphocyte spiperone-binding sites?

Bondy: There is no variation at all with regard to a circadian rhythm or circannual rhythms.

Felgenhauer: Have you studied the binding site during long relapses? Some of these patients may be in remission for several years.

Bondy: Our data extend over five or six years. Whenever a patient comes to the hospital the analysis is repeated, and they are always at the same level.

Höllt: Do you always relate your data to the number of cells, and do you get different recoveries of the cells with differences in the number of cells and types of lymphocytes?

Bondy: The reason we relate the density of the binding sites to the number of cells is because in no preparation of lymphocytes do you get rid of all the platelets, and the number of platelets in your lymphocyte preparation depends on whether the patients are treated or not. Patients who are being treated with neuroleptic drugs have a much greater degree of platelet contamination in the lymphocyte preparations, and if we were to relate the analysis to the total protein concentration we would get misleading results. Concerning your second question about the types of lymphocytes, there are some papers which state that the binding is predominantly on the B cells and less on the T cells. We did it ourselves a couple of times and obtained similar results, but for routine assays we don't separate B and T cells or the patients would need to provide too much blood.

Herz: Did you investigate the coupling of "your" receptor to cyclic AMP or to other second messenger systems?

Bondy: We tried it several times. On two occassions we obtained a decrease of the basic cyclic AMP levels with dopamine, and this could be inhibited by (+)-butaclamol and not by (−)-butaclamol. After these two results we never managed to repeat them again. There are some papers reporting that dopamine at higher concentrations increased cyclic AMP concentrations, but this certainly operates via the beta-receptors.

Riederer: What are the pharmacological properties of these binding sites, for example what do the saturation curves look like? And a second question, would it be better to take cell cultures of lymphocytes for these analyses, to exclude interfering platelets or other contaminations?

Bondy: The pharmacological properties are in fact very bad, and do not support the idea of a dopamine receptor. The binding curve is biphasic, which means that there is a kind of saturation, followed by an increase with increasing concentration. Displacement experiments with neuroleptics normally show values 100-fold higher than for central dopamine-receptors. As for your second question, we did try to do it with cultured transformed cells, and we tried to get this method established, because that way you get more cells. But it only works sometimes. Dick Epstein in Israel has been successful in doing the assay on transformed cells.

Korczyn: We are all bothered by the fact that the patients who were treated did not show any difference from those who were not treated, so the question is really one concerning the specificity of the spiperone-binding. Is it not inhibited in vitro by haloperidol or chlorpromazine? If it isn't then I don't know what is the significance of this spiperone-binding. If it does, then it's very difficult to understand why you do not show a therapeutic effect in vivo.

Bondy: In vitro it is inhibited. You have to be very careful with the assay: if you obtain blood from a patient who is under medication you need to wash the cells several times more than with untreated patients, otherwise you will have the drugs in the incubation medium and disturb your assay.

Danielczyk: Was the parkinsonian group free of psychiatric complications, hallucinations and so on?
Bondy: Yes, they were.
Wisser: As you know, in patients with pheochromocytoma the β2-receptors on leucocytes are down-regulated. Perhaps it would be an idea to analyse the leucocytes in patients with neuroblastoma with elevated concentrations of dopamine, and perhaps you could show that then the receptor density too is down-regulated.
Bondy: I should like additionally to mention a previous study we did: we studied the spiperone-binding to lymphocytes together with the dopamine concentrations in acute non-treated schizophrenic patients. There was no relationship between the spiperone-binding and the dopamine concentration, although the dopamine concentration was raised in acute schizophrenic patients.

Discussion of the paper by Kettler

Parkes: We recently looked at eldepryl—that's selegiline—in patients with narcolepsy, as a stimulatory drug, and could not really determine in these circumstances how much of the stimulant action of selegiline (30 mg/day by mouth) was due to conversion to amphetamine and metamphetamine and how much was due to inhibition of MAO. Your compound is not metabolised to amphetamine, has it any stimulant action?
Kettler: We have no stimulant action with that compound.
Parkes: So the entire stimulant action of selegiline is due to amphetamine?
Youdim: Deprenyl (selegiline) is itself a sympathomimetic amine, it increases the heart-rate. You see the amphetamine effect in animal experiments only at 10 mg/kg.

Discussion of the paper by Rolf and Brune

Da Prada: Do you think this decrease in the amino acids is due to a defect in the thrombocytes, or does it reflect a deficiency in their concentrations in the plasma? Do you have any idea where within the platelets they would be stored?
Rolf: I cannot really answer that because there are no known studies. It has not been established whether or not there are specialized organelles within the platelets. One might suppose that there could well be specific amino acid receptors because active uptake has been observed.
Riederer: Have you checked the patients' nutritional intake in comparison with that of the controls?

Rolf: These patients were at the clinic of Professor Fünfgeld in Bad Laasphe, and I have no information about their diets.

Rinne: I would like to inform you that *post mortem* analyses were done on patients at a very advanced stage of the disease, and we found a small but significant decrease of aspartate in the striatum, and a small decrease of glutamate which, however, failed to reach statistical significance. So in part this is in agreement with your findings. I would like to ask you about the patients with the "on-off" phenomena: were they having "end-of-dose-failure"? Or what kind of patients were they?

Rolf: These patients were scored 24 hours before our biological investigations, and we were told that they had a severe form of Parkinson's disease.

Felgenhauer: A very naïve question from an outsider: why are platelets used for these investigations? Is it because they are so readily available? If you are looking at genetic markers then this would be the most understandable explanation. One would expect that tissue cells too would show these effects, such as the increased binding. Have studies of this type been done? How about liver cells or muscle cells, to give just two examples.

Rolf: If you search the literature you will not find any reports on this. Platelets are used because they are assumed to be a suitable model, but platelets have not been very well etablished as suitable for studying glutamatergic processes.

Wesemann: May I extend Dr Felgenhauer's question a little, because it was rather general: If you find metabolic changes in receptor-binding in peripheral cells like lymphocytes or blood platelets, is it then correct to speak about a disorder just of the nervous system? Isn't it then more or less a systemic disease?

Brune: I don't think we have enough data to give a precise answer to that question. I have been reflecting that quite a few results we obtained in diseases of the nervous system could equally be seen in other tissues. This experiment, as demonstrated here in lymphocytes and platelets, should be done on other tissues as well, and the results should be used to determine whether those diseases which we usually call diseases of the central and peripheral nervous system may actually also be systemic diseases.

Korczyn: The literature is full of examples where peripheral models like these have been used for the sake of convenience. In Duchenne muscular dystrophy blood cells from patients were shown to differ from normal controls. But just recently when the gene for the disease was cloned and its product was demonstrated, it became evident that the abnormality was not in fact expressed in blood cells. I think this is

a very important lesson to us to be very careful to exclude artefacts (diet, drugs, hypertension etc.), which can all contribute indirectly to what happens to platelets or peripheral lymphocytes.
Brune: I would agree that we should try to control diets, but in this particular study the patients were from a hospital and the controls also received hospital diets, so I wouldn't think that diet would actually be a major factor for the observed differences.

Discussion of the paper by Lesch

Parkes: One of the most interesting contributions on prolactin secretion I've seen recently came from Tom Chase and his group, using the D 1-SKF-inhibitor. In six out of eight patients this resulted in an elevation and not a suppression of prolactin. Have you got any more data on this?
Lesch: No, I am sorry.
Markstein: I can confirm that the D 1 agonist can increase prolactin secretion, but in rats and primates this is a very short-lasting effect, and the result with SKF might equally well be attributed to a putative metabolite.
Rinne: I have one comment about this D 1 receptor stimulation and prolactin, because we too have measured prolactin secretion in a few patients treated with a D 1 agonist, and there was either no effect or a small decline like with D 2. It was completely different from that found by Chase using a different compound. So I was wondering whether this was really due to D 1-receptor stimulation. My question is, what is the behaviour of dopamine cells in the hypothalamus in Parkinson's disease? Are there the same sort of changes one sees in the striatum?
Lesch: Dopamine-containing cells in other brain regions are also affected in Parkinson's disease, but I think much less. This might influence the regulation of certain neuro-endocrine axes.
Rinne: You did not find any kind of basal changes in the hypothalamo-anterior pituitary system, for instance, and so in the early cases there is no effect at all, but maybe you could find such changes in the late phases. Dopamine is decreased in the hypothalamus only in advanced cases, but this is not nearly as prominent as in the nigrostriatal system.
Jellinger: Recently, in the *New England Journal of Medicine* the American authors stated that there was a considerable decrease of catecholamines in the adrenal medulla of advanced cases of Parkinson's disease. Peter Riederer and our group suggested the same thing eleven years earlier, and we found that there is also a considerable decrease

of tyrosine hydroxylase in the adrenal medulla. We further found a number of tyrosine hydroxylase immunoreactive neurons and probably fibres *en passage* in the supranuclear and paraventricular hypothalamic nucleus that might be taken as evidence of adrenergic and also noradrenergic innervation of these nuclei.

Lesch: Sure, but the regulation of neuro-endocrine axes is extremely complicated and involves neurotransmitters and peptidergic systems. I have tried to make it clear that this is only tentative and I am not certain whether neuro-endocrine methods or challenge tests are *really* helping us as aids in the diagnosis of receptor sensitivity, so we ought to be extremely careful on this issue.

Parkes: The concept of using neuro-endocrine tests as markers in neurology or psychiatry is badly founded and has no scientific basis. Prolactin release is pulsatile; 5% of the stored prolactin is released in a single pulse and it has a half-life of 18 minutes. Any system that does not look at the prolactin release over 24 hours and at 5-minute intervals is going to give a totally false impression of what is really happening. In a disorder which involves such features as stress, sleep disturbance and motor disturbances these tests are completely useless.

Lesch: As you say, such multiple factors need to be allowed for. Age, sex, and ovarian status especially where the lactotrophic axis is concerned, all these factors which are involved in the regulation of hormonal secretion, are very difficult to control.

Discussion of the paper by Ruß and Przuntek

Da Prada: How does the quaternary metabolite of MPTP get out of the glia? When plateles have been incubated with MPTP, the MPP^+ that is synthesized inside can no longer get through the membrane. There is a certain degree of blockade of the toxicity when you give a dopamine-uptake inhibitor, but the effect is not all that strong. My other point is this: I was unable to get any evidence of toxicity in the platelets, nor any accumulation within the mitochondria, but all these effects were seen by Singer *in vitro*. His concentrations were 10^{-3}, and I doubt whether one can attain such concentrations *in vivo*.

Ruß: The basic problem is that you can't get charged ionic species through membranes without active transport systems. It has been shown that MPTP can get through membranes, and there are investigations in which MPP^+ could be demonstrated in storage vesicles after lysis. The underlying mechanisms are, in my opinion, not at all clear. The same problem of course also applies to the uptake of MPP^+ into mitochondria, because they too have a double membrane. I don't know how the transport mechanism works there.

Youdim: We looked at the MAO-B-containing chromaffin cell which converts MPTP to MPP^+ and saw no toxicity. There must therefore be something special about the substantia nigra for toxicity to occur. MPTP is distributed throughout the brain, and we do not really understand the selective toxicity in the nigra. Two other neurotoxic substances are allylamine, which causes cardiovascular toxicity and MAO is involved because allylamine is a substrate, and there is a second compound, DSP4, which is a noradrenergic neurotoxin, and this too is metabolised via MAO-B.

Ruß: One possible explanation would of course be the concentration of these dopamine-uptake sites, leading to a relatively high concentration of MPP^+ in the corresponding neurons which causes their destruction. Such a high concentration of dopamine-uptake sites is not, for example, present in hepatocytes, although the toxic action on mitochondria is in principle identical in hepatocytes.

Herz: In rats MPTP does not produce a Parkinson-like syndrome. Could this have something to do with this special mode of action? Does melanin play any part in this?

Youdim: A recent paper in the *Journal of Neurochemistry* investigated the blood-brain-barrier in mice, rats and humans and showed that in the humans and the mice there was very little of the MAO-B, so that the MPTP could get across into the brain and there exert its neurotoxic actions, whereas in the rat microvessels, the endothelium which forms the blood-brain-barrier, there was a substantial amount of MAO-B so that the MPTP was metabolized there, and MPP^+ would not therefore get into the brain. This was put forward as a suggestion for MPTP having no effect on the rat, and might explain the species differences between mice, rats and humans. Another possible explanation might involve aldehydes: these are important because the first step in MPTP metabolism via the dehydrogenase results in the formation of an aldehyde which might already be toxic. Corsini blocked the dehydrogenase with diethyl-dithiocarbamate and was also able to block the toxicity.

Blau: The animal experiments are difficult to interpret, I wonder whether there are any clinical descriptions of the patients who suffered Parkinson-like symptoms because they accidentally took MPTP, and whether these symptoms were reversible as they were with the monkeys.

Ruß: The Parkinson syndrome that occurred in the drug-addicts in California who injected themselves with MPTP had a progressive character. On the other hand in the animal experiments which we carried out there was no such progression. This was already described once before in rhesus monkeys. This is certainly one of the basic problems when designing possible therapeutic experiments.

Herz: Reading your paper one gets the impression that in the marmosets the effects were relatively weak, or substantially reversible. On the other hand in the rhesus monkeys the picture was more severe and persistent. Is this a question of dosage or are we again seeing a species difference in various non-human primates?

Ruß: First of all we have to consider the age-effect: older monkeys react substantially more selectively to MPTP even at lower doses, and the main difference lies in the mode of application: our view is that the Parkinson-syndrome produced by the application of successive small doses is essentially more suitable as a therapeutic model.

Oertel: If you take marmosets and inject 10—15 mg MPTP over four days you get a severe Parkinson-syndrome, lasting 4—6 weeks, which takes a tremendous amount of animal care. This is then followed by a recovery leading to a partial Parkinson-syndrome, and this has the advantage that the animals are at least able to feed themselves, but they are then stable for about one or two years, and this gives you a certain period in which you are able to do therapeutic studies. If you use older monkeys, let's say 8-year-old marmosets, you can get them stable for 8—12 months at the bradykinetic stage with very low doses of MPTP, but it is a species problem.

Youdim: Going back to species differences, I should like to mention that MAO-B activity is not expressed in young rats and only to a limited extent in old rats: this might explain why rats are very insensitive to MPTP.

Streifler: With aging, the endothelial alignment in the blood-brain-barrier would be expected to become clogged with stomata, resulting in diminished permeability. We could therefore expect a little different behaviour in older experimental animals than in younger ones.

Ruß: Permeability would not be expected to play a decisive role in these investigations, because the substance which overcomes the blood-brain-barrier, MPTP, is highly lipophilic. For this reason age-related changes should not be relevant.

Korczyn: Proteins can easily cross the blood-brain-barrier in old age, so the blood-brain-barrier is not very strong. One ought to be very cautions about the use of the term "blood-brain-barrier" because it comprises a variety of barriers with different mechanisms of substrate transport.

Bogdahn: Has anybody tried to measure glutathione or other indices of repair capability in these animal species or in old or young animals to correlate these differences in susceptibility to MPTP?

Riederer: We looked at mice that had been treated with MPTP: there was a significant decrease in reduced glutathione while the oxidised glutathione went up.

Discussion of the paper by Konradi, Riederer and Heinsen

Herz: At the end you broached the subject of exotoxins: it is of course frightfully difficult to investigate experimentally, but perhaps we could get at the answer indirectly by looking at Parkinson's disease in different parts of the world, and at different cultural levels or communities. What do we know about the morbidity of Parkinson's disease?
Oertel: According to door-to-door investigations made in a certain county in Mississippi in the US, the prevalence of Parkinson's disease was found to be 95 per 100,000. The same group carried out a similar study in Red China and came up with an incidence of 63 per 100,000, and a recent study in San Marino done to essentially the same design gave a figure of 150 per 100,000, so there are geographical differences. Studies done at the NIH showed no differences in prevalence between whites and blacks, and the values quoted above from the Mississippi study may have something to do with the fact that 50% of the blacks had never seen a neurologist before.
Rinne: Do you think this is a cultural or a genetic difference?
Oertel: The study in Nigeria had only two patients in the sixties age-group, so I think it is very difficult to draw any conclusions.
Korczyn: In Israel we have been impressed by the fact that in the two ethnic groups the Ashkenazi (eastern European) and Sephardic (middle eastern and mediterranean) there is a wide discrepancy in the prevalence of Parkinson's disease: the Ashkenazis show a much higher prevalence. It is not clear, however, whether this is a genetic or an environmental difference.
Rinne: In Sweden there are some differences between the industrialized and non-industrialized areas. In Finland we have great areas with no industry, but there is no difference in the prevalence rate.

Discussion of the paper by Markstein and Vigouret

Poewe: I think you hinted in one of your slides that the differential activity of anti-parkinsonian drugs on akinetic as opposed to tremorous symptoms might have something to do with their D 1 or D 2 agonist properties. I wouldn't go along with that, because in the majority of oscillating "on-off" patients who have both akinesia and tremor, levodopa has a very strong effect on the tremor as well, which disappears in many of our patients. The same is seen with apomorphine, which, in some of the early studies, showed a better effect on tremor.

Horowski: I would like to remind you that there are other differences between L-dopa and dopamine agonists, not only the putative difference in D 1 activity. One of them is, of course, that in early parkinsonism L-dopa is being taken up and released by nerve-impulse flow. So it is buffered, it's very difficult to overdose it, much more so than compared to a dopamine agonist, and so what favours levodopa is tolerance as much as efficacy. Another important difference is that L-dopa can additionally be considered a "pro-drug" for noradrenaline, whereas the ergolines are α-adrenolytic compounds.

Discussion of the paper by Wesemann, Weiner, Nelzacka and Sontag

Horowski: I am very impressed by the amantadine data, however the concentration used was rather high. Would you say that *in vivo* the drug is maybe accumulating within the membrane?

Wesemann: We have some indications from a series of adamantane derivatives that the rather high concentrations which we need during an *in vitro* experiment can be reduced if you do an *in vivo* experiment. During an *in vivo* experiment with the 2-ethylamine derivative of amantadine you only beed 1/10 to 1/100 of the concentration that you need for *in vitro* experiments. We don't have many data about the accumulation of amantadine in *post mortem* brains. I know from the *post mortem* studies, that the range of amantadine is about the same order of magnitude as we find in the membranes after application *in vivo*, as measured by GC-mass spectrometry.

Kuschinsky: What is the effect of alcohol on the fluidity of dopamine- and serotonin-receptors?

Wesemann: It is known from rat and mouse experiments that the membrane fluidity is enhanced after single doses of alcohol. But fluidity is only one parameter. If you isolate the membranes from the chronically-treated rat and then add ethanol *in vitro*, you find only a small decrease in membrane fluidity as compared to membranes of naïve animals. The effect of membrane anisotropy is different with regard to the different types of receptor, and even if you have the same receptor type, the effect of membrane fluidity can vary with regard to the localization of the receptor. Membrane fluidity has an effect on the coupling either of the receptor with the second messenger system or on the coupling of the catalytic and regulatory unit.

Discussion of the paper by Oertel and Marsden

Korczyn: I would comment that transplantation as it stands is not justifiable because we have not yet accumulated sufficient data from basic scientific and animal studies. There is no doubt that any drug given to a Parkinson patient has a placebo effect, and I have no doubt whatsoever that the operation itself has a much stronger placebo effect than any drug. The self-selection of the patients and the special atmosphere surrounding the operation creates a very fertile ground for a significant placebo effect.

Rinne: It has been suggested that the therapeutic effect of the adrenal medulla transplantation is symptomatic only of the lesion, so why not have this operation without the actual transplantation? Perhaps this would give you the same result.

Oertel: One patient in Portugal *was* sham-operated and has not improved. I personally would never, at the present time, even consider sending a patient to have an adrenal medulla transplantation: I think it's unethical. I agree that there are no controlled studies of the effect of implantation with fetal human mesencephalon in non-human primate Parkinson's disease models. There is no increase of dopamine or of its metabolites in the CSF of patients with adrenal medulla transplants. In monkeys with fetal tissue transplants you may get an increase in certain parameters if the tissue is transplanted to a site where the implant is in contact with the CSF.

Rinne: How many embryos do you need for the fetal cell transplantation?

Oertel: I only know of estimates made on theoretical grounds by a group in Sweden: they think that they can cover a quarter of the putamen with one fetal brain.

Bogdahn: Has anybody tried to quantify nerve growth factors produced by adrenal medulla and by embryonic tissue?

Oertel: A group in Paris is characterizing a glycoprotein which keeps dopaminergic neurons alive. We don't know if there is any protein or peptide which will improve the health, or increase the ability of a diseased cell to survive.

Discussions on Saturday, May 7th, 1988

Discussion of the paper by Schneider

Brune: Dr Schneider, how would you define "early" and "late"?
Schneider: Yes, we have different ways of defining them: one definition is that "early" treatment means that levodopa would be started before the patient is in any way handicapped: you only see some minor signs of Parkinson's disease, but the patient is able to continue at work and doesn't need any help from other people. At the point where the patient starts having problems with his professional duties or aspects of his homelife, that would be another good time to start levodopa, but that's not late, that is also still "early".
Rinne: I would like to comment on these terms "early" and "late", because when we initiated long-term levodopa treatment in the middle sixties, and followed the same patient population treated for the whole of the rest of their lives. Afterwards we evaluated the results, and this enabled us to study the relationship of the patients' pre-treatment disabilities and their mortality and survival. It was very interesting to see that if you initiated levodopa treatment at Hoehn and Yahr stages one and two, you could offer 6–7 extra years to your patients, but if treatment was postponed to Hoehn and Yahr stages three, four and five, the outcome was much the same. This means that you have to initiate the dopaminergic treatment at least in Hoehn and Yahr stages one and two, because if you postpone it until stage three then you lose the capacity of extra years on levodopa for your patients. As far as I am concerned the Hoehn and Yahr staging is very useful because it enables you to tell what is "early" and what is "late".

Discussion of the paper by Gerlach and Przuntek

Danielczyk: The older the patients the worse is their renal function. Do you have any data correlating the excretion of this Retard-formulation with age? Because the incidence of side-effects rises very steeply with age in old patients, and this is to some extent governed by the deterioration in their renal function. How do you measure renal function?
Gerlach: These are only the first pharmacokinetic data based on plasma values. We have not so far measured renal excretion rates or done renal function studies.
Poewe: We now have three years' experience with patients treated with HBS and CR 4. I would just like to make a few comments on that

in the light of what we have heard about these slow-release preparations. We found that improvements in mobility which we might have expected from the pharmacokinetics of these substances failed to materialize. For example, in our patients there were only insignificant alterations in dose-frequency and the mean dosage-intervals, compared to the rapidly released L-dopa. Independently of this there was a high proportion of treated patients who showed an improvement in oscillation. One problem, and one which I believe may also be expected with the preparation just presented here, is that there were very pronounced restless movements which erupted explosively in our patients, particularly in the afternoon. I don't believe that we can reduce these peak-dose restless movements with this mode of treatment, which is more likely to lead to biphasic dyskinesias and naturally also to dystonias coupled to on-off phases. What I particularly liked about the preparation you presented was that it produced a plasma level relatively rapidly, because we found that with the hitherto available slow-release preparations i.e. CR 4 and HBS we needed to give the patients a "kick-start", particularly in the morning to get them going at all.

Gerlach: I think that there is an essential difference in the mode of action of HBS and our slow-release formulation. HBS is outstanding *in vitro* and releases L-dopa over a period of hours. HBS-madopar unfortunately has a decisive disadvantage, and that it that its absorption is very dependent on gut motility. This was shown originally in normal healthy subjects, who exhibited mean gastric dwell-times of HBS capsules which varied between 10 minutes and 10 hours. The slow-release formulation which we are using works in a fundamentally different fashion: it simultaneously liberates a whole lot of micropellets inside the stomach. We achieve the differential release of the L-dopa by differential covering of the micropellets with gastric juice-resistant coatings.

Przuntek: In using slow-release preparations it is absolutely essential to determine absorption and pharmacokinetics. There are two possible ways to achieve a "kick-start": a) to use a fixed combination of a quick-acting and a slow-release preparation of L-dopa; or b) to use large doses of a slow-release preparation. We too have observed the accumulation effect described by Poewe, again particularly in the afternoon, and we can explain it on the basis of the steep rise in plasma L-dopa levels, especially when we give equal doses at equal intervals. We believe that the best effects on fluctuations in mobility are achieved when the morning level rises relatively rapidly, declines slightly towards noon, and then rises again gently in the afternoon. Using the form of L-dopa administration proposed by us it is possible to extend the intervals between doses, because we can maintain therapeutic levels over prolonged periods. We can also document this sustained effect with the motor performance series.

Discussion of the paper by Birkmayer

Przuntek: Professor Birkmayer, people who don't believe in the iron theory say that iron cannot pass through the blood-brain-barrier. During the Austrian Parkinson Society meeting Kornhuber Jr told us that in a group of patients with Parkinson's disease the blood-brain-barrier was more permeable than in age-matched subjects. Do you think we ought to determine the blood-brain-barrier permeability before we apply the iron treatment, to see who would benefit? Could you explain the abbreviated effective period of the iron preparations on the basis that iron therapy restores the normal impermeability of the blood-brain-barrier?

Birkmayer: That could be the case.

Przuntek: Youdim told us yesterday that, at least in animal experiments, the administration of iron could eventually lead to toxic effects. Should we be concerned about such possible toxic side effects of the long-term administration of iron?

Birkmayer: Don't ask me! Internists phone me up and say: "Are you crazy? You're giving these patients iron! Don't you realise that iron accumulation leads to haemosiderosis?" Well, it doesn't do anything of the kind at these tiny doses: 15 mg, where's the harm in that? We have been using iron treatment for over two years now, without any side effects worth mentioning.

Danielczyk: I have only very limited experience with the iron treatment. Up to now we have treated only 15 old and multisymptomatic patients with Parkinson's disease in an open-ended study. We have seen no side effects except a slight, and indeed desirable increase in blood-pressure, and increased tremor. I would therefore not give iron to patients with tremor. Subjectively the patients all reported improvement, but objectively the improvements were very slight and mostly transient. Up to a week ago I would have been of the opinion that iron was probably ineffective. But then I had the opportunity to examine a 61-year-old patient who had been treated for seven weeks with twice-weekly iron injections, and since 11 years with amantadine and madopar. This patient experienced a substantial improvement on the iron treatment, even with a reduction of L-dopa to a half and of amantadine to one third of previous doses. The patient is also psychologically improved, and he puts it all down to these iron injections. He feels an improvement 2–3 hours after each, the following day is the best, and then the effect slowly wears off after four days. This patient too, who had a low blood pressure of 100 and 110 mm Hg, showed an increase in blood pressure after the iron injections. I ask myself whether a peripheral sympathetic effect may not be the decisive factor in this patient's improvement.

Discussion of the paper by Parkes

Schneider: What was your experience of the interruption of amantadine therapy in patients who had received this drug together with L-dopa for about 7,8 or 10 years?
Parkes: It was as dramatic as the withdrawal of any antiparkinsonian drug—on the other hand they may show not the slightest difference whatsoever, and I do not know why this is. Let me tell you one clinical story about one of the patients we included in an original amantadine trial 19 years ago. This patient has continued on amantadine alone, 600 mg daily, for 19 years with no toxicity and with continued excellent antiparkinsonian response. But that is only one patient amongst 30 or 40 in an initial trial; the others have nowhere near achieved that kind of response, or have died.
Danielczyk: I believe amantadine is usually given in too low dosage, and so its effectiveness is underrated. In my experience the average effective dose lies in the region of 500–600 mg, and side-effects are slight. It is only the older Parkinson patients who in advanced stages show a pharmaco-toxic psychosis which responds to a reduction in dosage. It is a pity that we know so little about the mode of action of these preparations.
Parkes: Yes, I agree completely.
Wesemann: I agree that the basic research has not come up with an explanation for the mode of action of amantadine, but to give only two hints, first it was shown that amantadine enhanced the electrically-stimulated release of dopamine and 5-hydroxytryptamine, and second, Osborn demonstrated that amantadine and amantadine drivatives have some stimulating effect on the second messenger system.
Przuntek: Do you know any long term trial that shows that amantadine works for longer than two years?
Parkes: There is a little French study for slightly longer periods.
Przuntek: I would like to ask Wesemann this: you showed that the membrane fluidity is increased with ethanol, and if you then administer it for a longer period the fluidity decreases again. You also showed an increased membrane fluidity with amantadine. What happens if you administer amantadine in a long-term trial?

Wesemann: Well, the "long-term trial" lasted only two weeks! Acutely you get a rather high increase in membrane fluidity, and this is reduced in a long-term trial, i.e. within two weeks. But there is a decrease nevertheless. But there is another effect—yesterday I mentioned the experiment with imipramine, and the effect with amantadine is similar—if you compare naïve animals with animals that were treated with amantadine or amantadine derivatives, the fluidising effect is less if you isolate membranes from the treated animals and add more amantadine *in vitro* than with membranes isolated from the untreated animals. That means amantadine pretreatment must somehow have altered the chemistry of the membranes, but in what way we do not know.
Korczyn: We tested the effect of amantadine in several animal models of Parkinson's disease made in rats by treatment with tetrabenazine, reserpine, haloperidol or oxotremorin. Amantadine had much the same effect in all of these, whereas with either levodopa or anticholinergics there were different effects, depending on the mode of action. This led us to suggest that amantadine acted quite differently from either the dopaminergic or the cholinergic systems. It is extremely important to find out the mode of action of amantadine, because this may give us an approach to the development of even more effective.

Discussion of the paper by Rinne

Da Prada: Dr Rinne, you are one of the few people using both dopamine agonists and L-dopa plus deprenyl. How would you compare these two modes of treatment?
Rinne: We have now followed this patient population receiving levodopa plus lisuride and deprenyl for nearly three years, and with this combination you need to use much less levodopa—only about 200 mg on the average—and get an improvement in antiparkinsonian activity that is equal to that obtained with levodopa combined with lisuride or to high doses of levodopa alone. But we have done it for too short a time to compare the possible incidences of late complications, though so far it is at least as good. So I suggest that if we consider the protective effect of deprenyl on these toxic radicals and so on, it might be better in future to initiate the early treatment with deprenyl, L-dopa and a dopamine agonist.
Mertens: I would like you to clarify your strategy, because earlier you told us that first you use amantadine, and now you are talking about a three-drug combination: how do you manage with all four?
Rinne: I initiate anti-parkinson treatment with amantadine only in elderly Parkinson patients who don't tolerate anticholinergic drugs or

levodopa so well. But I mentioned earlier, in connection with our findings on mortality and survival, that it is not wise to postpone the start of dopaminergic treatment too long, certainly not as long as Hoehn and Yahr stage three. So we start the dopaminergic treatment in stages one or two, but not by using levodopa alone because after 3–5 years this would lead to complications such as motor fluctuations, but by using this type of combined treatment you can offer the patients several extra years of a better quality of life.

Korczyn: With prolonged treatment, "peak of dose" akinesia and the "on-off" phenomena are the major, perhaps the most significant problems, and you have shown, and we too have seen, that with D 2-agonists we have much less of a problem. If we take your presentation today together with Markstein's of yesterday relating the "on-off" and the fluctuations to D 1-receptor stimulation, how would this fit in with the suggestion that we should add a D 1 agonist to our armamentarium?

Rinne: I agree that it is important to analyse the relative roles of D 1- and D 2-receptors as far as therapeutic responses to treatment are concerned. I also agree that the balance between D 1 and D 2 stimulation is relevant to the development of fluctuations: in our *post mortem* brain studies, for instance, there was a significant relationship between changes in the D1-receptor in the striatum and peak-dose dyskinesia.

Danielczyk: In advanced Parkinson's disease we are forced to use combination treatments, and in practice we use the dopaminergic agonists bromocriptine, lisuride and terguride. In the last four years we have found terguride to be a very valuable substance because of its effectiveness and safety and its lower incidence of side-effects, particularly the appearance of pharmaco-toxic psychoses. This preparation enables us to reach higher dose levels and a therapeutic effect relatively rapidly. It is currently our most certainly effective dopamine agonist for the treatment of patients aged 75–80.

Rinne: I agree: this new dopamine agonist may prove to be very useful.

Parkes: To follow on that, which dopamine agonist should we be using?

Rinne: I suggest one with both a D 1 and a D 2 action.

Baas: If you add an agonist to the levodopa after the appearance of fluctuations, we know that we can reduce them. Do you get more fluctuations in patients where you add the dopamine agonist to levodopa monotherapy or in the group pretreated on long-term combination therapy?

Rinne: It was not our purpose to have this kind of additional treatment in the end-phase, but our earlier experience was that the addition of a dopamine agonist in patients with fluctuations only reduced them, but as far as I know you can never completely eliminate end-of-dose failure.

Riederer: If you start with amantadine, L-dopa, dopamine agonists and deprenyl in the very early phase of Parkinson's disease, are you not very soon going to run into trouble with psychosis?

Rinne: I said earlier, I never have four drugs including amantadine, but I have initiated early combined treatment with deprenyl, lisuride and levodopa and not had any more psychiatric problems than with high doses of levodopa alone.

Discussion of the paper by Kempster and Stern

Oertel: I can confirm most of your data, but one thing puzzles me: two of our patients with severe orthostatic hypotension received domperidone to block the peripheral D 2-receptor, which is supposed to be related to hypotensive action, and they still had severe orthostatic hypotension: can you explain that?

Kempster: I can't give you a valid explanation.

Przuntek: It is not excluded that apomorphine may work centrally.

Scholz: Could you say something about central nervous side-effects with apomorphine, such as hallucinations?

Kempster: The only thing I can say is that they didn't happen. Among our 19 patients who had motor fluctuations, 10 had a history of neuropsychiatric problems, hallucinations or confusion on L-dopa or ergolid agonist medication. Apomorphine produced an equivalent motor response without any of these neuropsychiatric side-effects.

Scholz: What was the duration of treatment?

Kempster: The mean treatment period is 14 months.

Korczyn: Would you now regard this as the treatment of choice for patients with "off"-periods? What factors would dictate your selection of patients for this treatment?

Kempster: We chose patients who showed benefit from conventional L-dopa treatment but with high-amplitude fluctuations: people who don't respond to L-dopa are not improved by apomorphine.

Deuschl: We generally confirm what you say, but had some patients who only responded when the dose reached 7 mg. What was the highest dose you gave, and what would you recommend?

Kempster: We're talking about a single-dose response: the usual dose that gave satisfactory movement response was between two and three milligrams, but we certainly had some patients who needed six or seven.

Horowski: What about your use of domperidone?
Kempster: When the patients first started treatment we gave them 20 mg three times a day, and nearly all of them tolerated this dopamine receptor stimulation and it was almost always possible to cut it down; a few patients were able to stop it completely without becoming nauseated, but most patients still needed some domperidone.
Rinne: Our experience with lisuride rather than apomorphine is similar, but I think that for patients with fluctuation one ought first to try oral treatment unless they are so incapacitated that you can no longer manage them by the oral treatment. Then one can go to the more complicated pump treatment.
Kempster: One advantage of that treatment is that the patients were most satisfied with it because it gave them greater control over what had previously been erratic and unpredictable fluctuations. Those who got the greatest benefit were the ones who learnt how to manipulate the pulse controls of the pump, according to the requirements of their motor function.
Suchy: I gather that you give a very low background dose and the patient is taught that if he feels an "off"-period coming on he can give himself an additional bolus dose. Could you go into more detail about how the dosage is worked out and what the maximum daily dose would be?
Kempster: The mean dosage-rate was 0.04 mg/kg per infusion, giving about 4 mg/hour. The bolus doses under the patient's own control were 2—5 mg, and as I was saying, many patients might give themselves 20—30 bolus doses a day.
Suchy: You quoted the maximum effective dose given acutely in an "off"-period as being 7 mg. Is the single effective dose lower if you are receiving a basal infusion?
Kempster: I can't say, because we have not studied it comprehensively.
Riederer: What percentage of patients on D 1- plus D 2-receptor stimulation were non-responders compared to patients with purely D 2 stimulation?
Kempster: Again, I can't say: none of our patients were on D 2 stimulation as well as apomorphine.
Kuschinsky: In rats, low doses of apomorphine produce hypo- and akinesia. Did any of your patients under any conditions of apomorphine administration show a worsening of their Parkinson's disease?
Kempster: We found, with single dose studies, that the motor response to apomorphine in individual patients was very like their motor response to L-dopa. So one didn't particularly see any worsening with apomorphine, just as one didn't with L-dopa. One could pre-

dict a patient's motor response to apomorphine by looking at his motor response to L-dopa.
Schneider: Did you notice any diurnal variations in the patients' response?
Kempster: Although we have not looked at that in a comprehensive way, we found no evidence: "off"-periods were reversible at the usual bolus dose with respect to the time of day.
Przuntek: I agree with Rinne that the pump treatment is very complicated. We have treated more than 50 patients with the lisuride pump, and in 6 of them who were severely disabled Parkinson patients, we changed to apomorphine, but they showed a deterioration, so that we had to go back to lisuride infusion combined with levodopa therapy. As Bittkau will show in the next paper, it is important to select the right match between lisuride and levodopa, and furthermore it is possible to improve the combined treatment by using a slow-release levodopa.

Discussion of the paper by Bittkau, Klewin, Suchy and Przuntek

Rinne: With regard to the psychiatric side-effects, our experience paralleled yours: using a lisuride infusion, a confusional state could appear with a rapid increase in the rate of infusion, but we have now been using this treatment for about two years and you can avoid psychiatric problems by giving a reduced dose.
Bittkau: That's been our own experience, so we tend to give lower doses: 0.6—0.8 mg/day. If the patients tolerate this, and should need any more, we can always go up.
Rinne: When you give apomorphine you should really also give domperidone to control nausea: has this been a problem?
Bittkau: In the first few days I think nausea can be a problem.
Suchy: Where patients have been treated with dopamine agonists previously, nausea is never a problem. Where they have not, then nausea may be troublesome, but it can be minimised by giving domperidone in the first few days. At one centre domperidone was given continuously, but at another it was only given during the first week, and the patients were able to continue without further nausea.
Bittkau: We have a patient who does not tolerate oral lisuride even at low doses but who tolerates quite high doses if it is subcutaneously infused. We must ask ourselves whether it is the fluctuating levels that are the trouble, or whether we have here stimulation of the gastrointestinal tract as well as of the area postrema.
Riederer: What is the percentage of non-responders to D 1- plus D 2-stimulated patients compared to those treated with pure D 2-agonists?

Suchy: Patients treated with the lisuride infusions were concomitantly receiving levodopa, so you could say they were getting both D 1 and D 2-agonists. There were theoretically no non-responders: I say theoretically, because some have to be excluded on account of psychiatric side effects, but even those had an improved motor disability at the time of exclusion. Of those on long-term treatment, in a small number—two or three out of about 120—their motor fluctuations failed to respond to continuous stimulation. Our problem is to define the therapeutic window and to keep the continuous dose as low as possible.

Rinne: In your opinion, why are there more psychiatric side-effects on continuous lisuride than on apomorphine? What is the mechanism for these clear-cut differences between your results and those from London?

Horowski: Of course there are pharmacological differences between lisuride and apomorphine, but I do not wish to speculate until we have done a clear-cut comparative study. However, in the two studies quite different conditions were used: we started infusing very high doses, and we now infuse for over 24 hours; the London group were able to treat patients with discontinuous administration of apomorphine. And then, were the patients in the two places strictly comparable? To answer Riederer, the Rome group infused levodopa for one day and lisuride in the same way the next day in the same 16 patients. Ten of the 16 responded in a similar way, the other six had a less brilliant or negligible response to lisuride. However, on chronic treatment with some levodopa added these patients also responded to chronic subcutaneous infusions: that's as near as I can come to an answer.

Korczyn: The problem of psychiatric side-effects raised by Rinne needs looking at more carefully, because I don't think we should put all such side-effects into one bag: the patients vary, their treatment varies. Patients with an instrumental impairment because of old age or Alzheimer's disease or a multi-infarct dementia may be more vulnerable to intoxication with CNS-active drugs, and the confusional state may manifest itself in many ways, including paranoid ideation. But with lisuride you may get primary phenomena which are not related to confusion. We are also using lisuride for other indications e.g. dystonia, where it may be highly effective in some patients. One of our patients with focal dystonia in fact developed paranoid ideation although he was relatively young and not in any sense demented. This is probably related to the fact that lisuride is an ergot derivative so one can see effects very much like those one can sometimes see with LSD, which is of course different from what one would expect from patients treated with apomorphine which has a "cleaner" effect on dopamine-receptors.

Riederer: I wanted to ask Kempster whether all the patients on apomorphine responded.

Kempster: All patients who had a clear-cut periodic response to L-dopa responded to apomorphine. Patients who remained incapacitated after L-dopa didn't do any better on apomorphine. The patients I presented were selected, because they had isolated motor function disturbances that responded well to L-dopa, and they universally reacted equally well to apomorphine, with a more stable and predictable pattern of motor response.

Oertel: Were the patients who failed to respond to L-dopa and to apomorphine not of the idiopathic type?

Kempster: They had earlier shown a perceptible response to L-dopa and had no evidence of other neurodegenerative conditions, but when tested at times of peak motor improvement with L-dopa they remained really disabled and reacted no better to apomorphine.

Parkes: Do you believe that in true idiopathic Parkinson's disease we ever see total failure of dopamine-agonist or L-dopa response? I don't, but do you?

Rinne: In the initial stages of idiopathic Parkinson's disease you always get a therapeutic response, but of course at the endstage there are many patients who no longer respond: with the progression of the disease they have lost their ability to respond.

Suchy: I was surprised that the apomorphine patients additionally needed levodopa. If apomorphine is such a good D 1- plus D 2-agonist, why do the patients need levodopa as well?

Kempster: They nearly all needed L-dopa. There were two patients whose motor function we could start up in the morning with a dose of L-dopa and whose motor function would then be sustained on apomorphine for the rest of the day. When we tried that on many of the other patients at the doses we were using they eventually went off again. In several patients where I was trying to replace L-dopa completely with apomorphine, I had to increase the dose to such levels that they got nauseated in spite of the domperidone they were getting, and in any case, you cannot infuse too large volumes into subcutaneous tissue. We didn't push on with that approach because the combination seemed to be perfectly satisfactory.

Round Table Discussion

Przuntek: At the end of the symposium we should try to reach some conclusion, and prepare a summary, and answer the question that forms the theme of the symposium: do we really have any chance of an early diagnosis and preventive therapy for Parkinson's disease? So I should like to ask the participants whether they think that early diagnosis and preventive treatment will become possible within the next few years. Dr Steinberg, what is your opinion?

Steinberg: Yes, I think so: if we sharpen up our methods and shorten them by factor analysis of the various test items in such a way that the test session only takes about 30 minutes, then it should be possible.

Przuntek: The cost of early diagnosis must be rather low so that everybody can get tested, and the examinations should only take a short time. Dr Rinne, do you believe that this will be possible?

Rinne: It is difficult to see how one could achieve this for neuropsychological testing. We have to ask ourselves what would be the role of cognitive evoked potentials, and how sensitive are they in giving us hints of the disease before the appearance of clear-cut Parkinson's disease symptoms.

Przuntek: Dr Scholz, do you have any experience?

Scholz: As Dr Maurer told us on Friday, it is possible to improve the evaluation of psychological symptoms with these technical measurements. The only significant visually evoked potential data do not concern the whole system, but only the retina.

Przuntek: Dr Kraus, Dr Scholz, is there any possibility of improving early diagnosis by kinesiological or electrophysiological means?

Kraus: With the help of a test-battery I think we *could* get a contribution to an early diagnosis. Using our results as a basis, we can expect that by concentrating on the tests of motor performance which give the best results as a first approach, we should be able to simplify and improve the method.

Scholz: I think it is an important tool for the early diagnosis. The second point concerns tremor: I find it interesting that we can diagnose tremor and even try to make a differential diagnosis with these stress experiments.

Korczyn: These methods may be sensitive but they lack specificity, so we may find a high false positive rate.

Przuntek: That is true: we may screen out a collection of patients including, for example, bot depression and Parkinson's disease: we would then have to separate them by more specific methods.

Korczyn: But maybe not patients with progressive supranuclear palsy (PSP) or olivo-ponto-cerebellar ataxia (OPCA) and so on, which constitute about 20–25% of the total population presenting with movement disorders of this type.

Przuntek: Dr Leenders, can you separate out these subtypes of Parkinson's syndrome?

Leenders: It should be possible, by a combination of several tracers, to distinguish between idiopathic Parkinson's disease and other degenerative conditions, mainly on the basis that in idiopathic Parkinson's disease the receptor population is essentially intact, particularly in untreated cases. That is the basis of the effectiveness of agonist therapy. If you should find a decrease in receptor population as well as impaired pre-synaptic function then I think we have good reason to suspect that something else is going on.

Przuntek: Dr Korczyn, do you believe there is any chance of earlier diagnosis by quantifying the investigations of vegetative or of sensory functions?

Korczyn: Probably not; as I said in my talk, the importance of detecting autonomic disturbances is to exclude other diseases such as multisystem atrophy, which may have autonomic manifestations, and so I don't think this is necessarily a very good method. One possibility, however, is that we might be able to get tissue diagnosis by examining peripheral autonomic ganglia. We do not at present know how specific or how sensitive this is, but there is a possibility that Lewy bodies may exist in autonomic ganglia, and if this were true then it might be important.

Mertens: I have to admit that what I found most exciting was the possibility of finding receptors on lymphocytes, and if this could be confirmed elsewhere I would gladly follow this up.

Przuntek: Mrs Bondy told me that she needs four hours per patient to perform this test: this seems very time-consuming.

Oertel: You have to do 16,000 pipetting steps to do four patients in a day. That is very labour intensive.

Nevertheless we decided at this meeting to introduce the technique at Innsbruck hospital, and then we shall see if it is reproducible. We also determine iron and investigate patients with other motor disturbances such as PSP and OPCA, and I hope that in two or three years' time we may have an answer.

Riederer: I would suggest that several other groups of patients ought to be tested to find out the specificity of this trial. I must say that I am not yet convinced that lymphocyte spiperone-binding sites are a useful tool for diagnosing patients with Parkinson's disease: perhaps it works in schizophrenia, but even that still has to be confirmed.

Przuntek: I totally agree: at the moment I too am not convinced by this test, but we have to check it out and to do that we will have to select other patient groups such as those with OPCA and with essential tremor, for example.

Gibb: Can I go back to a point of pathology: there are few situations where one can make the pathological diagnosis, i.e. identify Lewy bodies before death. As we have heard, Lewy bodies occur in peripheral sympathetic and parasympathetic tissues, also in sympathetic ganglia and in myenteric ganglion cells. Duvoisin and Yahr in 1965 reported the case of a man with what appeared to be intermittent claudication, and who was operated for it by the removal of the lumbar sympathetic ganglion. That ganglion contained Lewy bodies. Well, of course the patient did not have intermittent claudication of the vascular type, he had dystonic feet. That was a clear example of the early diagnosis of Parkinson's disease. There are other examples where Lewy bodies have been identified in peripheral tissues, e.g. in the oesophagus, the colon and the rectum. The frequency of myenteric ganglion cells and Lewy bodies would not make rectal biopsy a plausible diagnostic approach, but perhaps if one could surgically remove a sympathetic ganglion at the time of some other operations in suspected cases of Parkinson's disease this might be useful.

Danielczyk: We had this unexpected finding of high MAO-B values in thrombocytes. I never considered the possibility of using this as a criterion for making differential diagnoses, because these were all patients with advanced Parkinson's disease, but I wanted to ask Riederer whether one could not also investigate this in incipient Parkinson's syndrome.

Riederer: That should be tried, of course, but we have one problem: patients we see with dementia of the Alzheimer type show the same thing, and with Parkinson's disease patients we are not sure whether this significant increase of MAO-B activity may not be drug-induced. We really need to study some untreated patients to find out whether this effect is truly significant in drug-naïve patients. If it is not, then the MAO-B increase in platelets is rather a test for dementia of the Alzheimer type than for Parkinson's disease.

Korczyn: I want to complement what Gibb said or to reply to him. In my opinion the occurrence of Lewy bodies comes quite late in the sequence of degenerative processes within the cells, and it is quite possible that by refining the methods—looking at ubiquitin perhaps, or whatever else may be easy to find—it might be possible to spot the degenerative changes at a much earlier stage. Perhaps Jellinger or Gibb may care to comment, but if you could do the diagnosis of Parkinson's disease by, say, a rectal biopsy, then that might be a most sensitive and specific test.

Gibb: I totally agree: the potential of ubiquitin must be investigated further—although one has to remember that myenteric ganglion cells are not very frequent in the rectum, and really you need to remove the whole rectum to get sufficient results, scarcely a diagnostic possibility! I should say, however, that Forno reported a number of patients with incidental Lewy bodies within the substantia nigra but without Parkinson's disease, and she found that most of them also had Lewy bodies in the sympathetic ganglia. So I think that they occur there relatively early in the disease, or else there is a wide variation between cases.

Herz: For the last two days we have been hearing a lot about the long-term side-effects of anti-parkinsonian drugs. I don't really understand why an early diagnosis is necessary.

Przuntek: That's a very good point. We observe Parkinson's disease when more than 80% of the dopaminergic systems have been destroyed, and we hope that if we can make an earlier diagnosis we might be able to diminish this degenerative process, for example by protecting the degenerating brain with MAO-B inhibitors. At the moment we still have to treat the patients with MAO-B inhibitors plus L-dopa plus dopaminergic drugs. We hope that we shall be able to identify useful substances using the MPTP-model, substances which might be suitable for the prevention of Parkinson's disease.

Rinne: I believe that we *are* able to identify these pre-synaptic changes in the brain more than ten years before the clinical features become apparent. I believe that there is one *in vivo* method for studying pre-synaptic dopaminergic neurons, and that is positron emission tomography.

Leenders: Yes I agree, an *in vivo* method to detect the functional state of the dopaminergic nigrostriatal system is only possible at the moment with those radiolabel tracer methods, but you can't start screening the whole population by that method.

Rinne: Only suspect cases.

Leenders: If you see a functional impairment of 60% in a suspected case, all it tells you is that this particular functional brain system is impaired, it does not yet amount to a diagnosis of Parkinson's disease.

Rinne: I have seen some PET-scans showing some unilateral decreases of pre-synaptic dopaminergic neurons, and it will be interesting to follow these cases: maybe in ten years' time they will develop Parkinson's disease, as Firnau in Canada was able to show.

Kuschinsky: Might it not be an easy and pragmatic approach to use a test-dose of a neuroleptic drug in a pre-parkinsonian phase in cases where there is an index of suspicion?

Scholz: Even in patients who already have Parkinson's disease who are treated with high doses of neuroleptics the response is evident in only 50% of the cases, so I doubt whether that would be a sufficiently sensitive test.

Gibb: It seems quite likely to me that the elderly patient who is given a small dose of a phenothiazine e.g. chlorpromazine, and shows Parkinson-like symptoms may have presymptomatic Parkinson's disease ot at least a degree of nigro-striatal deficiency. On the pathological side there is some speculative evidence that the duration of the presymptomatic phase for the average patient developing the first symptoms at the age of 60 is something like 20—30 years, so that the first pathological changes in the substantia nigra may start as early as between the ages of 25 and 35. So there may be quite an extended presymptomatic phase when PET-scanning or some other approach could identify those patients at risk. Another possibility might perhaps be to screen the population for the appearance of some kind of pre-morbid personality and to treat the introspectives, or whoever is identified to have to "pre-parkinsonian personality", with selegiline or whatever else may turn out to be most useful.

Mertens: I don't believe that we are yet at a stage when PET-scanning will permit us to accumulate sufficient experience, although I think that one or two significant individual cases will lead us to fresh insights. In my opinion at our present state of play with Parkinson's disease, if we are to have any hope of success, there are in the first place two main directions to follow: the first is psychological testing for bradyphrenia, while the second is the pharmacological-kinesiological investigation of the latent presence of tremor and akinesia. I hope the psychologists will learn to recognise bradyphrenia at an early stage. Dementia is apparently only a chance finding which is not typical of Parkinson's disease. So the point is to isolate the phenomenon of bradyphrenia, and this should be possible using tachistoscopic or other psychological testing methods at an early stage of bradyphrenia. Of course this is not completely specific. But if one additionally finds an increased tendency to tremor, which you would have to test pharmacologically with cholinergic drugs, and further, a tendency to akinesia, which might be picked up pharmacologically with short-acting dopamine agonists, then it must eventually become possible, with some such test battery of two or three methods, successfully to approach the early diagnosis of Parkinson's disease.

Parkes: In real life my difficulty lies in separating early benign central tremor from early Parkinson's disease, and that is why I think I must often go wrong. Yesterday we saw a very nice example of the use of physiological techniques in making that distinction, and so I think tremor recording is of great value.

Horowsky: I would propose instead a combination of a dopamine-depleting compound such as tetrabenazine plus α-methyl-p-tyrosine, since the limiting enzyme is tyrosine hydroxylase. This enzyme's activity decreases with the progression of the disease, but there is some compensation in the appearance of hyperactivity. This could therefore be a sensitive test in combination with some suitable motor function test.

Przuntek: Parkes seems to have the answer ...

Parkes: We clearly need to be able to clone the gene for the expression of tyrosine hydroxylase, and possibly dopa decarboxylase as well!

Danielczyk: I am disappointed to report that my psychological colleages—and they are very experienced—tell me that it is not possible to make a differential diagnostic distinction between depression and an incipient Parkinson-syndrome, even with a very extensive battery of tests.

Beckmann: I wanted to warn you against taking the early vegetative signs too seriously. You have to consider that a practising physician will find exactly the same vegetative signs in 30—40% of the elderly patients that pass through his consulting room: I would be inclined to say rather more than 40% than less. I would also caution against these challenge methods: in schizophrenic patients they regularly indicate a Parkinson-syndrome in 25—30% of the patients, and they can't all be going to have Parkinson's disease! So I would counsel you not to pin your hopes on such test-combinations as possible methods for early diagnosis.

Mertens: Of course to a certain degree Beckmann is right, nevertheless I believe that if we combine akinesia-provocation with resting-tremor-provocation then the possibility of detecting a latent Parkinsonism will be much greater. I further believe that just as with curare in the diagnosis of myasthenia gravis, so there ought to be an age-related threshold dose for these challenging drugs in Parkinson's disease.

Beckmann: However, I think that once bradyphrenia has appeared we should be able to observe some abnormalities in the motor function.

Da Prada: In latent Parkinson's disease there is a deficiency of dopamine in the brain, 10% or 20% or 30% or even more. I think we simply need to find a biochemical approach to stimulating the biosynthesis or inhibiting the degradation of dopamine. The concentrations of the metabolites would be different and perhaps measurable in urine or plasma.

Riederer: In my opinion such determinations, whether in the CSF or in the plasma, or also in the urine are entirely nonspecific. It would be more to the point to develop an adrenal test. It has been reported that the adrenal function of patients with Parkinson's disease is different. I am not sure how specific *that* is, but perhaps we could check out other peripheral dopamine-dependent systems.

Da Prada: In Parkinson patients the endogenous plasma dopa level is increased after peripheral decarboxylase inhibition, and this must have something to do with the peripheral sympathetic system. Maybe the increase in plasma dopa in Parkinsonian patients is less rapid or smaller in extent.

Rinne: If you really have idiopathic Parkinson's disease there is always a significant decrease in CSF-HVA. Of course this is not specific, but if you have neuroleptic-induced parkinsonism there is always a significant *increase,* quite the opposite. So in some of those rare cases where you are not sure whether it is idiopathic or drug-induced you can find out just by measuring CSF-HVA.

Felgenhauer: I would like to ask my fellow physicians, supposing we had some yardstick for measuring preclinical Parkinson's disease, which of the clinicians sitting here in this room would treat these patients?

Rinne: Yes I would, with deprenyl.

Przuntek: Yes, we have to try that, to see how well it works. I would like to mention that as far as the CSF-HVA in patients with Parkinson's disease: we have seen patients who were not getting neuroleptics who had levels within the control range.

Felgenhauer: I would like to extend the scope of my original question: can you tell me any other degenerative disease which can be treated and cured just by treating the symptoms? Any disease at all, I'd like to hear one!

Przuntek: Wilson's disease.

Felgenhauer: Well of course all you are doing is removing the copper deposits, that's completely different, I am sorry, but I don't count that.

Rinne: It is very important to try to find the primary cause of Parkinson's disease, because this would let you design a more rational treatment approach. It is very good that the Parkinson patient's symptoms are focussed on the degeneration of the nigrostriatal dopamine neurons, because it gives us a very good way of treating the patients. Admittedly we are only treating the symptoms, but all the same, we can improve the quality of our patients' lives for more than ten years, so for those patients it is a very good therapeutic system.

Przuntek: I think that Wilson's disease is not a bad example of our problem: if we can show very early on that they have this disease, then we can protect the patients with chelating compounds and so prevent their further deterioration.

Birkmayer: I agree that the best pharmacological test is certainly the challenge with α-methyl-*p*-tyrosine, because this compound inhibits tyrosine hydroxylase activity in a very short time. By correlation of the dose which elicits parkinsonian symptoms you could decide whether the patient had latent Parkinson's disease.

Oertel: The European Multicenter Trial of deprenyl plus L-dopa has been going on for four years now, is there any information available about how it is progressing?

Przuntek: It is a double-blind study, so no information can be released until it has been completed.

Gibb: One other weak ray of light on the pathological side: from biopsied peripheral autonomic tissue. It is a link with Hallervorden-Spatz disease, because the pathological levels of the disorders are very similar and so there might well be shared pathogenetic factors. In Hallervorden-Spatz disease abnormalities have been described in Bernmayer's histiocytes and peripheral blood lymphocytes at the ultrastructural level. So if anyone is looking for a project it might be worth looking at the ultrastructural level in Parkinson's disease.

Steinberg: As a psychologist I want to make this point about the role of psychology, that as psychologists we can only try to explain those aspects of Parkinson's disease which are accessible to psychological methods. As long as there are no psychological cognitive deficits which are exclusive to Parkinson's disease only, then it becomes a question of cross-validation or multivariate procedures, a question we have to pass on to other approaches such as those which have been mentioned in our symposium, for example aspects of motor performance deterioration, so that we can piece together the whole picture of Parkinson's disease and of its diagnosis.

Przuntek: Summing up, we must try to find a combination of psychometric, motor performance, biochemical and neuro-imaging tests to detect Parkinson's disease in its very early stages.

Now we most turn our attention to treatment, and discuss the possibilities as well as the problems that we face of developing strategies of preventive and long-term therapy as well as additional treatments in Parkinson's disease.

Danielczyk: If we ask ourselves when to start treatment of Parkinson's disease, the answer depends very much on the age of the patient. With a patient aged 55 I would not hesitate a single week before initiating immediate treatment, but with a 75-year old patient I would hold back and not necessarily rush into therapy.
Birkmayer: I agree with that: with a 75-year old there is more sensitivity to the drugs, and therefore we have to limit our engagement in these cases.
Korczyn: No less important in Parkinson's disease than making an early diagnosis is to discover the factors that influence the natural progression of the disease, because we are all of us aware that there are wide differences among patients in the rate at which the disease advances, and this will have to be taken into account when one treats patients aggressively in the early stages. If there is any therapy that holds the progression of the disease in check then obviously this will have side-effects. We must, therefore, try to develop criteria which can indicate to us which patients are likely to progress rapidly and which more slowly so that we can match our treatment to these different rates of progression.
Parkes: How can we prevent Parkinson's disease? In the animal MPTP studies, MPP^+-reuptake blockers like amphetamine and mazindol are as effective as selegiline in stopping or preventing toxicity. At the two last meetings of the International Narcolepsy Symposia we took a vote from people using large quantities of amphetamine on patients as to how often they see Parkinson's disease developing after 20 years' treatment with amphetamine. Parkinsonism and progressive parkinsonism is not at all uncommon in narcoleptic subjects taking amphetamine, and I have also seen it in a single patient taking mazindol. So at the level of reuptake blockers we are certainly not preventing either disease onset or disease progression.
Youdim: In my opinion Parkinson's disease is a neurotoxin-dependent disorder. As I suggested yesterday, it could just be the formation of radicals. In London they have developed a method for actually monitoring radical formation in vivo, and it would be very interesting to apply this method to seeing whether such phenomena can also be occurring in the parkinsonian condition. So I agree with the view expressed by Oleh Hornykiewicz that the MPTP-model, although it is a nice model, does not quite fit in with what we know about the classical Parkinson, and the progressive neurodegeneration could be associated with the enzyme MAO and the formation of radicals by hydrogen peroxide.
Brune: Parkinson's disease is a disease of advanced age. We may not believe there is a causal relationship, but still, we have patients with

disturbances of perfusion. So we not only have to look for depression, we also have to look for disturbances of cerebral perfusion, and what is the treatment for that?

Parkes: Although of course Parkinson's disease is much commoner in old people, it is not desperately uncommon in younger people: Neal Quinn from our department published a series of 60 young-onset Parkinson patients, so I'm not sure the basic premise is right. Have we any evidence about cerebral blood-flow from PET-scanning, Leenders?

Leenders: The studies performed by the St Louis group, by the Vancouver group and by myself, on patients with idiopathic Parkinson's disease, have shown global decreases in cerebral blood-flow in line with global decreases of energy metabolism. There was no focal sign of ischemia, indicating that there is no primary vascular problem responsible for the dopaminergic deficit. Of course Parkinson's disease is common and vascular disease is common, so there might be an overlap in some cases, but in typical cases the perfusion studies do not specifically tend to implicate cerebrovascular disease.

Brune: Well I accept that Parkinson's disease may manifest itself in patients aged less than 40 or a little bit later, but in all departments we more frequently see patients with Parkinson's disease who are 60 and over, and when I talk about disturbances of cerebral blood-flow I mean those patients and not the ones with very early manifestations. And in those patients, when you have symptoms indicating disturbed blood-flow, what do you do? Have you any experience with the combination of anti-parkinsonian drugs and drugs designed to improve perfusion?

Korczyn: I know of no drug that improves perfusion in such a situation, but we should take the question seriously. These days we shy away from the term "arteriosclerotic parkinsonism", and rightly, but perhaps we are not 100% justified in our reluctance to use that term. Our patients with Parkinson's disease who developed dementia had very significant risk factors: vascular risk factors such as diabetes and hypertension, indicating that vascular effects may contribute to the development of dementia. We have also looked at patients who have undergone cerebral CT, and we took those that had holes in the basal ganglia due to infarcts, and of those quite a significant number were diagnosed as having Parkinson's disease. The probably do not have idiopathic Parkinson's disease although the diagnoses are based on clinical manifestations such as gait and the response to dopaminergic drugs; however it points up the possibility that we do have to take vascular factors into account in some of the patients presenting with Parkinson's disease.

Gibb: We all commonly see patients with strokes who have fragments of the Parkinson syndrome, but whether there are clinical manifestations correlated with dopamine deficiency is entirely another matter. I *have* to say that there is not a single clinically and pathologically studied patient who has been shown to have a true parkinsonian dopamine-deficiency disorder as a result of vascular disease.

Horowski: There is another interaction between Parkinson's disease and cardio-vascular disease: it is my impression, and I want to have the clinicians confirm this, that there is an inverse relationship between Parkinson's disease and hypertension.

Parkes: That is almost certainly correct. The parkinsonian patients lose weight, a fact not stressed in the literature. It is not uncommon for a patient to lose 20 kg during the course of Parkinson's disease, and I suspect that this may be a partial explanation for your observation.

Danielczyk: Many patients start with idiopathic Parkinson's disease, and since they survive so much longer than before, something like 12 years in our patients, they reach an age where cerebrovascular complication rise in a linear fashion, so that between the ages of 75 and 85 encephalomalacial foci appear. At first we were rather disappointed with the CT-scans because these foci often failed to show up, but now with the MRI-investigations we observe much more than before how much more frequent these secondary and tertiary diseases occur in patients with Parkinson's disease.

Schneider: I wanted to remind you that there is a long-standing investigation of the incidence of arteriosclerosis in parkinsonian patients compared with a normal population, which showed that the parkinsonians did not have an increased incidence. As a second disease it then has adverse effects on the symptomatology, whether it is the brain-organic psycho-syndrome or the motor function disorder.

Przuntek: Professor Youdim, do we really have a chance of preventive treatment in Parkinson's disease?

Youdim: Rinne mentioned selegiline earlier on: when I lecture on Parkinson's disease, I tell my students that if I thought, from some early-warning technique, that I was at risk of contracting Parkinson's disease, then I would sprinkle selegiline over my cornflakes every morning. I say this in all seriousness, because I think it is the only drug we have available. Yesterday I suggested that iron-chelators might prove valuable, and we now have very good evidence that iron does play a role in cytotoxicity in cases where radicals are formed and produce cell degeneration. Paraquat is the most frequently mentioned toxic substance: it has a similar structure to MPTP or MPP^+ and desferrioxamine is now very successfully used as an effective drug in treating paraquat poisoning. Maybe we ought to be looking for similar kinds of chelating-agents that can cross the blood-brain-barrier and I am aware of several centres that are following this up. If we have any indication that iron is increased in any brain region, that would be further evidence for the aetiology of the disease being neurodegenerative due to radical formation. Now, if I may, I'd like to come back to what Parkes said about amphetamine: amphetamine or reuptake blockers would not prevent the formation of radicals, so I think we should go back a step, to MAO. I'd like to ask Beckmann, looking at it another way, how many patients on antidepressive drugs or MAO-inhibitors ended up in parkinsonism?

Korczyn: But amphetamine is an MAO-inhibitor.

Youdim: No, it is not really an MAO-inhibitor, or only a poor one: it is in the first place a much better releaser of dopamine.

Beckmann: To answer your question, I have no figures, but it is becoming more and more difficult to give patients long-term MAO-inhibitor treatment. Once we have a better tolerated MAO-inhibitor it might be more worth doing again.

Youdim: Phenelzine has been used for many years both in Europe and in the US and I think it is one of the safest drugs.

Rinne: I would like us all to remember that while we cannot stop the progression of the disease by using levodopa, we can, by initiating its use early, increase survival and reduce mortality. For this reason it is a very important treatment for the patients if you can offer them extra years by giving them levodopa in the early phase.

Leenders: It may be a naïve question, but I had the impression that the iron concentration in the brain is increased, so why does Birkmayer advocate giving patients iron?

Youdim: Maybe he should answer why be believes the iron therapy should work, but these are two different issues: we don't even know why iron should accumulate in the substantia nigra. Iron probably is accumulated as ferritin, the protein which keeps iron in an active form, and it may be that when iron is released in an active form that is when the damage occurs. That is exactly how paraquat works: it releases the iron from the ferritin. The free iron is involved in the production of radicals which cause the damage to the cells. Maybe something along these lines is occurring in Parkinson's disease, that somehow—perhaps by some toxic agent—the iron is released from accumulated deposits, but these speculative ideas need further studies for confirmation.

Da Prada: You obviously must not accumulate iron in the periphery. Can you give any iron-chelator? Would that not be very difficult?

Youdim: Desferrioxamine is given, and it does not seem to produce that many side-effects when it is used in treating paraquat toxicity. It is a very certain treatment. We should refer to the work of Barbeau with the MPTP-model, and we have done similar studies in Haifa. He showed that iron made the MPTP-model more sensitive, and desferrioxamine actually blocked MPTP toxicity.
Korczyn: I must say I am beginning to get quite confused! Most of us agree that the early treatment of Parkinson's disease should consist of levodopa with a peripheral dopa decarboxylase inhibitor, so that's two drugs. Then some of us use anti-radicals and advocate radical scavengers like vitamin E, and MAO-inhibitors, so that's four. Then Da Prada says COMT-inhibitors should be added to those, that's five, and Konradi suggested reuptake-blockers could prevent toxic effects of drugs, that makes six. Then data are accumulating that indicate that giving patients D 2-agonists would retard the rate of loss of dopamine cells. Felten in Rochester has shown that in a very nice study. Then Rinne and Markstein say that you have to add D 1-agonists as well, so that's eight, to say nothing of the classical adjuvants, I think for example that amantadine should given very early on, and of course we all treat patients with antimuscarinic drugs. So there you have it, we begin by treating our patients with at least ten drugs, and we have not even mentioned iron plus iron-chelators: where is all this going to end?
Przuntek: If I had to make a choice at this moment, I would go along with Youdim and if had an early diagnosis I too would sprinkle MAO-B inhibitors on my cornflakes!

Now, if there are no more comments, I should like to conclude this symposium with the words of Professor Birkmayer. He quoted Popper, "Science is always on its way . . ." and say to you: "we are on *our* way". I hope that we can come together again within the next three years to look at how far we have progressed with the methods and approaches we have been discussing during the last two days. Many people asked me: "why have you organised this meeting one month before the big Parkinson meeting in Jerusalem?" There are two reasons: the first reason might be that we had to discuss a very circumscribed theme. The second reason is this: all of you know that the route of the pilgrims from Ireland to Jerusalem passed through Würzburg, and before the pilgrims went on to Jerusalem they sat here in the wine-bars and discussed their problems, and hoped that all their problems would disappear in Jerusalem. Therefore we are glad that Dr Korczyn took part in our symposium. He was heard all our questions and all our problems, and we hope that when we arrive—

not next year but next month—in Jerusalem, he will have all the answers ready for us!

Now I would like to thank all our participants who attended this meeting. I hope that you enjoyed this symposium and this wonderful city.

Finally, I must say a very heartfelt "thank you" to Mr. Wanzlik, who organised this meeting in such an excellent fashion, and furthermore I would like to express, on all our behalf, our sincere thanks to Hoffmann-La Roche for their very generous sponsorship.

Subject Index

J. A. Obeso, R. Horowski, C. D. Marsden (eds.)

Continuous Dopaminergic Stimulation in Parkinson's Disease

(Journal of Neural Transmission / Supplementum 27)
1988. 58 partly coloured figures. XII, 255 pages. ISBN 3-211-82034-5
Soft cover DM 114,–, öS 800,–*.

The subject of this book is the outcome of a workshop held in Alicante/Spain in 1986 on the topic of "Continuous dopaminergic stimulation in Parkinson's disease" and consequently deals with a new continuous s.c. infusion method using the dopamine agonist lisuride for treatment of fluctuations in motor performance following long-term treatment with L-DOPA in Parkinson's disease which occur in a majority of cases. For galenical reasons, intravenous L-DOPA treatment can be given only for a short period, so it became necessary to devise new ways of achieving an easily controllable and constant dopaminergic stimulation, both for further research and treatment.

M. B. H. Youdim, M. Da Prada, R. Amrein (eds.)

The Cheese Effect and New Reversible MAO-A Inhibitors

(Journal of Neural Transmission / Supplementum 26)
1988. 39 figures. VII, 136 pages. ISBN 3-211-82031-0
Soft cover DM 70,–, öS 490,–*

This collection of papers covers aspects of the interaction between moclobemide, a new reversible MAOI which preferentially inhibits MAO type A and tyramine. In vitro and pharmacokinetic interaction studies and studies in man are described. Moclobemide is shown to be an effective antidepressant and the "cheese effect", when tyramine-rich food is consumed during moclobemide treatment, is very unlikely to occur and thus diet restrictions are unnecessary.

* *10% price reduction for subscribers to "Journal of Neural Transmission"*

Springer-Verlag Wien New York

Moelkerbastei 5, A-1010 Wien
Heidelberger Platz 3, D-1000 Berlin 33
175 Fifth Avenue, New York, NY 10010, USA
37-3, Hongo 3-chome, Bunkyo-ku, Tokyo 113, Japan

MAO-B-Inhibitor Selegiline (R-(-)-Deprenyl)

A New Therapeutic Concept in the Treatment of Parkinson's Disease